THIEME ORGANIC CHEMISTRY MONOGRAPH SERIES

Analytical Chemistry of Carbohydrates

Related Thieme Titles of Interest

Organic Chemistry Monograph Series

A. Hirsch
The Chemistry of the Fullerenes

T. Eicher and S. Hauptmann
The Chemistry of Heterocycles

J. Lehmann
Carbohydrates Structure and Biology

Foundations of Organic Chemistry Series

P. J. Kocieński
Protecting Groups

H. B. Kagan
Asymmetric Synthesis (in preparation)

M. Hesse, H. Meier and B. Zeeh
Spectroscopic Methods in Organic Chemistry

Analytical Chemistry of Carbohydrates

Heimo Scherz

German Research Institute
of Food Chemistry

Günther Bonn

Leopold-Franzens-University
of Innsbruck

Georg Thieme Verlag Stuttgart • New York 1998

Prof. Dr. Heimo Scherz
German Research Institute of Food Chemistry
Lichtenbergstraße 4
85748 Garching
Germany

Univ. Prof. Dr. Günther Bonn
Institute of Analytical Chemistry and Radiochemistry
Leopold-Franzens-University of Innsbruck
Innrain 52a
6020 Innsbruck
Austria

Die Deutsche Bibliothek – Cataloging-in-Publication Data

Scherz, Heimo:
Analytical chemistry of carbohydrates: Thieme organic chemistry monograph series
Stuttgart; New York: Thieme, 1998
NE: Bonn, Günther

© 1998 Georg Thieme Verlag, Rüdigerstraße 14, D-70469 Stuttgart
Printed in Germany by Gutmann & Co GmbH, D-74388 Talheim

Georg Thieme Verlag, Stuttgart   Thieme Medical Publishers, Inc., New York
ISBN 3-13-102351-1   ISBN 0-86577-666-0   1 2 3 3 4 5 6

Foreword

Carbohydrates are some of the most important chemical compounds of the biological world and are the key substances of life since they are the primary products of photosynthesis. They occur in nearly all living systems either as high molecular skeletal substances, as high or low molecular energy resource compounds as well as key intermediates in human, animal and plant metabolisms. In combination with proteins or lipids they give highly physiologically active glycoproteins (e.g., blood group substances) or glycolipids.

A lot of commercial products consist of carbohydrates or contain carbohydrates, particularly most foods, and especially those of plant origin (cereals, pulses, fruits and vegetables), or are products of daily use such as paper and textiles. It should also be noted that sugar itself with the chemical name sucrose is the most wide spread chemical in the world produced in the highest purity.

Indispensable for the quality control of carbohydrate containing products such as foods, medical or pharmaceutical products, medical diagnostic applications (e.g., in the field of diabetes) and for the field of research in organic chemistry, biochemistry and physiological chemistry are methods for qualitative and quantitative analysis of carbohydrates, which has been a part of analytical chemistry from its scientific beginning. In the last four decades considerable progress has occurred in this field dealing preferably with the new chromatographic and electrophoretic methods such as thin-layer chromatography, gas chromatography, high pressure lipid chromatography (HPLC) and high voltage capillary electrophoresis, but also with the development of new enzymatic procedures.

This monograph presents an overview of the classical methods such as different photometric procedures as well as the modern enzymatic, chromatographic and electrophoretic methods and should be helpful to the broad spectrum of researchers who are involved in carbohydrate analysis.

Heimo Scherz   Günther Bonn
January 1998

Acknowledgments

The authors would like to thank:

- Christel Hoffmann for her excellent cooperation preparing the manuscript.
- Dip. Ing. Dr. Rainer Gagstetter for computer drawing the chemical formulas.
- Prof. Dr. Andreas Zemann for his contributions toward the chapter 'Electrophoresis'.
- Dr. Ray Boucher for his extensive help at the publishing office.
- Dr. Torren Peakman and Lindsey Sturdy for English polishing.
- Margarethe Podlesch and Sabine Koschnitzke for preparing the layout and the drawings (supplementary figures).

Furthermore, the authors would like to express their thanks to the following publishers for their kind permission to include some materials from their publications into this book:

– Elsevier Science B. V. (Amsterdam, Netherlands):
  Journal of Chromatography; Carbohydrate Research.
– American Chemical Society (Washington DC, USA):
  Analytical Chemistry.
– Springer Verlag GmbH & Co. KG (Berlin, Heidelberg, Germany):
  Zeitschrift für Lebensmittel-Untersuchung und -Forschung; Fresenius' Zeitschrift für Analytische Chemie.
– Springer Verlag Austria (Vienna, Austria):
  Mikrochimica Acta.
– Wiley VCH Publishing GmbH (Weinheim, Germany):
  Die Nahrung; Chromatographia.
– Wissenschaftliche Verlagsgesellschaft mbH (Stuttgart, Germany):
  Deutsche Lebensmittelrundschau.
– National Academy of Science (Washington DC, USA):
  Proceedings of the National Academy of Science; Biochemistry.
– Bundesamt für Gesundheit (Bern, Switzerland):
  Mitteilungen aus dem Gebiete der Lebensmitteluntersuchung und Hygiene.
– Academic Press (Orlando, USA):
  Analytical Biochemistry.
– f. a. Dionex GmbH.

Contents

Introduction

Carbohydrates are, in terms of their quantity, the largest group of organic compounds on earth and are implicated in every form of living activity in plants, animals and microorganisms. They are one of the most important food constituents (e.g., starch), the raw materials for human clothing, and the raw material for paper and for many biotechnological processes, such as the production of beer and antibiotics. It is obvious that methods are necessary to identify the individual species and to determine them quantitatively, either alone or as a mixture of more than one component in the presence of other compounds.

Earlier methods of analysis of carbohydrates deal with gravimetric, titrimetric and colorimetric procedures and are the so-named "classical methods" which enable only the qualitative and quantitative analyses of different groups of carbohydrates but not the individual compounds. The major breakthrough began fifty years ago with the development of chromatographic, electrophoretic and enzymatic methods. These procedures enable the qualitative and quantitative analysis of single species in complicated mixtures of carbohydrate compounds.

Chemical Structures of Carbohydrates

Carbohydrates are classified according to their chemical structure into three main groups of compounds namely (a) monosaccharides, (b) oligosaccharides and (c) polysaccharides.

Monosaccharides

Monosaccharides are organic compounds with the molecular formula $C_n(H_2O)_n$ (the origin of the name carbohydrate) and are vicinal polyhydroxy aldehydes and polyhydroxy ketones.

The simplest monosaccharide is the three-carbon compound glyceraldehyde. According to the asymmetric carbon atom, it has a chiral center and occurs as an enantiomeric pair namely in the D- and L-forms as shown in Figure 1.

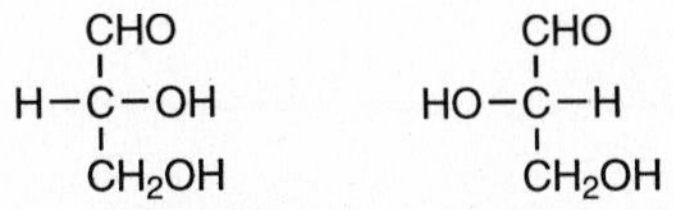

$$
\begin{array}{cc}
\text{CHO} & \text{CHO} \\
| & | \\
\text{H}-\text{C}-\text{OH} & \text{HO}-\text{C}-\text{H} \\
| & | \\
\text{CH}_2\text{OH} & \text{CH}_2\text{OH} \\
\text{D-Glyceraldehyde} & \text{L-Glyceraldehyde}
\end{array}
$$

Figure 1

A fictive reaction, namely an aldol condensation with "formaldehyde", leads to the enlargement of the molecule by one carbon atom yielding two diastereoisomeric tetroses with a new chiral center. A repetition of this yields four possible pentoses and then eight possible hexoses as shown in Figure 2.

```
           CHO
          ┣OH
          CH₂OH
```

D-Glyceraldehyde
(D-glycero-)

```
    CHO              CHO
   ┣OH          HO┫
   ┣OH              ┣OH
   CH₂OH            CH₂OH
```

D-Erythrose   D-Threose
(D-erythro-)   (D-threo-)

```
   CHO         CHO          CHO         CHO
  ┣OH      HO┫            ┣OH      HO┫
  ┣OH          ┣OH    HO┫         HO┫
  ┣OH          ┣OH         ┣OH         ┣OH
  CH₂OH        CH₂OH       CH₂OH       CH₂OH
```

D-Ribose   D-Arabinose   D-Xylose   D-Lyxose
(D-ribo-)   (D-arabino-)   (D-xylo-)   (D-lyxo-)
[D-rib]   [D-arab]   [D-xyl]   [D-lyx]

```
CHO       CHO       CHO       CHO       CHO       CHO       CHO       CHO
┣OH    HO┫        ┣OH     HO┫        ┣OH     HO┫        ┣OH     HO┫
┣OH       ┣OH    HO┫       HO┫        ┣OH       ┣OH    HO┫       HO┫
┣OH       ┣OH       ┣OH       ┣OH    HO┫       HO┫       HO┫       HO┫
┣OH       ┣OH       ┣OH       ┣OH       ┣OH       ┣OH       ┣OH       ┣OH
CH₂OH     CH₂OH     CH₂OH     CH₂OH     CH₂OH     CH₂OH     CH₂OH     CH₂OH
```

D-Allose   D-Altrose   D-Glucose   D-Mannose   D-Gulose   D-Idose   D-Galactose   D-Talose
(D-allo-)   (D-altro-)   (D-gluco-)   (D-manno-)   (D-gulo-)   (D-ido-)   (D-galacto-)   (D-talo-)
[D-all]   [D-alt]   [D-glc] or [glu] [D-man]   [D-gul]   [D-id]   [D-gal]   [D-tal]

( ) Parenthesis: abbreviation for the description of the structure in monosaccharide
[ ] Parenthesis: abbreviation for the description of the structure in oligo- and polysaccharides

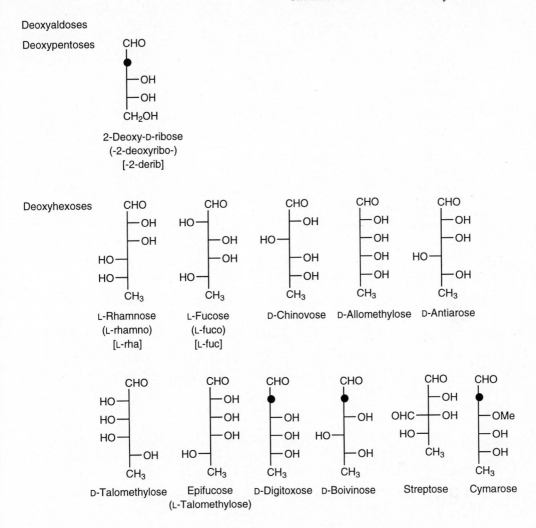

Figure 2. Structure of the Monoaldoses (Fischer Projection) and Naturally Occurring Deoxyaldoses

When the carbonyl group is located in the middle of the molecule such compounds are called ketoses. The number of diastereoisomers is reduced to half those of the aldoses with the same number of carbons. Their structures are illustrated in Figure 3. Figure 4 shows the structures of further natural monosaccharides and their derivatives.

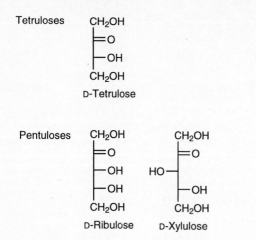

Tetruloses

D-Tetrulose

Pentuloses

D-Ribulose   D-Xylulose

Hexuloses

D-Psicose   D-Fructose   D-Sorbose   D-Tagatose
              (D-fructo)
              [D-fru]

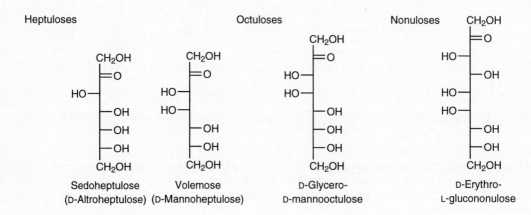

Heptuloses   Octuloses   Nonuloses

Sedoheptulose   Volemose   D-Glycero-   D-Erythro-
(D-Altroheptulose) (D-Mannoheptulose) D-mannooctulose   L-glucononulose

Figure 3.   Structure of Monoketoses (Fischer Projection)

Uronic Acids (Fischer Projection)

CHO	CHO	CHO	CHO	CHO
—OH	HO—	—OH	HO—	—OH
HO—	HO—	HO—	HO—	HO—
—OH	—OH	HO—	—OH	—OH
—OH	—OH	—OH	HO—	HO—
CO$_2$H	CO$_2$H	CO$_2$H	CO$_2$H	CO$_2$H
D-Glucuronic acid	D-Mannuronic acid	D-Galacturonic acid	L-Guluronic acid	L-Iduronic acid

Sugar Alcohols Pentitols (Fischer Projection)

CH$_2$OH	CH$_2$OH
—OH	HO—
—OH	—OH
—OH	—OH
CH$_2$OH	CH$_2$OH
D-Ribitol (Adonitol)	D-Arabitol

Hexitols (Fischer Projection)

CH$_2$OH	CH$_2$OH	CH$_2$OH
—OH	HO—	—OH
HO—	HO—	HO—
—OH	—OH	HO—
—OH	—OH	—OH
CH$_2$OH	CH$_2$OH	CH$_2$OH
D-Sorbitol D-Glucitol	D-Mannitol	Galactitol Dulcitol

Heptitols (Fischer Projection)

CH$_2$OH
HO—
HO—
HO—
—OH
—OH
CH$_2$OH
Volemitol

Cyclitols

D-Inositol L-Inositol Quebrachitol *myo*-Inositol

scyllo-Inositol Mytilitol Quercitol Viburnitol Conduritol

Amino Sugars (Fischer Projection)

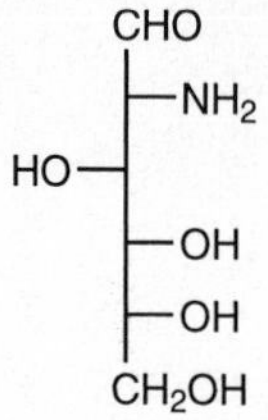

CHO
|—NH₂
HO—|
|—OH
|—OH
CH₂OH

D-Glucosamine
D-Chitosamine

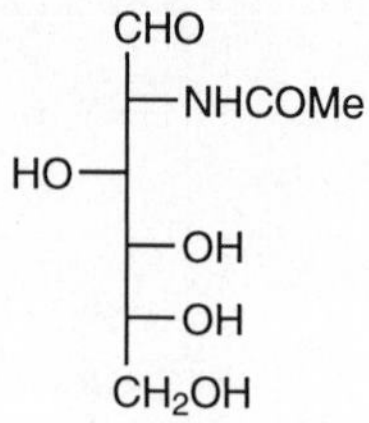

CHO
|—NHCOMe
HO—|
|—OH
|—OH
CH₂OH

N-Acetyl-
D-glucosamine

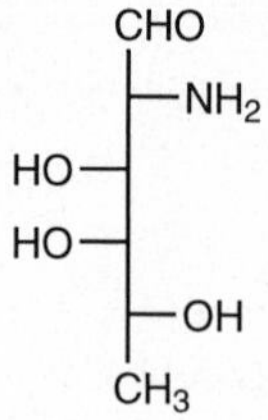

CHO
|—NH₂
HO—|
HO—|
|—OH
CH₃

D-Galactosamine

CHO
|—NHCOMe
HO—|
HO—|
|—OH
CH₃

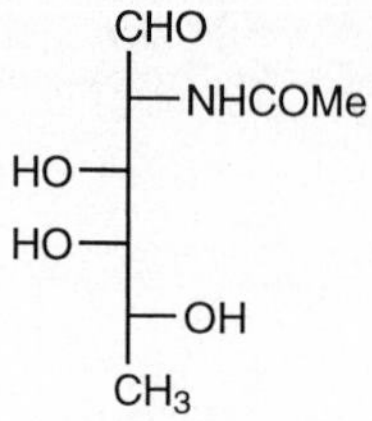

N-Acetyl-
D-galactosamine

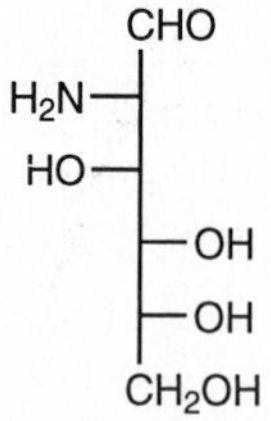

CHO
H₂N—|
HO—|
|—OH
|—OH
CH₂OH

D-Mannosamine

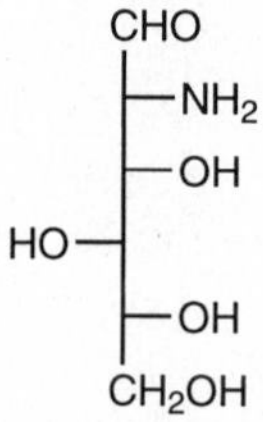

CHO
|—NH₂
|—OH
HO—|
|—OH
CH₂OH

D-Gulosamine

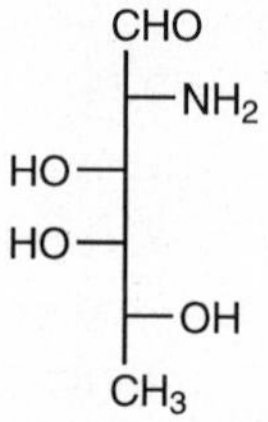

CHO
|—NH₂
HO—|
HO—|
|—OH
CH₃

D-Fucosamine

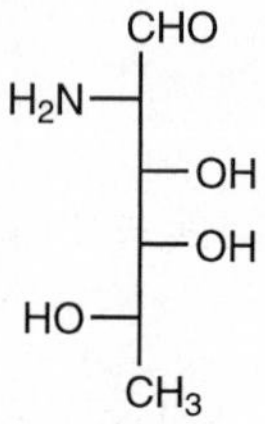

CHO
H₂N—|
|—OH
|—OH
HO—|
CH₃

L-Fucosamine

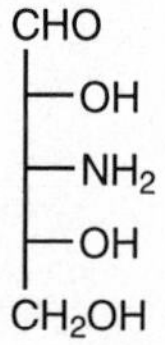

CHO
|—OH
|—NH₂
|—OH
CH₂OH

3-Amino-
3-deoxy-D-ribose

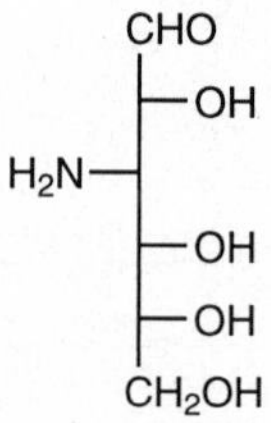

CHO
|—OH
H₂N—|
|—OH
|—OH
CH₂OH

Kanosamine

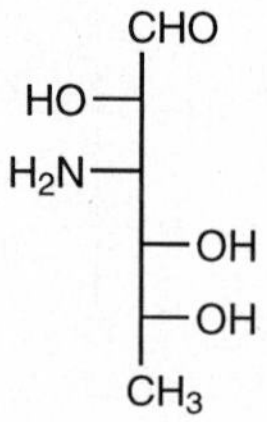

CHO
HO—|
H₂N—|
|—OH
|—OH
CH₃

Mycosamine

CHO
|—OH
Me₂N—|
|—OH
|—OH
CH₃

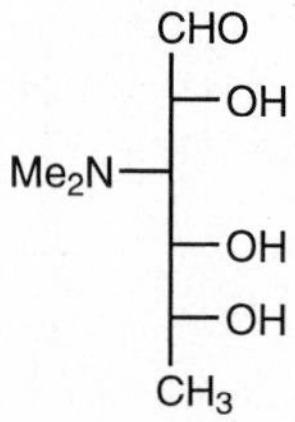

Mycaminose

Amino Sugar Carbonic Acids (Fischer Projection)

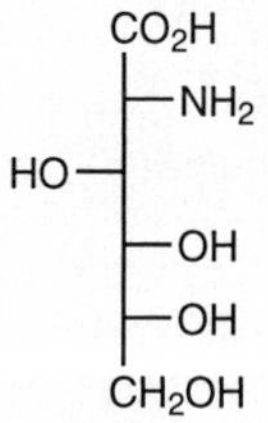

CO₂H
|—NH₂
HO—|
|—OH
|—OH
CH₂OH

D-Glucosaminic
acid

CHO
|—NH₂
HO—|
HO—|
|—OH
CO₂H

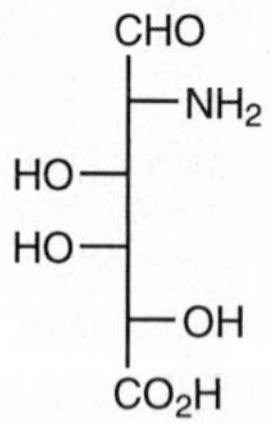

D-Galactosamine-
uronic acid

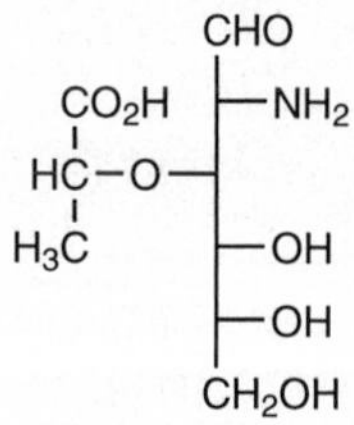

CHO
|—NH₂
CO₂H
|
HC—O—|
|
H₃C |—OH
|—OH
CH₂OH

Muramic acid

CO₂H
‖O
●
|—OH
RCONH—|
HO—|
|—OH
|—OH
CH₂OH

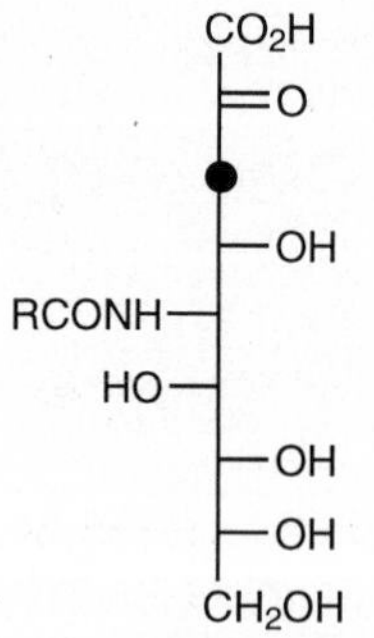

R=Me, CH₂OH
etc.
Sialic acids

Amino Cyclitols

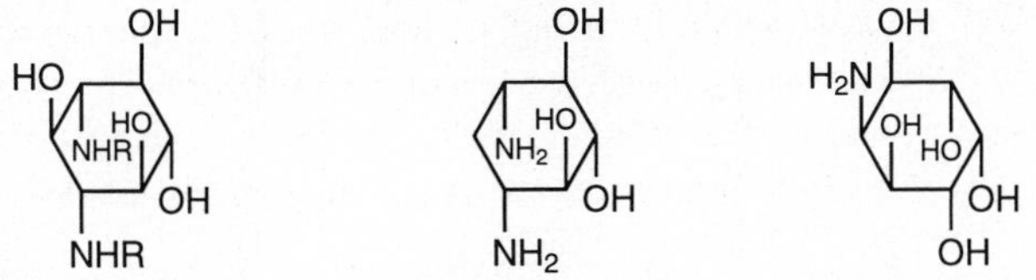

Streptamine R = H
Streptidin R = −C(NH)NH₂

2-Deoxystreptamine

neo-Inosamine

Figure 4. Structure of Sugar Derivatives in Biological Systems

In reality, the open-chain configuration of aldoses and ketoses as represented in Figures 1–3 exists only as minute quantities in solution. The main forms are the cyclic pyranosides and furanosides derived from the hypothetical "acetal" of the open-chain form by dehydration. This "hemiacetal" formation provides a new chiral center at C-1, forming two additional diastereoisomers which are called the α- and β-form of such aldoses.

In the Haworth projection of the structures, the hydroxyl group at C-1 of the D-sugars is below the pyranoside or furanoside plane in the α-form and above in the β-form as shown in Figure 5.

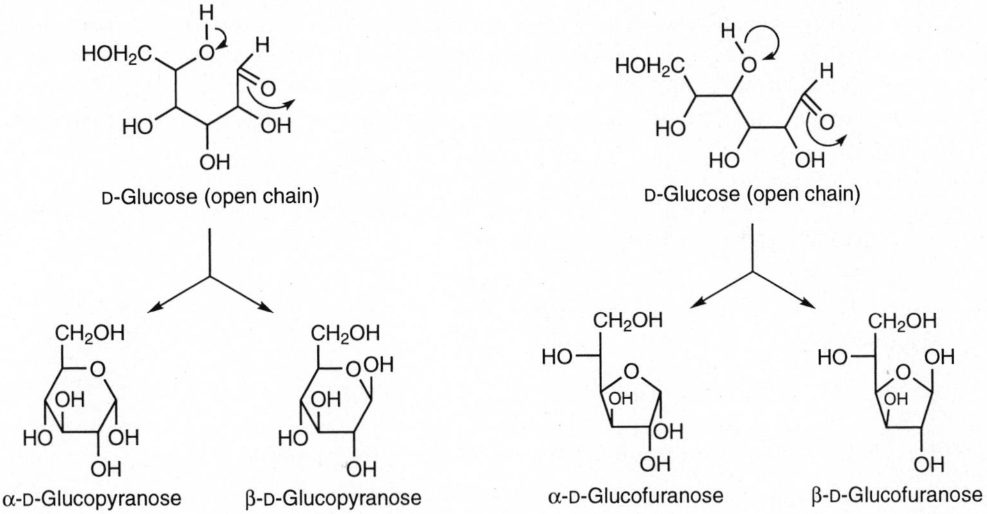

D-Glucose (open chain)

D-Glucose (open chain)

α-D-Glucopyranose

β-D-Glucopyranose

α-D-Glucofuranose

β-D-Glucofuranose

Figure 5. Formation of the Pyranoside and Furanoside Hemiacetals from the Free Aldehyde Form of Reducing Monosaccaride (Example: D-Glucose)

The configuration of these pyranosides and furanosides is not planar. The preferred structure of the pyranose compounds is the chair conformation which also exists in two different forms due to the positions of the single hydroxyl groups, these being either axial or equatorial according to their thermodynamic stabilities.

Molecules with the C-4 carbon in the upper position of the chair conformer are named as having $^4C_1$ conformation, whilst molecules with the C-4 carbon in the lower position of the conformer are said to have $^1C_4$ conformation as shown in the case of D-glucose (see Figure 6). For most pyranoses, the $^4C_1$ conformation is preferred due to lower interaction between the single hydroxyl groups.

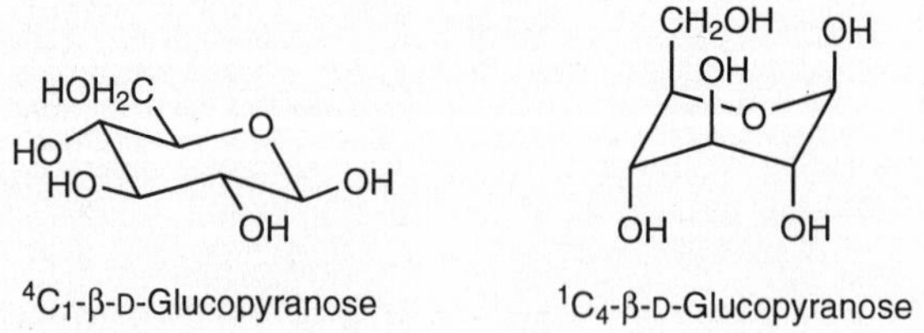

$^4C_1$-β-D-Glucopyranose $^1C_4$-β-D-Glucopyranose

Figure 6.   Conformation of β-D-Glucopyranose

Optical Rotation

Sugars are optically active compounds and their specific constants [α] are dependent on the position of the hydroxyl group at C-1 and on the ring form (pyranose or furanose). In solid form only one of the configurations exists, but upon dissolving, e.g. in water, the freshly prepared solution of the isomer shows a change in the optical rotation until equilibrium is reached. This effect is called "mutarotation" and is caused by the formation in solution of the other cyclic species of the sugar and of the free aldehyde form (the latter occurring only in very small amounts). This is shown in the simple case of D-glucose, where almost only two structures, namely those with the α- and β-pyranoside configuration, exist in an aqueous medium (see Figure 7 and Section 1.7.).

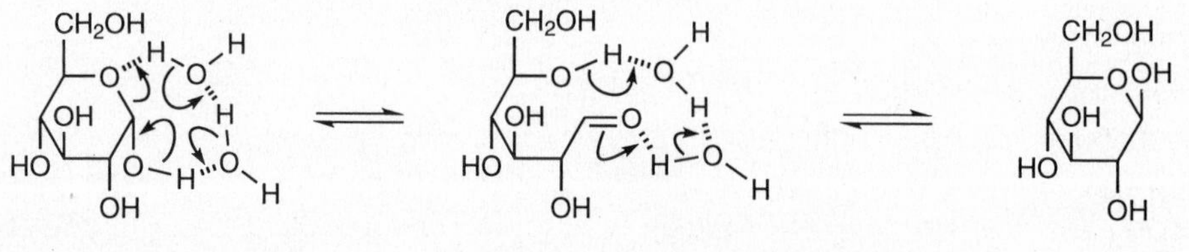

Figure 7.   Mutarotation of Reducing Sugars in Aqueous Solution (Example: D-Glucose)

Occurrence of Monosaccharides

Only a limited number of monosaccharides are present in living systems such as plants, animals and microorganisms as shown in Tables 1 and 2. In Table 3, the most important naturally occurring monosaccharide derivatives are listed. The structures of the last group of compounds are shown in Figure 4.

Table 1. Occurrence of Monosaccharides in Biological Systems

Name of the Monoaldose	Occurrence
Pentoses	
L-Arabinose	Units of many plant polysaccharides such as plant gums (tragacanth), woods and cereals; constituent of glycosides
D-Arabinose	Constituent of aloe glycosides and bacterial polysaccharides
D-Xylose	Units of many plant polysaccharides such as the xylans of cereals and hemicelluloses of many hardwoods and conifers
D-Ribose	Constituent of nucleic acids (RNA), nucleosides, nucleotides and some of the important coenzymes such as NAD; small amounts of free sugars in muscles of slaughtered animals
D-Lyxose	Small amounts as units of yeast nucleic acid; free form in human heart muscle
L-Lyxose	Constituent of the antibiotic Curamycin
D-Apiose (branched pentose)	Constituent of parsley and celery glycosides; in the bark of *Hevea brasiliensis*
Hexoses	
D-Glucose	Widespread in all plants and animals and microorganisms; one of the most important organic compounds in biological systems; constituent of many polysaccharides like starch and cellulose
D-Galactose	As the free compound, only in small amounts, in wine and fermented milk products such as cheese and yogurt; constituent of many oligosaccharides, glycosides and polysaccharides
D-Mannose	As the free compound, only in traces, in fruits such as apples, peaches, etc. Unit of many polysaccharides such as the mannans or galactomannans
D-Talose	Constituent of the antibiotic Hygromycin B
Deoxymonoaldoses	
Deoxypentoses	
2-Deoxy-D-ribose	Constituent of Deoxyribonucleic acid (DNA)
Deoxyhexoses	
L-Rhamnose (6-Deoxy-L-mannose)	As the free compound, only in small amounts, in wine and in the extract of the leaves of poison ivy (*Rhus toxidendron L.*); constituent of many glycosides and plant polysaccharides
D-Rhamnose (6-Deoxy-D-mannose)	Unit of capsular polysaccharides of gram negative bacteria strain GS
L-Fucose (6-Deoxy-L-galactose)	Only as a component of oligosaccharides, especially those of human milk, of glycosides of some plants, units of polysaccharides of some seaweeds like fucoidan and some glycoproteins
D-Chinovose (D-Epirhamnose) (6-Deoxy-D-glucose)	Glycoside of the bark of Chinchona
D-Allomethylose (6-Deoxy-D-allose)	Cardiac glycosides from *Gomphoro carpus fructicosus*, *Digitalis canariensis var. isabelliana*

Table 1. (Continued)

Name of the Monoaldose	Occurrence
Antiarose (6-Deoxy-D-gulose)	Component of the steroid glycoside in the sap of the upas tree (*Antiaris toxicaria*) and as the glycoside of *Digitalis canariensis var. isabelliana*
Talomethylose (6-Deoxy-D-talose)	Unit of the capsular polysaccharide of gram negative bacteria strain GC
Epifucose (6-Deoxy-L-talose)	Unit of the cell wall polysaccharide of a strain of *Actinomyces bovis* and in the K-antigen of *Pseudomonas pseudomallei* ; component of cardiac glycosides
Digitoxose (2,6-Dideoxy-D-allose)	Component of the *Digitalis* glycosides
Boivinose (2,6-Dideoxy-D-xylohexose)	Component of *Strophantus* glycosides
Streptose (5-Deoxy-3-*C*-formyl-L-lyxose)	Component of the antibiotic Streptomycin
Cymarose (Digitoxose-3-*O*-methyl ether 2,6-Dideoxy-3-*O*-methyl-D-allose)	Component of the glycosides of *Strophantus* and other plants

Table 2. Occurrence of Some Monoketoses in Biological Systems

Name of the Monoketose	Occurrence
Pentuloses	
D-Ribulose (D-Erythropentulose)	As the free compound in seaweed, leaves of sugar beet and leaves of barley germ; the phosphorylated compound is an early product of photosynthesis
Hexuloses	
D-Fructose	As the free compound in many fruits, plants and in honey; unit of many oligosaccharides such as sucrose and polysaccharides such as inulin
D-Psicose	As the free compound in the residue of fermented molasses
D-Tagatose	Unit of the gum exudate of *Sterculia setigera*
L-Sorbose	Unit of pectin from the skin of the fruit of *Passiflora edulis* (passion fruit)
Heptuloses	
Sedoheptulose (D-Altroheptulose)	Free compound as an intermediate of photosynthesis in all green plants; preparation from Sedum leaves
Volemose (D-Mannoheptulose)	Free compound in avocado
Octuloses	
D-Glycero-D-mannooctulose	Free compound in Californian avocado
Nonuloses	
D-Erythro-L-glucononulose	Free compound in avocado

Table 3. Occurrence of Monosaccharide Derivatives in Biological Systems

Name of the Monosaccharide Derivative	Occurrence
Uronic Acids	
D-Glucuronic acid	Unit of the polysaccharides of connective tissue, of heparin, of hyaluronic acid, of plant polysaccharides such as gum arabic and wood hemicelluloses: Compound of glycosides, especially of the glucuronides in urine
D-Galacturonic acid	Unit of plant polysaccharides such as pectin, plant gums such as tragacanth or karya gum
D-Mannuronic acid	Unit of the polysaccharide Algin
L-Guluronic acid	Unit of the polysaccharide Algin
L-Iduronic acid	Unit of the mammalian polysaccharides chondroitin sulfuric acid and heparin
Sugar Alcohols	
Pentitols	
Adonitol, Ribitol	Free compound in *Adonis vernalis*; in the roots of *Bapleurum sp.*; component of vitamin B_2 (lactoflavin)
D-Arabitol, D-Lyxitol	Free compound in mushrooms such as *Boletus bovinus* and in some lichens (e.g., *Ramalina scopulorum*)
Hexitols	
D-Glucitol, D-Sorbitol	Free compound in several fruits such as apples, pears, plums and rowan berries
D-Mannitol	Free compound in several mushrooms such as chanterelle and in marine algae such as in all brown algae
Galactitol, Dulcitol	Free compound widespread in plants; greater amounts in Torula yeasts and in Madagascar manna
Heptitols	
Volemitol (D-Glycero-D-taloheptitol)	Free compound in several plants and lichens and in the mushroom *Lactarius volemus*
Perseitol (L-Glycero-D-mannoheptitol)	Free compound in seeds of avocados
Cyclitols	
D-Inositol	Free compound in the wood of Californian sugar pine
L-Inositol	Free compound in the latex of *Hevea brasiliensis*
Quebrachitol (2-*O*-Methyl-L-inositol)	Free compound in the latex of *Hevea brasiliensis* and Quebracho barks
myo-Inositol (*meso*-Inositol)	As the free compound it is widespread in plants, animals and microorganisms; in plants mostly as the ester of phosphoric acid (phytic acid)
Scyllitol, *scyllo*-Inositol, Cocositol	Free compound in several plants such as acorns and in fish such as shark and ray

Table 3.   (Continued)

Name of the Monosaccharide Derivative	Occurrence
Mytilitol, *C*-Methyl-*scyllo*-inositol	Free compound in sphincter of mussels
D-Quercitol (1-Deoxy-*muco*-inositol)	Free compound in the bark of oak trees and in acorns
L-Viburnitol, *vibo*-Quercitol (1-Deoxy-*myo*-inositol)	Free compound in the leaves of spurge and in *Viburnum Sinus L.*
Conduritol	Free compound in the bark of the condurango tree
Amino Sugars	
D-Glucosamine, Chitosamine (2-Amino-2-deoxy-D-glucose) and N-Acetyl-D-glucosamine (2-Acetamido-2-deoxy-D-glucose)	As the *N*-acetyl derivative unit of several oligo- and polysaccharides such as milk oligosaccharides, glycoproteins, chitin and several glycosaminoglycans
D-Galactosamine (2-Amino-2-deoxy-D-galactose) and N-Acetyl-D-galactosamine (2-Acetamido-2-deoxy-D-galactose)	The acetylated compound is a widespread unit of animal glycosaminoglycans in connective tissues (chondroitin sulfate, dermatan sulfate); component of many glycoproteins and of bacterial immunopolysaccharides
D-Mannosamine (2-Amino-2-deoxy-D-mannose)	Degradation product of sialic acids
D-Gulosamine 2-Amino-2-deoxy-D-gulose	Cleavage product of the Streptomyces antibiotics Streptothricin and Streptolin B
D-Fucosamine (2-Amino-2-deoxy-D-fucose)	Unit of the lipopolysaccharides of some strains of *Chromobacterium violaceum*
L-Fucosamine (2-Amino-2-deoxy-L-fucose)	Unit of the capsular polysaccharides of *Pneumococcus* Type V
3-Amino-3-deoxy-β-D-ribose	Component of the antibiotic Puromycin
Kanosamine (3-Amino-3-deoxy-D-glucose)	Component of the antibiotic Kanamycin
Mycosamine (3-Amino-3,6-dideoxy-D-mannose)	Component of the antibiotics Nystatin, Amphotericin and Pimaricin (Natamycin)
Mycaminose (3,6-Dideoxy-3-dimethylamino-D-glucose)	Component of the antibiotic Carbomycin
D-Fructosamine (1-Deoxy-1-Amino-D-fructose)	As the free compound in human and mammalian blood serum
Amino Sugar Carbonic Acids	
D-Glucosaminic acid (2-Amino-2-deoxy-D-gluconic acid)	Metabolism product of D-glucosamine with *Pseudomonas fluorescence*
D-Galactosamineuronic acid (2-Amino-2-deoxy-D-galacturonic acid)	As the *N*-acetyl compounds: Component of the Vi-Antigen of *Escherichia coli*, etc.
Muramic acid [2-Amino-2-deoxy-3-*O*-(α-D-carboxyethyl)-D-glucose]	Component of cell walls and spores of many bacteria

Table 3.   (Continued)

Name of the Monosaccharide Derivative	Occurrence
Sialic acids: Mono- and diacylneuraminic acids Neuraminic acid: (D-glycero-D-galacto-5-amino-3,5-dideoxy-2-ketononic acid)	Widespread component of gangliosides, lipopoly-saccharides, milk oligosaccharides and glycoproteins in human and animal organisms
Amino Cyclitols	
Streptamine (1,3-Diamino-1,3-dideoxy-*scyllo*-inositol)	Degradation product of Streptidin a component of the antibiotic Streptomycin
neo-Inosamine	Component of the antibiotic Hygromycin
2-Deoxystreptamine (1,3-Diamino-1,2,3-trideoxy-*scyllo*-inositol)	Component of the antibiotics Kanamycin and Neomycin

Oligosaccharides

The hemiacetal group at the C-1 position (aldose) or at the C-2 position (ketose) of monosaccharides has a slightly acidic character and is able to form ether-like compounds called "glycosides" with other hydroxyl-bearing compounds, such as aliphatic or alicyclic alcohols or phenols, with elimination of water. Acids catalyze the reaction (see Figure 8).

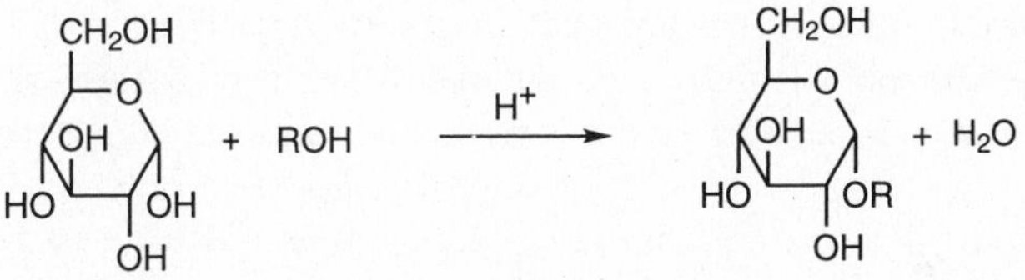

Figure 8.   Formation of a Glycoside by Reaction of a Monosaccharide (Example: α-D-Glucose) with an Alcohol or Phenol

In the same manner, the hydroxyl group at C-1 can react with primary, secondary and also with glycosidic hydroxyl groups of other monosaccharide molecules with elimination of water to give dimeric or oligomeric compounds connected by such glycosidic linkages (or external ether bridges). If the number of individual monosaccharide units is only a few (up to 10 monosaccharides) such compounds are named as "oligosaccharides". The position of the glycosidic hydroxyl group in the molecule itself (α or β; at C-1 or C-2) and the position of the glycosidic bond to the carbon of the neighboring monosaccharide determine the chemical structure of the oligosaccharide. If the glycosidic linkage is established between one hemiacetal carbon atom and one of the non-hemiacetal carbon atoms of the neighboring monosaccharide, a reducing oligosaccharide is formed which is denoted as a "glycosylglycose". If the glycosidic linkage is located between the two hemiacetal carbons of two monosaccharides, nonreducing sugars are formed named "glyosylglycoside". The examples given in Figure 9 show maltose for the first and sucrose for the second case of glycosidic linkage.

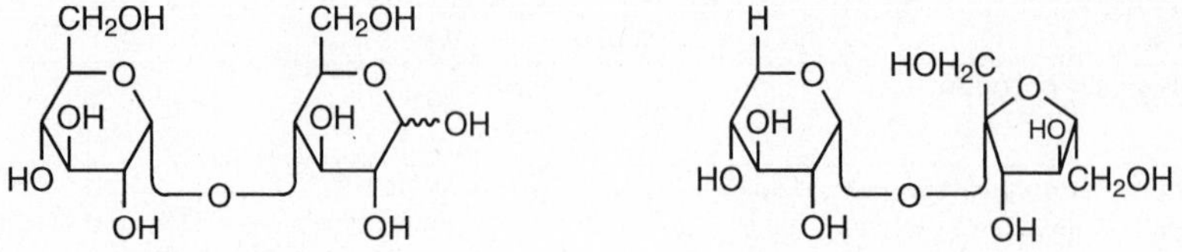

O-α-D-Glucopyranosyl-
(1 → 4)-D-glucopyranose = Maltose

O-β-D-Fructofuranosyl-
α-D-glucopyranoside = Sucrose

Figure 9.   Examples of Glycosylglycose and Glycosylglycoside Structures of Oligosaccharides

The structure of an oligosaccharide is abbreviated by using three-letter symbols for the individual monosaccharide, the letter α and β for the respective anomeric form and the suffix *f* for the furanose or *p* for the pyranose configuration. The positions of the glycosidic linkages are described by numbers in parentheses. For example:

Maltose: *O*-α-D-glucopyranosyl-(1→4)-D-glucopyranose = *O*-α-D-Glc*p*-(1→4)-D-Glc*p*.

Sucrose: *O*-β-D-fructofuranosyl-(2→1)-α-D-glucopyranoside = *O*-β-D-Fru*f*-(2→1)-α-D-Glc*p*.

Branched oligosaccharides often occur as hydrolysis products of polysaccharides and are abbreviated as follows:

O-α-D-glucopyranosyl-(1→4)-*O*-[α-D-glucopyranosyl-(1→6)]-D-glucopyranose
= *O*-α-D-Glc*p*-(1→4)-D-Glc*p*-(1→6)-*O*-α-Glc*p*

Oligosaccharides occur in living systems as free compounds as well as bonded to many other plant constituents, such as phenols, steroids, saponins and flavor components, as their corresponding glycosides. A number of such oligosaccharides occurring in plants and in microorganisms are listed in Table 4.

The most important naturally occurring compounds of this group, which are used preferentially in food stuffs, are sucrose and lactose. Sucrose is the constituent of many roots (sugar beet, carrots) and fruits (plums, apricots) and the pure compound is the most used sweetening ingredient of households in the world. The lactose *O*-β-D-Gal*p*-(1→4)-D-Glc*p* is the main carbohydrate constituent of milk and its occurrence in food products always indicates the use of milk or milk products in their preparation.

Figure 10 shows the whole stereochemical structure of sucrose which is one of the major pure chemicals produced in the world.

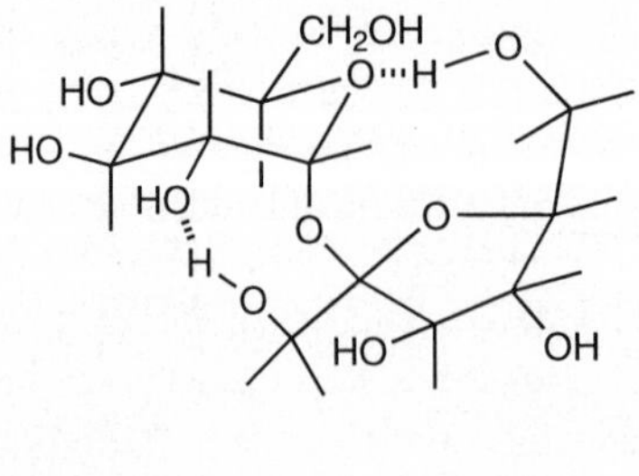

Figure 10.   Conformational Structure of Sucrose

Table 4.   Structure and Occurrence of Important Oligosaccharides in Biological Systems

Name	Structure	Occurrence
Disaccharides		
Cellobiose	*O*-β-D-Glc*p*-(1→4)-D-Glc*p*	Unit of cellulose
Gentiobiose	*O*-β-D-Glc*p*-(1→6)-D-Glc*p*	Sugar component of glycosides such as amygdalin and crocin
Isomaltose	*O*-α-D-Glc*p*-(1→6)-D-Glc*p*	Unit of amylopectin; hydrolysis product of starch
Maltose	*O*-α-D-Glc*p*-(1→4)-D-Glc*p*	As the free compound in malt, beer and, in small amounts, of some vegetables and fruits and in honey; main unit of starch
Nigerose	*O*-α-D-Glc*p*-(1→3)-D-Glc*p*	As the free compound in honey and beer; unit of the polysaccharide nigeran
Laminaribiose	*O*-β-D-Glc*p*-(1→3)-D-Glc*p*	As the free compound in honey; unit of the polysaccharide laminaran and of the glucan of yeast
Sophorose	*O*-β-D-Glc*p*-(1→2)-D-Glc*p*	Sugar component of glycosides of *Sophora Japonica L* and of the glycoside of Stevioside
Kojibiose	*O*-α-D-Glc*p*-(1→2)-D-Glc*p*	Free compound in honey
Trehalose	*O*-α-D-Glc*p*-(1→1)α-D-Glc*p*	Free compound in mushrooms and in the blood of several insects and of grasshoppers
Neotrehalose	*O*-α-D-Glc*p*-(1→1)-β-D-Glc*p*	Free compound in the extract of Koji
Sucrose	*O*-β-D-Fru*f*-(2→1)α-D-Glc*p*	Free compound in sugar cane, sugar beet and widespread in plants, especially in fruits
Turanose	*O*-α-D-Glc*p*-(1→3)-D-Fru*p(f)*	Free compound in honey
Maltulose	*O*-α-D-Glc*p*-(1→4)-D-Fru*f*	Conversion product of maltose; free compound in malt, beer and honey
Palatinose	*O*-α-D-Glc*p*-(1→6)-D-Fru*f*	Free compound in the microbiological conversion product of sucrose
Lactose	*O*-β-D-Gal*p*-(1→4)-D-Glc*p*	Free compound in milk and milk products
Lactulose	*O*-β-D-Gal*p*-(1→4)-D-Fru*f*	Conversion product of lactose; in sterilized milk
Rutinose	*O*-α-L-Rha*p*-(1→6)-D-Glc*p*	Sugar component of glycosides (e.g., hesperidin)
Neohesperidose	*O*-α-L-Rha-(1→2)-D-Glc*p*	Sugar component of glycosides (e.g., naringin, neohesperidin)
Melibiose	*O*-α-D-Gal-(1→6)-D-Glc*p*	Degradation product of raffinose; as the free compound in cocoa beans
Scillabiose	*O*-β-D-Glc*p*-(1→4)-L-Rha*p*	Sugar component of the glycosides of *Scilla maritima*
Robinobiose	*O*-α-L-Rha*p*-(1→6)-D-Gal*p*	Sugar component of robinin
Digilanidobiose	*O*-β-D-Glc*p*-(1→4)-D-Digitoxose	Sugar component of cardiac glycosides (Lantanoside, purpurea glycoside)
Strophanthobiose	*O*-β-D-Glc*p*-(1→4)-D-Cymarose	Sugar component of K-strophanthin
Mannobiose	*O*-β-D-Man*p*-(1→4)-D-Man*p*	Unit of the guaran polysaccharide

Table 4.   (Continued)

Name	Structure	Occurrence
Lycobiose	*O*-4-β-D-Glc*p*-(1→4)-D-Gal*p*	Sugar component of the glycoside tomatin
Primverose	*O*-β-D-Xyl*p*-(1→6)-D-Glc*p*	As the free compound in carob fruits; sugar component of the glycoside of *Primula officinalis*
Vicianose	*O*-α-L-Arab*p*-(1→6)-D-Glc*p*	Sugar component of the glycoside vicianin, gein and violutoside
Solabiose	*O*-β-D-Glc*p*-(1→3)-β-D-Gal*p*	Sugar component of the glycoside β-solanin
Trisaccharides		
Maltotriose	*O*-α-D-Glc*p*-(1→4)-*O*-α-D-Glc*p*-(1→4)-D-Glc*p*	As the free compound in the degradation product of amylose (starch)
Panose	*O*-α-D-Glc*p*-(1→6)-*O*-α-D-Glc*p*-(1→4)-D-Glc*p*	As the free compound in honey, in the degradation product of amylopectin
Isopanose	*O*-α-D-Glc*p*-(1→4)-*O*-α-D-Glc*p*-(1→6)-D-Glc*p*	As the free compound in honey
Isomaltotriose Dextrantriose	*O*-α-D-Glc*p*-(1→6)-*O*-α-D-Glc*p*-(1→6)-D-Glc*p*	As the free compound in honey
Centose	*O*-α-D-Glc*p*-(1→4)-*O*-α-D-Glc*p*-(1→2)-D-Glc*p*	As the free compound in honey
1-Kestose Isokestose	*O*-α-D-Glc*p*-(1→2)-β-*O*-D-Fru*f*-(1→2)-β-D-Fru*f*	As free compounds in honey and some fruits; reaction of sucrose with yeast saccharase
6-Kestose	*O*-α-D-Glc*p*-(1→2)-β-*O*-D-Fru*f*-(6→2)-β-D-Fru*f*	Free compound in honey
Melezitose	*O*-α-D-Glc*p*-(1→3)-*O*-β-D-Fru*f*-(2→1)-α-D-Glc*p*	Free compound in the nectar of many plants, in excudates of lime and pine; in manna and honey
Raffinose	α-*O*-D-Gal*p*-(1→6)-*O*-α-D-Glc*p*-(1→2)-*O*-β-D-Fru*f*	As the free compound widely distributed in plants, particularly in leguminose seeds such as soja beans, mung beans, etc.
Manninotriose	*O*-α-D-Gal*p*-(1→6)-*O*-α-D-Gal*p*-(1→6)-D-Glc*p*	As the free compound in Manna; degradation product of stachyose
Erlose	*O*-α-D-Glc*p*-(1→4)-α-D-Glc*p*-(1→2)-*O*-β-Fru*f*	As the free compound in honey
Umbilliferose	*O*-α-D-Gal*p*-(1→2)-*O*-α-D-Glc*p*-(1→2)-β-D-Fru*f*	As the free compound in the roots of Umbilliferae
Gentianose	*O*-β-D-Glc*p*-(1→6)-*O*-α-D-Glc*p*-(1→2)-β-D-Fru*f*	As the free compound in the rhizomes of Gentiana varieties
Planteose	*O*-β-D-Gal*p*-(1→6)-*O*-β-D-Fru*f*-(2-1)-α-D-Glc*p*	As the free compound in seeds of Plantago varieties
Lycotriose I	*O*-D-β-Glc*p*-(1→2)-β-D-Glc*p*-(1→4)-α-D-Gal*p*	Sugar component of the glycoside tomatin of wild tomatoes
Solatriose	*O*-β-D-Glc*p*-(1→3)-*O*-D-β-Gal*p*-(1→2)-α-L-Rha*p*	Sugar component of the glycoside solanin

Table 4. (Continued)

Name	Structure	Occurrence
Strophanthotriose	*O*-β-D-Glc*p*-(1→6)-D-β-Glc*p*-(1→4)-D-Cymarose	Sugar component of h-strophanthoside
Tetrasaccharides		
Maltotetraose	*O*-α-D-Glc*p*-(1→4)-*O*-α-D-Glc*p*-(1→4)-*O*-α-D-Glc*p*-(1→4)-D-Glc*p*	As the free compound in starch hydrolysates
Stachyose	*O*-α-D-Gal*p*-(1→6)-*O*-α-D-Gal*p*-(1→6)-*O*-α-D-Glc*p*-(1→2)-β-D-Fru*f*	As the free compound in seeds of pulses such as soya beans; in rhizomes of the Japanese artichoke
Lychnose	*O*-α-D-Gal*p*-(1→6)-*O*-α-D-Glc*p*-(1→2)-*O*-β-D-Fru*f*-(1→1)-α-D-Gal*p*	As the free compound in rhizomes of *Lychnis dioica*
Pentasaccharides		
Verbascose	*O*-α-D-Gal*p*-(1→6)-*O*-α-D-Gal*p*-(1→6)-*O*-α-D-Gal-*O*-α-D-(1→6)-Glc*p*-(1→2)-β-D-Fru*f*	As the free compound in several seeds of pulses such as cow peas, winged beans and lima beans; in the rhizomes of the wool plant
Hexasaccharides		
Ajugose	[*O*-α-D-Gal*p*-(1→6)]₄-*O*-α-D-Glc*p*-(1→2)-β-D-Fru*f*	As the free compound in the roots of *Verbasum thapsiforme Schrad* and *Ajuga nipponensis*

Polysaccharides

An enhancement in the number of units of oligosaccharides over 10 gradually causes a change in the physical and chemical properties of these compounds which are then designated as polysaccharides. Their molecular weights are between 10 000 and $1–10 \times 10^6$ Daltons. This group of compounds is present in nearly all living systems. They are the main constituents of plant cells such as cellulose, pectin and starch but also occur in fauna such as the chitin of the shells of insects and crustaceans, and glycogen in the liver of mammals. From a chemical point of view, homopolysaccharides containing only one monosaccharide unit, such as D-glucose, D-mannose, D-fructose, D-galactose, etc., can be distinguished from heteropolysaccharides which contain more than one monosaccharide unit.

Polysaccharides can occur as long unbranched chains like cellulose and agar or as branched chains as in most plant exudates such as tragacanth and gum arabic. Their structures influence to a high degree their solubility in water and the viscosities of their solutions. Table 5 gives a short overview of several important polysaccharides in plants, mammals and microorganisms.

Table 5.   Occurrence of Important Polysaccharides in Biological Systems

Name	Constituent and Structure	Occurrence
Homopolysaccharides		
Cellulose	Unit: D-Glucose; β-1,4-glucan; linear chain	Cell wall constituent of nearly all plants
Starch	Mixture of Amylose and Amylopectin	Grains of cereals (wheat, rye), in tubers (potato) and some roots (maniok)
Amylose	Unit: D-Glucose; α-1,4-glucan; linear chain	
Amylopectin	Unit: D-Glucose; α-1,4 resp.; 1,6-glucan; branched	
Glycogen	Unit: D-Glucose; α-1,4; α-1,6-glucan; branched molecule	In animal cells, particularly in liver cells
Lichenin	Unit: D-Glucose; β-1,3; β-1,4-glucan; linear chain	Extract of Islandic moose; in oats
Pustulan	Unit: D-Glucose; β-1,6-glucan; linear chain	In lichens such as *Umbilicaria*
Laminaran	Unit: D-Glucose; β-1,3-glucan; linear chain; some -1,6- linkages	In seaweeds such as *Laminaria digitata*
Yeast glucan	Unit: D-Glucose; β-1,3; β-1,6-glucan; linear chain	Cell membranes of yeast
Nigeran	Unit: D-Glucose; α-1,3; α-1,4 alternating bonds; linear chain	Fermentation product of *Aspergillus niger*
Curdlan	Unit: D-Glucose; β-1,3-glucan; linear chain	Fermentation product of *Alcaligenes faecalis*
Scleroglucan	Unit: D-Glucose; β-1,3-glucan; linear chain, some β-1,6-side molecules	Fermentation product of *Sclerotium glucanicum*
Pullulan	Unit: D-Glucose; α-1,4; α-1,6-glucan; linear chain	Fermentation product of *Aureobasidium pullulans*
Dextran	Unit: D-Glucose; α-1,6-glucan; linear chain; some α-1,3-, α-1,4- and α-1,2-molecules	Fermentation product from *Leuconostoc mesenteroides*
Chitin	Unit: N-Acetylglucosamin; β-1,4-glycan; linear chain	Shells of insects, shellfish and fungi
Inulin	Unit: D-Fructose; some D-Glucose units; β-2,1-fructan linear chain	Component of salsify, Jerusalem artichoke, roots of chicory
Levan	Unit: D-Fructose; β-2,6-fructosan; linear chain	Component of several grass varieties: Fermentation product of *Bac. mesentericus* and *Aerobacter levanicum* with sucrose
Stone nut mannan	Unit: D-Mannose; β-1,4-mannan; linear chain	Shells of stone nuts
Yeast mannan (Yeast gum)	Unit: D-Mannose; α-1,6-mannan; linear chain with α-1,2 branches	Cell of baker's yeast (*Saccharomyces cerevisiae*)
Lupine galactan	Unit: D-Galactose; β-1,4-galactan; linear chain	Seed of white lupins
Snail galactan	Unit: D-Galactose, L-galactose; 1,3; 1,6-galactan; branched	Mucine of *Helix pomatia*

Table 5. (Continued)

Name	Constituent and Structure	Occurrence
Bovine lung galactan	Unit: D-Galactose, L-galactose; 1,3; 1,6-galactan	Component of bovine lung tissue
Esparto grass xylan	Unit: D-Xylose; β-1,4-chain with some β-1,3 branches	Component of Esparto grass

Heteropolysaccharides

Name	Constituent and Structure	Occurrence
Pectin	Units: D-Galacturonic acid, L-rhamnose; α-1,4 galacturonan chain interupted with β-1,2-L-rhamnose molecules	In nearly all higher plants; constituent of the primary cell wall
Guar	Units: D-Mannose, D-galactose β-1,4-mannan; linear chain; α-1,6-galactose as side molecules	In seeds of the Indian legume *Cyamopsis tetragonolobus*
Locust bean gum Carob gum	Units: D-Mannose, D-galactose β-1,4-mannan; α-1,6-galactose as side molecules	In seeds of the carob tree *Ceratonia siliqua L.*
Larch gum	Units: D-Galactose, L-arabinose; β-1,3-galactan, α-1,6-branches with galactan, araban and arabinogalactan chains	In compressed wood of Laric species
Tamarind gum	Units: D-Glucose, D-galactose, D-xylose, L-arabinose; β-1,4-glucan chain; α-1,6-D-xylose and L-arabinose side molecules	In seeds of the Tamarind tree (*Tamarindus indicans*)
Algin	Units: D-Mannuronic acid, L-guluronic acid; linear chain of β-1,4-D-mannuronic acid and α-1,4-L-guluronic acid	In the seaweed *Laminaria digitata*, *Macrocystis pyrifera*, etc.
Agar	Units: D-Galactose, 3,6-anhydro-L-galactose; linear chain alternating β-1,4-D-galactose and α-1,3-3,6-anhydro-L-galactose	In the seaweed *Gelidium sp.*, *Gracilaria sp.* and *Pterocladia sp.*
Carrageenan	Units: D-Galactose sulfuric acid esters, 3,6-D-anhydrogalactose; linear chain; alternating β-1,4-D-galactose sulfuric acids and α-1,3-3,6-anhydro-D-galactose	In the seaweed of *Condrus sp.*, *Gigartina sp. a. Eucheuma sp.*
Gum arabic	Units: L-Rhamnose, D-galactose, L-arabinose, D-glucuronic acid; strongly branched β-1,3-galactan	Exudate of Acacia trees (Sudan, Senegal)
Tragacanth	Units: L-Rhamnose, L-fucose, D-xylose, L-arabinose, D-galactose, D-galacturonic acid; strongly branched α-1,4-D-galacturonan	Exudate of *Astragalus gummifer* (Turkey, Iran)
Karaya gum	Units: L-Rhamnose, D-galactose, D-galacturonic acid, D-glucuronic acid (partly acetylated); α-1,4-D-galacturonan interrupted by β-1,2-L-rhamnose units; side chains: D-galactose, D-glucuronic acid	Exudate of *Sterculia sp.* (*Sterculic urens*)
Ghatti gum	Units: L-Arabinose, D-galactose, D-mannose, D-xylose, D-glucuronic acid traces of L-rhamnose; consists of more than one β-1,6-galactan chains with several side chains with other units	Exudate of the tree *Anogeissus latifolia* (India, Sri Lanka)

Table 5.   (Continued)

Name	Constituent and Structure	Occurrence
Xanthan	Units: D-Glucose, D-mannose, D-glucuronic acid, pyruvic acid, acetic acid, β-1,4-D-glucan; α-1,3-attached short side chains of D-mannose and D-glucuronic acid; partly esterified with acetic acid and pyruvic acid	Fermentation product of *Xanthomonas campestris*
Gellan	Units: D-Glucose, L-rhamnose, D-glucuronic acid; chains of β-1,4-units of D-glucose and D-glucuronic acids are connected with α-1,3-L-rhamnose	Fermantation product of *Auromonas (Pseudomonas) elodea*
Hyaluronic acid	Units: D-Glucuronic acid, N-acetyl-glucosamine; linear chain of β-1,3-D-glucuronic acid and β-1,4-N-acetylglucosamine	Constituent of connective tissue, eye-glass corpus, fluid of ankles, skin
Chondroitinic acid	Units: D-Glucuronic acid, L-iduronic acid, D-galactosamine; sulfuric acid; linear chains of β-1,3-uronic acids with β-1,4-galactosamine; distinct positions of the sulfuric acid group	Constituent of connective tissue, skin and cartile
Heparin	Units: D-glucuronic acid, L-iduronic acid, glucosamine, N-acetylglucosamine, sulfuric acid; linear chain of β-1,4-D-glucuronic acid connected with α-1,4-L-iduronic acid and α-1,4-N-acetyl-glucosamine and α-1,4-glucosamine; sulfuric acid groups in the 2- and 6-position	Constituent of human and animal muscles, blood, liver lung and kidney; connected to proteins

Physical Properties of Mono-, Oligo- and Polysaccharides

Due to the large number of hydroxyl groups, carbohydrates are strongly hydrophilic. The monosaccharides and oligosaccharides are soluble in water (partly to an extreme degree) and in aprotic polar organic solvents such as pyridine, dimethylformamide, dimethyl sulfoxide and morpholine, partly soluble in lower alcohols like methanol or ethanol, and insoluble in hydrocarbons, chloroalkanes and ethers.

In the case of polysaccharides, their solubilities are predominately dependent, as remarked before, on their chemical structures. Chain molecules like cellulose are totally insoluble in water due to a high number of intra- and intermolecular hydrogen bridges which enable, to a large extent, the formation of crystalline subunits. Polysaccharides with branched structures are partly soluble in water often yielding colloidal solutions with high viscosities like guar or xanthan. Aqueous solutions of some polysaccharides grow stiff to a gel by including water molecules in a three-dimensional network as in the case of pectin or agar.

The monosaccharides and the lower oligosaccharides have distinct melting points. Some of them have two melting points, the first one as the hydrate and the second as the water-free substance (for example, D-galactose monohydrate: mp 118–120 °C; water-free α- or β-D-galactose: mp 167 °C). At the melting point of water-free substances most of the mono- and oligosaccharides undergo partial decomposition. Polysaccharides decompose at high temperature, without melting, yielding a large number of volatile and nonvolatile compounds.

Chemical Reactions

The primary and secondary hydroxyl groups as well as the aldehyde or keto group of carbohydrates are responsible for their chemical reactivities.

The hydroxyl groups react with inorganic and organic acids as well as their reactive derivatives, such as acid chlorides, giving the corresponding esters. Some of them are very important in living systems such as monosaccharide phosphates (e.g., glucose-6-phosphate, glucose-1-phosphate, fructose-1,6-diphosphate). A strong alkali enables the ionization of primary and secondary hydroxyl groups of carbohydrates which react with alkyl halides or sulfates to form the corresponding ethers. This reaction is important for the elucidation of the structures of oligo- and polysaccharides. The free hemiacetal hydroxyl group of the carbohydrate molecule is more acidic and can be transferred into its corresponding ethers the so-called "glycosides", more easily.

Reaction of the *cis*-vicinal hydroxyl group with ketones forms mostly 1,3-dioxalanes under the influence of acids (sulfuric acid, 4-toluenesulfonic acid, Lewis acids) and dehydrating agents since aldehydes prefer 1,3-dioxanes. These compounds are alkali stable, thermally stable and are also not attacked by oxidizing or reducing reagents. The 1,3-dioxalane and 1,3-dioxane groups are important protecting groups in the field of organic syntheses with carbohydrates. The more commonly used ketones and aldehydes are acetone, cyclohexanone and benzaldehyde, and the isopropylidene monosaccharides, obtained by reaction of these sugars with acetone, belong to the most used intermediate products in the synthetic pathways of sugars.

Some inorganic multivalent acids like boric acid, arsenous acid, molybdates or vanadates give complexes with the vicinal hydroxyl groups. In the case of boric acid, the complexation leads to a strong enhancement of the acidity of the system; the sugar molecules become charged and move in an electric field. Periodic acid and lead tetraacetate cleave the C–C linkages of vicinal polyhydroxy compounds yielding carbonyl group bearing fragments.

The carbonyl group represented as an aldehyde or keto group in carbohydrates is responsible for most of their chemical reactivity, including:

(a)   Reductive properties of monosaccharides and oligosaccharides with free glycosidic hydroxyl groups, especially in alkaline medium.

(b)   Nucleophilic reactions with nitrogen compounds containing a free electron pair, such as amines, hydrazines and hydroxylamines giving the corresponding aminoglycosides, hydrazones, osazones and oximes.

The reaction with amines and amino acids, the so-called "Maillard Reaction" leads to a large number of different compounds like furans, pyrroles, pyranones, pyrones and pyrazines, and plays an important role in the field of flavor in foods.

(c)   Treatment with strong acids causes dehydration leading to reactive furan derivatives like ω-hydroxymethylfurfural. These reactions are the base of many photometric methods for the determination of carbohydrates.

(d)   Treatment with strong alkali leading to enolization of carbonyl groups due to the adjacent hydroxyl groups. Beside the strong reductive capacities of these compounds, these intermediates undergo successive reactions like retro-aldol scission and benzilic acid rearrangements.

(e)   Reduction of the reducing sugars with elementary hydrogen under catalysis or with sodium borohydride in weak alkaline medium giving the corresponding sugar alcohols which are very stable compounds.

(f)   Oxidation with halogens such as bromine and iodine as well as derivatives like *N*-bromosuccinimide giving the corresponding aldonic acids. More stronger oxidative reagents like nitric acid additionally attack the primary and secondary hydroxyl groups yielding aldaric acids or keto-aldonic acids.

(g)   Addition of hydrocyanic acid or alkali cyanides to the free carbonyl group leading to nitriles of the two epimeric aldonic acids with one additional carbon atom. Saponification of the nitriles to their free acids and reduction of their lactones in neutral medium with sodium borohydride or sodium amalgam yields the corresponding reducing sugars. This reaction developed by Fischer and Kiliani in 1886 is one of the most used methods for increasing the number of carbons in sugars.

(h)   Enzymatic reactions. A great number of enzymes deal with reactions with carbohydrates involving, amongst others, hydrolysis, phosphorylation, dehydrogenation and oxidation. Some of these reactions are the basis of quantitative determinations of single sugars in a mixture due to their great specificities.

A great number of these reactions are used for the qualitative and quantitative analysis of carbohydrates and are described in detail in the following chapters.

Glycoproteins

These compounds are combinations of proteins with carbohydrates connected by covalent glycosylic linkages. They are classified into two main groups according to whether the glycosylic linkage is attached to the hydroxyl group or the amino group of a protein. The carbohydrate moiety can be either a monosaccharide, oligosaccharide or polysaccharide such as the long-chain highly charged glycosaminoglycans. The latter group of compounds consists of a central core of protein to which the glycosaminoglycan chains are linked; they are designated as proteoglycans in the literature. The linkages between the protein and the carbohydrate parts occur only between a few amino acids and sugars as shown in Table 6.

Table 6.   Commonly Occurring Types of Linkage Between Protein and the Carbohydrate Components in Glycoproteins

Amino Acid	Type of Glycosidic Linkage	Corresponding Monosaccharide
L-Serine	-O-	N-2-Acetylamido-2-deoxy-D-galactose
		N-2-Acetylamido-2-deoxy-D-xylose
		D-Xylose
L-Threonine	-O-	N-2-Acetylamido-2-deoxy-D-galactose
5-Hydroxy-L-lysine	-O-	D-Galactose
L-Asparagine	-NH-	N-2-Acetylamido-2-deoxyglucose

In the carbohydrate part, the individual monosaccharides are connected together by glycosidic linkages. In many cases, they are branched oligosaccharides with L-fucose, D-galactose, *N*-2-acetylamido-2-deoxy-D-galactose or sialic acids as nonreducing terminals (e.g., blood-group specific glycoproteins). Glycoproteins are widespread compounds in animals, plants, microorganisms and viruses and have a number of different functions in these biological systems. One of the most important functions of the carbohydrate part is as the recognition region for the specific cell membrane receptors.

Glycolipids

These physiologically active compounds are also widely distributed in animals, plants and microorganisms. They are components of semipermeable cell membranes in which the carbohydrates are connected by glycosidic linkages to lipid molecules. Cerebrosides and gangliosides are examples of such compounds and are found in the membranes of nerve and brain cells. Their lipid component is the amino alcohol sphingosin, D-*erythro*-2-amino-1,3-dihydroxy-4-*trans*-octadecene, to which the oligosaccharide part is connected by an oxygen glycosidic linkage. The oligosaccharides contain mainly D-galactose, *N*-2-acetylamido-2-deoxy-D-galactose, D-glucose and sialic acids. The amino groups are substituted by fatty acid residues, as shown for one example in Figure 11.

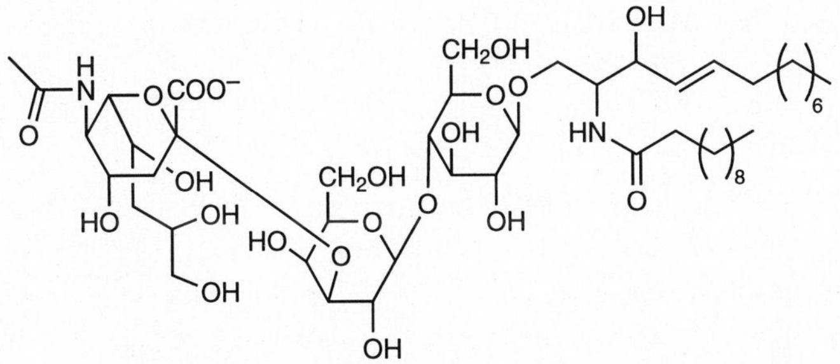

Figure 11. Structure of a Gangloiside of Milk (Belitz, Grosch)

General References

The Carbohydrates, Vol. IA, IB, IIA, IIB; Wander, J.; Pigman, W.; Horton, D.; Herp, A., Eds.; Academic: New York, 1970–1980.

Belitz, H.-D.; Grosch, W. In *Lehrbuch der Lebensmittelchemie*; Springer: Berlin, 1992.

The Polysaccharides, Vol. 1–3; Aspinall, G., Ed.; Academic: New York, 1983.

Buddecke, E. In *Grundriß der Biochemie*; Walter de Gruyter: Berlin, 1980.

Rauen, H. M. In *Biochemisches Taschenbuch*; Springer: Berlin, 1964.

Lehman, J. In *Chemie der Kohlenhydrate*; Thieme: Stuttgart, 1976.

Stephan, A. M.; Merrifield, E. H. In *Encyclopedia of Analytical Science*, Vol. 1; Townshend, A., Ed.; Academic: London, 1995; p 451.

1. Analytical Methods without Separation

1.1. Reduction Methods

Treatment with aqueous alkali converts the cyclic molecules of the monosaccharides into the open-chain configuration by ionizing the hydroxyl group at C-1. Under the influence of the vicinal hydroxyl groups, these aldehydes are converted into the so-called "enediol compounds"[1-3] (see Figure 1).These enediols are strong reducing compounds and react with various reagents to form precipitates or intense colors. In the case of oligosaccharides, it depends on the position of the external ether bridges as to whether the sugar is reducing or nonreducing. When one of the hydroxyl groups at the C-1 position is free, such enediol groups can be formed in an alkaline medium and the same reactions can be applied as for monosaccharides. Nonreducing sugars with no free hydroxyl group at the C-1 atom must be hydrolyzed by acids or enzymes before applying any reactions for reducing sugars.

1.1.1. Reaction with Cu(II) Ions

Complexes of Cu(II) ions are converted into Cu(I) ions by heating with reducing sugars in an aqueous alkaline medium. The deep blue solutions turn into a reddish brown color due to the precipitation of Cu_2O. This formal "Fehling reaction" is one of the oldest identification reactions for sugars and is applied in practice even today due to its simplicity and cheapness. Some of the procedures given in collections of official methods are based on this reaction (see General References).

The pathways of this chemical reaction are complex and consist of two main routes which occur simultaneously

(a)   Oxidation of the enediol compounds by Cu(II) ions.

(b)   Occurrence of the so-called "retro-aldol scission" and oxidation of the enediols of the lower membered hydroxy aldehydes also by Cu(II) ions.

The reaction was studied in detail for D-glucose[4] and Table 1 summarizes the yields of the main products obtained. The reaction scheme shown in Figure 1, which is based on identified reaction products[4], can be assumed.

The first step is the ionization of the hydroxyl group at C-1 which occurs because the hydroxyl group at this position has a stronger acidic character than the other hydroxyl groups of D-glucose[1,2]. Following this, a ring opening of the cyclic hexose molecule occurs after which two reaction pathways are possible. In the case of Path A, the enediol compounds are formed via an intermediate pseudocyclic carbanion. This scheme is generally accepted today from spectroscopic and chromatographic evidence[3,5]. The enediols are strongly reducing compounds and are oxidized very quickly by the Cu(II) complexes. When neighboring carbon atoms bear hydroxyl groups, the enediol group is cleaved into two carboxyl groups. If as in the case of D-glucose, the enediol group is located between C-1 and C-2, the C_5-aldonic acids (arabonic acid, ribonic acid) and formic acid are the oxidation products. C_4-Aldonic acids (erythronic acid, threonic acid) are also found, probably due to isomerization of the enediol group towards the middle of the molecule between C-2 and C-3. Between these two enediols an equilibrium must occur which causes an epimerization of the hydroxyl group at C-3 as equal amounts of arabonic acid and ribonic acid are found among the reaction products.

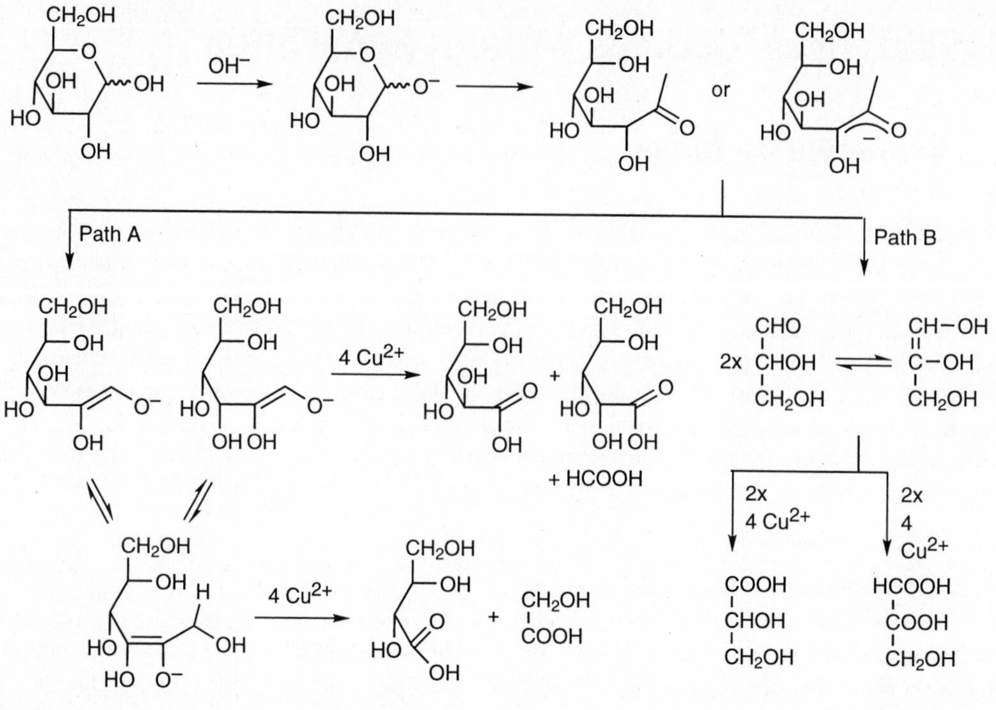

Path A   Path B

Figure 1.   Proposed Scheme for the Reaction between D-Glucose and Cu(II) Ions in an Alkaline Medium (Scherz)

Table 1.   Quantitative Determination of the Compounds Which are Formed by the Reaction of D-(+)-Glucose with Cu(II) Ions in an Alkaline Medium[4]

Compound	Yield[a,b] (mmol)				
	A	B	C	D	E
Glycolic acid	0.22	0.51	0.54	0.54	0.39
Glyceric acid	0.06	0.17	0.19	0.34	0.35
Erythronic acid	0.14	0.30	0.28	0.14	0.17
Threonic acid	0.00	0.03	0.04	0.02	0.02
3-Deoxypentonic acid	0.00	0.08	0.11	0.10	0.10
Ribonic acid	0.05	0.12	0.09	0.07	0.08
Arabonic acid	0.08	0.18	0.14	0.09	0.09
Mannonic acid	-	-	-	0.04	0.09
Gluconic acid	-	-	-	0.01	0.01
Formic acid[c]	-	0.68	-	-	-
Relative Cu(II)–D-(+)-Glucose (in Mol)	3.93	4.26	4.60	5.00	5.00

[a]Reaction conditions: concentration of D-(+)-glucose 1 mmol.
[b]pH values and reaction times: A: pH 9.8, 5 min; B: pH 9.8, 15 min; C: pH 9.8, 60 min; D: pH 12.6, 5 min; E: pH 12.6, 15 min.
[c]The concentration of formic acid could only be determined using the reaction conditions given for B.

The second pathway involving retro-aldol scission[6] leads to the formation of two triose molecules from the hexose molecule and the oxidation products of these enediols are glyceric acid, glycolic acid and formic acid.

For carrying out analyses, solutions of the samples and reagents are mixed and heated for an exact period of time. The Cu_2O precipitate is then quantitatively determined either by gravimetric, titrimetric or photometric methods.

The most important procedures for this type of reaction are described in the following sections.

Table 2. List of the Most Important Procedures for Determination of Reducing Sugars with Cu(II) Complexes in Alkaline Solution

Name	Complexing Compound	Procedure for Production of the Reagent
1. Fehling	Potassium sodium tartrate (Seignette Salt)	$CuSO_4 \cdot 5H_2O$ (69.28 g) is dissolved in distilled water and the solution is made up to 1 L (Fehling solution A); potassium sodium tartrate (173 g) and solid KOH (125 g) are dissolved in distilled water and made up to 500 mL (Fehling solution B)[8]
2. Soxhlet	Potassium sodium tartrate	$CuSO_4$ solution with the same concentration as above; potassium sodium tartrate (173 g) and KOH (51.6 g, AOAC: 50 g) are dissolved in distilled water and made up to 500 mL[7,8]
3. Luff–Schoorl	Sodium citrate	$CuSO_4 \cdot 5H_2O$ (25 g) is dissolved in 100 mL of distilled water (solution I); citric acid (50 g) and $Na_2CO_3 \cdot 10H_2O$ (388 g cryst) is dissolved in 400 mL distilled water (solution II); solution I is added in to solution II and made up with distilled water to 1 L[9,10]
4. Potterat–Eschmann	Ethylenediamine tetraacetate disodium salt (Komplexon III, Titriplex III)	$CuSO_4 \cdot 5H_2O$ (25 g) is dissolved in 100 mL of distilled water (solution I); $Na_2CO_3 \cdot 10H_2O$ (286 g) is dissolved in 500 mL of warm distilled water and Komplexon III (38 g) is added (solution II); solution I is added into solution II and made up to 1 L with distilled water[11]

1.1.1.a. Meissl Procedure resp. Munson–Walker[7]

The reagent involved is a Cu(II)–potassium sodium tartrate complex prepared according to the method of Soxhlet[8] (see Table 2, entry 2).

In a 400-mL beaker, mix 25 mL of each of the $CuSO_4$ and the potassium sodium tartrate solutions. To the mixture, add 50 mL of an aq solution of the sugar containing sample. Then make the mixture up with distilled water to 100 mL and heat on a net of asbestos (original procedure) allowing reflux to be reached within two min. Maintain reflux for 2 min while covering the beaker with a watch glass. Then filter the solution through a Gooch crucible (original procedure) or through a sintered porcelain funnel. The Cu_2O remaining is determined gravimetrically and from this result the quantity of the sugar is estimated empirically.

1.1.1.b. Luff–Schoorl Procedure[9,10]

The reduced alkalinity of the reagent means that an extension of the reflux time (10 min) is required in order to obtain constant results. Nevertheless, this reagent has the following advantages over classical Fehling's mixtures:

(a)　　The solution of the reagent is stable and can be kept for extended periods of time.

(b)   Only polyhydroxy aldehydes and polyhydroxy ketones react with this mixture; unsubstituted aldehydes do not interfere.

(c)   The extended heating time leads to equal reduction equivalents for glucose and fructose.

The main disadvantage of this reagent is its sensitivity towards variation in pH. For the composition of this reagent see Table 2, entry 3.

Pipet 25 mL of the Luff–Schoorl reagent together with a solution of the sugar sample (containing 10–60 mg of the reducing sugar) into a 300-mL Erlmeyer beaker and make up the mixture to 50 mL. After adding boiling chips the solution should reach reflux within 2 min. After refluxing for 10 min, cool the solution and determine the amount of precipitated Cu_2O. It is important to keep to the reflux time exactly.

1.1.1.c. Potterat–Eschmann Procedure[11]

Cu(II) ions are complexed in an alkaline solution by ethylenediaminetetraacetic acid (EDTA); such solutions are stable. For composition of the reagent see Table 2, entry 4

Mix 10 mL of the reagent and 10 mL of the sample in a special filtration flask (see Figure 2) and, after connecting to a reflux condenser, heat the solution to reflux within 2 min and keep at reflux for 10 min exactly. Then, cool the solution thoroughly and filter through a sintered plate. The amount of Cu_2O is determined by complexometric titration.

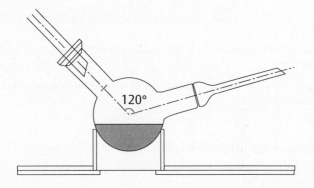

Figure 2.   Filtration Flask used for the Determination of Reducing Sugars (Potterat and Eschmann[11]; with permission)

Methods of Determination of Cu_2O

Gravimetric Procedure

After washing the precipitated Cu_2O with EtOH and Et_2O, it is dried at 100 °C or annealed to give CuO.

Titrimetric Procedures
Procedure A

Reaction of ammonium iron(III) sulfate to Fe^{2+} takes place according to the following scheme. The Fe^{2+} ions formed are then titrated with $KMnO_4$.

$$Cu_2O + 2\,Fe^{3+} + 2\,H^+ \longrightarrow 2\,Cu^{2+} + 2\,Fe^{2+} + H_2O$$

Determination

Dissolve the precipitated Cu_2O in to an acidic solution of ammonium iron(III) sulfate [120 g $(NH_4)Fe(SO_4)_2$ and 100 mL of concd H_2SO_4 are made up with distilled water to 1 L] and titrate with $KMnO_4$ (0.1 M)[12]. Application of the Zimmermann–Reinhard procedure improves recognition of the change of color[13]. In the AOAC method the Cu_2O is dissolved in a neutral aq solution of $(NH_4)Fe(SO_4)_2$ and the sulfuric acid is added shortly before titration using *o*-phenanthroline as the indicator[14].

Procedure B: Iodometric Titrations

Oxidation occurs with iodine according to the following scheme[15]

$$Cu_2O + I_2 + 2\,H^+ \longrightarrow Cu^{2+} + H_2O + 2I^-$$

Determination

Acidify the cooled solution of the Luff–Schoorl reaction with 50 mL of acetic acid (0.4 M) and 25 mL of M KI–I_2 (0.1 M) solution and after shaking intensively, carefully add 55 mL of HCl (0.75 M) and shake the suspension again until the Cu_2O is totally dissolved. Excess iodine is titrated with sodium thiosulfate (0.1 M). Addition of solid $NaHCO_3$ diminishes the influence of air by creating a CO_2 atmosphere.

Another way is performed in following most of the official procedures. In these cases, excess Cu(II) is converted into Cu_2I_2 by addition of KI in an acidic solution and then the formed iodine is titrated with sodium thiosulfate.

Official procedure from § 35 LMBG Germany (determination of sugar in fruit juices)[10]:

After performing the Luff–Schoorl reaction, add 10 mL of an aq solution of KI (30 g/100 mL) and 25 mL of sulfuric acid (w = 25%) to the cooled reaction mixture. After standing for a while, the liberated iodine is titrated with sodium thiosulfate (0.1 M) using soluble starch as the indicator.

A blank is run in the same manner and the difference between both volumes is equivalent to the amount of Cu_2O which corresponds to the amount of sugars. Table 3 shows a correlation between the volumes of the consumed 0.1 M sodium thiosulfate and the corresponding amounts of invert sugar. In the second column the difference between a single mL of 0.1 M sodium thiosulfate and the amounts of the corresponding invert sugars are listed; it shows a derivation from a linear correlation.

The excess Cu(II) ions after the reaction can also be determined by an ion-selective electrode[16].

1.1.1.d. Lane–Eynon Procedure

This is recommended in some cases such as the determination of an invert sugar in industrial sugar products[17,18].

Determination

Pour 20 mL of Fehling's solutions A and B, prepared according to the method of Soxhlet (Table 2, entry 2) into a 500 mL flask and dilute with 15 mL of distilled water; from a burette add a measured volume of the standard solution of invert sugar (deficit) and heat the mixture within 2.5 min to reflux and maintain reflux for exactly 2 min. Then add 2–3 drops of a solution of methylene blue and continue the titration until the blue color disappears. The same procedure is then carried out with the samples to be analyzed and, from the relationship between both volumes, the amount of invert sugar in the samples is calculated. In the presence of sucrose, a correction according to Table 4 is necessary.

Table 3.   Table for the Estimation of the Amount of Reducing Sugars[10]

Consumption of Sodium Thiosulfate (c = 0.1 M) (mL)	Reducing Sugar as Invert Sugar (mg)	Difference (mg)[a]
1	2.4	2.4
2	4.8	2.4
3	7.2	2.5
4	9.7	2.5
5	12.2	2.5
6	14.7	2.5
7	17.2	2.6
8	19.8	2.6
9	22.4	2.6
10	25.0	2.6
11	27.6	2.7
12	30.3	2.7
13	33.0	2.7
14	35.7	2.8
15	38.5	2.8
16	41.3	2.9
17	44.2	2.9
18	47.1	2.9
19	50.0	3.0
20	53.0	3.0
21	56.0	3.0
22	59.1	3.1
23	62.2	3.1

[a]Differences between single volumes of consumed sodium thiosulfate and the amount of invert sugars (in mg).

Table 4.   Correction Factors for the Presence of Sucrose in the Sample

Sucrose[a] (g)	Correction factor (f)	Sucrose[a] (g)	Correction factor (f)
0.0	1.000	5.5	0.910
0.5	0.982	6.0	0.904
1.0	0.971	6.5	0.898
1.5	0.962	7.0	0.893
2.0	0.954	7.5	0.888
2.5	0.946	8.0	0.883
3.0	0.939	8.5	0.878
3.5	0.932	9.0	0.874
4.0	0.926	9.5	0.869
4.5	0.920	10.0	0.864
5.0	0.915		

[a]Amount of sucrose present in the refluxing mixture.

Photometric Procedures

For determination of sugar concentrations with in the microgram range, several methods have been developed based on Cu(II) reduction to Cu_2O. These methods require phosphormolybdic and phosphotungstic acids which are converted into molybdic and tungstic blue, respectively. These methods are described in a separate section (see Section 1.1.4.a.).

1.1.1.e. Special Modifications

Reaction of Cu(II) in Weak Acetic Acid Medium (Barfoed Reaction)[19–21]

In the range of pH 6.0 only monosaccharides and not disaccharides, especially those with 1,4-linkages, react with Cu(II) ions to form of Cu_2O. Under these conditions it is possible to determine monosaccharides such as glucose in the presence of reducing disaccharides such as maltose and lactose.

1.1.2. Reaction with Hexacyanoferrate(III)

In alkaline solution, hexacyanoferrate(III) ions oxidize monosaccharides resulting in the formation of hexacyanoferrate(II) ions. This reaction is the basis of several procedures used for the determination of reducing sugars. In general, a known excess of reagent is reacted with the reducing sugar in the sample and the amount of the remaining reagent is determined by iodometric titration according to the reaction previously described in Section 1.1.1. and the equation below.

$$2 \, [Fe(CN)_6]^{3-} + 2 \, I^- \longrightarrow I_2 + 2 \, [Fe(CN)_6]^{4-}$$

Hexacyanoferrate(II) is removed from the equilibrium by precipitation with Zn ions. The difference between the titrimetric volumes required for the blank and for the sample is equal to the amount of sugar. Several macro- and micromethods are based on this principle, such as the determination of glucose in blood (Hagedorn–Jensen procedure)[22,23]. In the microgram range the excess hexacyanoferrate(III) can be determined by photometry at the wavelength of 420 nm where hexacyanoferrate(II) has no absorption.

Determination
Photometric Procedure[24]
Reagent: 13.2 g of potassium hexacyanoferrate(III) and 50 g of sodium carbonate (water free) are dissolved in distilled water and the solution made up to 1 L. The reagent is stable for several months if stored in the dark.
Mix 5 mL of an aq solution of the sample with 5–15 mL of the reagent and heat the mixture at 80 °C for 30 min, then cool the solutions to 20–25 °C and make up to a set volume with distilled water. Then measure the absorbance at 418–420 nm. A parallel reaction is performed with a blank and the difference between the results obtained from the blank and the sample corresponds to the amount of reducing sugars. The empirical relation is estimated by using defined standards.

1.1.3. Reaction with Iodine

In a weakly alkaline medium aldoses are oxidized to aldonates according to following scheme. Ketoses and nonreducing sugars remain unaffected by this reaction.

$$I_2 + 2\ OH^- \longrightarrow IO^- + I^- + H_2O$$

$$IO^- + RCHO + OH^- \longrightarrow RCOO^- + I^- + H_2O$$

An important criteria for this reaction is the exact observance of the pH value otherwise this reaction extends to other hydroxy compounds. The disadvantage of the original procedure of Willstätter and Schudel[25], where NaOH (0.1 M) was used for pH regulation, has been overcome by using a sodium carbonate/sodium bicarbonate buffer[26].

Determination
Reagent A: mixture of equal volumes of sodium carbonate and sodium bicarbonate (0.2 M).
Reagent B: iodine solution (0.1 M).
Reagent C: sodium thiosulfate (0.1 M).
Mix 25 mL of the sample (10–200 mg of the reducing sugar) with 100 mL of buffer (reagent A) and with an exact known volume of reagent B. After storing for 90–120 min in the dark, acidify the solution with 12 mL of 25% sulfuric acid and titrate the excess iodine with reagent C. In parallel, a blank solution is treated in the same manner and the difference in sodium thiosulfate volumes corresponds to the amount of iodine necessary to oxidize the aldoses (e.g., 1 mL 0.1 M I_2 solution = 9.005 mg D-glucose).
Recent procedures recommend magnesium oxide for the stabilization of the pH[27] and *N*-bromosuccinimide instead of iodine as the oxidation reagent[28].

1.1.4. Reaction with Phosphomolybdic and Phosphotungstic Acid

The following modifications exist for this reaction:
(a)   Formation of Cu_2O by reaction between reducing sugars and alkaline Cu(II)–tartrate complexes followed by reaction of Cu_2O with phosphomolybdic acid to molybdene blue.
(b)   Direct reaction between phosphomolybdic or phosphotungstic acid and reducing sugar.
Both of these modifications result in the formation of intensely colored compounds which allow the sugar to be determined photometrically.

1.1.4.a. Reaction between Reducing Sugars and Alkaline Cu(II)–Tartrate Complexes
The best known procedures for this type of reaction are those of Folin–Wu[29,30] and Nelson–Somogyi[31,32].

Determination
Folin–Wu Procedure
Reagent A: 100 mL of a 2% solution of Na_2CO_3 in NaOH (0.1 M) is mixed with 2 mL of a 0.5% aq solution of $CuSO_4$ in a 1% aq solution of potassium tartrate freshly prepared.

Reagent B: 35 g of MoO_3 and 5 g of sodium tungstate are refluxed in 10% NaOH (200 mL) for 20–40 min and then diluted with distilled water to 350 mL. To the mixture is added 25% phosphoric acid (125 mL) and it is then made up with distilled water to 500 mL.

To 2 mL of the sample (0.1–2.5 mg of reducing sugar) pipet 2 ml of reagent A heat in a boiling water bath for exactly 6 min and then cool with ice water. After waiting for 10 min, add 3 mL reagent B and, after a further 10 min, make up the solution to a set volume with 13% aq phosphoric acid. The absorbance of the intense blue color is measured at 700–750 or 550 nm.

Proteins and other reducing compounds interfere with this reaction. Phosphoric acid can be substituted by a complex of molybdate and arsenic acid. This procedure created by Somogyi and Nelson[31,32] can be used for the determination of sugars in an autoanalyzer[33].

1.1.4.b. Reaction between Reducing Sugar and Phosphomolybdic Acid without Interaction of Cu(II) ions

This method also leads to the formation of molybdene blue. In an acidic solution (H_3PO_4) the reaction detects mono- and oligosaccharides, while in a neutral solution only monosaccharides are active[34,35].

Determination
Reaction in Neutral Solution[35,36]
Reagent A: 7.5% aq solution of ammonium molybdate.
Reagent B: potassium dihydrogen phosphate (0.02 M).
To 1–10 mL of the sample (0.5–10 mg of reducing sugar) pipet 10 mL of reagent A and 5 mL of reagent B and heat the mixture in a boiling water bath for 15 min (ketoses) or 30 min (aldoses). After cooling in ice water and making the mixture up to a set volume (equal to that of the standard) the absorbance is measured at 640–660 nm.

The acid procedure uses 10% phosphomolybdic acid and phosphoric acid (1 M). This reagent is not specific and reacts with many other substances like phenols, uric acid, ascorbic acid and reducing steroids, etc.

1.1.5. Reaction with 3,5-Dinitrosalicylic Acid and Other Aromatic Nitro Compounds

1.1.5.a. Reaction with 3,5-Dinitrosalicylic Acid[37–40]
Heating reducing sugars in an alkaline medium with dinitrosalicylic acid leads to the formation of an intense color with an absorption maxima in the range of 500–550 nm. The reaction is not stoichiometric unless several reaction products are formed. The colored product is believed to be 3-amino-5-nitrosalicylic acid (Figure 3).

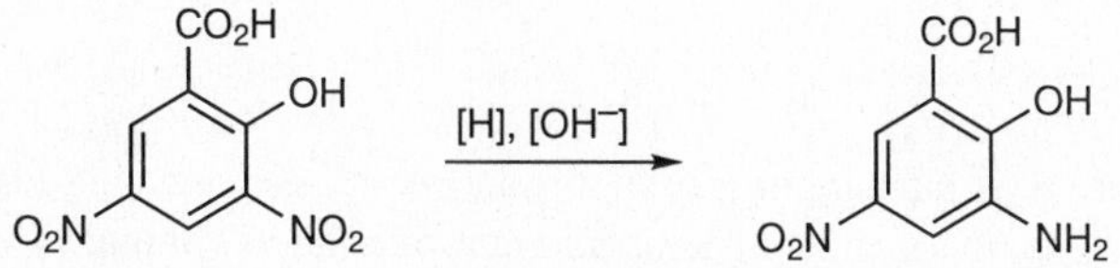

Figure 3. Reaction between Reducing Sugars and 3,5-Dinitrosalicylic Acid (Hofstettler and Borel)

Determination

Reagent A: 1 g of cryst phenol and 10% NaOH (2.2 mL) are mixed together, made up to 10 mL with distilled water and $NaHSO_3$ (1 g) is added; Potassium sodium tartrate (25 g) is then dissolved in 4.5% NaOH (30 mL) and mixed with a 1% aq solution of 3,5-dinitrosalicylic acid (88 mL). Both solutions are mixed together stored for 2 days and then filtered. The reagent is stable for several months if stored in the dark.

Reagent B: 50% aq potassium sodium tartrate.

To 3–4 mL of the sample (0.1–0.6 mg sugar) pipet 3 mL of reagent A, 2 mL NaOH (6 M) and 1 mL of reagent B. Make the mixture up to 10 mL and then heat for 10 min in a boiling water bath. After cooling measure the absorbance at 500 or 543 nm[38].

In a similar way other nitro compounds have been used for the determination of reducing sugars.

1.1.5.b. 3,4-Dinitrobenzoic Acid[41]

On reaction of reducing sugars with 3,4-dinitrobenzoic acid, in an aqueous sodium carbonate solution, a red-violet color forms which is identical to that of 4-hydroxyamino-3-nitrobenzoic acid (λ_{max} = 548 nm) (Figure 4). Further reduction leads to 4-amino-3-nitrobenzoic acid.

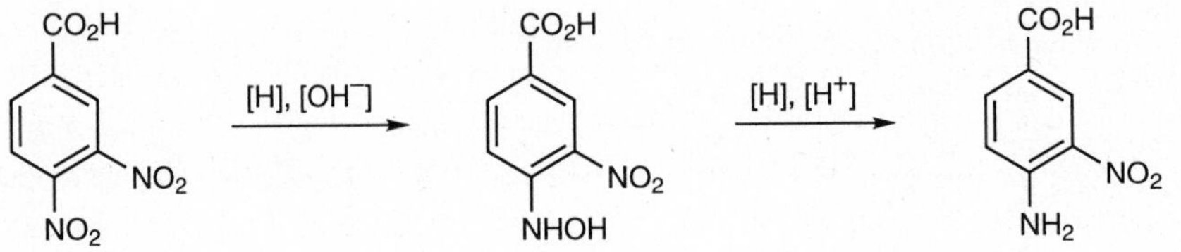

Figure 4.   Reaction between Reducing Sugars and 3,4-Dinitribenzoic Acid (Borel and Deuel)

1.1.5.c. Reaction with 3,6-Dinitrophthalic Acid[42–44]

In a sodium carbonate environment in the presence of thiosulfate this reagent reacts to give the deep red color of 3,3′-azobis(6-nitrophthalic acid) (λ_{max} = 450 nm).

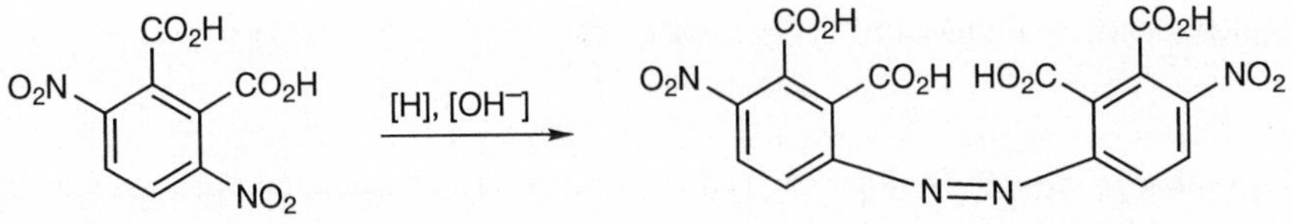

Figure 5.   Reaction between Reducing Sugars and 3,6-Dinitrophthalic Acid (Momose and Inaba)

1.1.5.d. Reaction with Picric Acid[45,46]

Reaction of this reagent with reducing sugars in a sodium carbonate medium leads to a deep red color (λ_{max} = 502 nm). Ketoses react at room temperature while aldoses require heating. Other reducing compounds such as ascorbic acid and substances with active methylene groups interfere with this reaction.

1.1.5.e. Reaction with 1,2-Dinitrobenzene[44,47,48]

Heating 1,2-dinitrobenzene in an alkaline medium with reducing sugars leads to the formation of an unstable deep red violet compound which is supposed to be the acid form of 1-nitroso-2-nitrobenzene (Figure 6). Substances such as uric acid, hydroxylamine and hydrazine compounds interfere with this reaction.

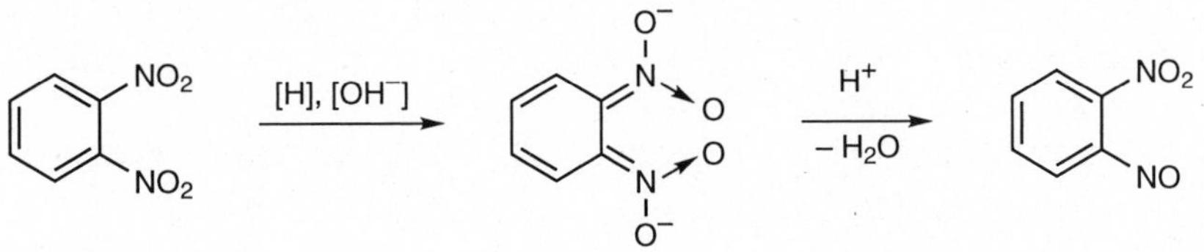

Figure 6.   Reaction between Reducing Sugars and 1,2-Dinitrobenzene (Péronnet and Hugonnet)

1.1.6. Reaction with Tetrazolium Salts

The colorless tetrazolium ions are converted by reducing sugars in an alkaline medium to deeply colored water insoluble formazans[49–51] which can be extracted with organic solvents from the aqueous phase. The reactant with the simplest form from of this group of compounds is triphenyltetrazolium chloride (TTC), which yields the deep red triphenylformazan (see Figure 7).

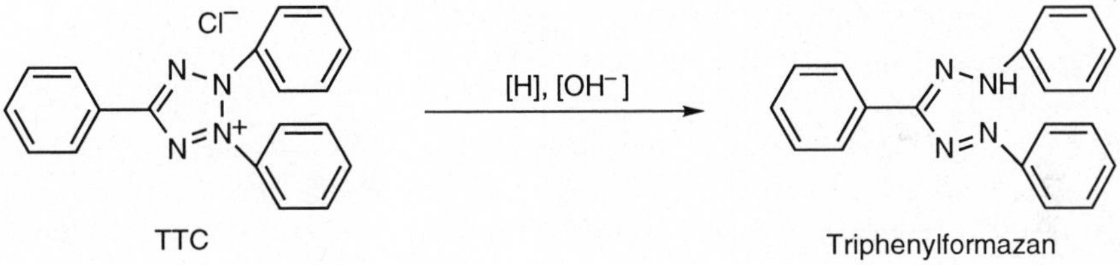

TTC   Triphenylformazan

Figure 7.   Reaction between Reducing Sugars and Triphenyltetrazolium Ions in an Alkaline Medium (Fischer and Dörfel)

Several procedures for the photometric determination of reducing sugars use this compound[49,52]. The reaction has the disadvantage of needing high alkali concentrations and the formazan itself is sensitive towards light therefore, all procedures have to be carried out in the dark.

The best results are obtained with the following tetrazolium compounds (see Figure 8): 2,2'-(4,4'-stilbenylene)bis(3,5-diphenyl-3H-tetrazolium 2-chloride) (STC)[53], 2,2'-diphenyl-5,5'-di(p-anisyl)-3,3'-(3,3'-dimethoxy-4,4'-diphenylene)ditetrazolium dichloride (ATC)[54,55] and 2,5-diphenyl-3-(4-styryl-phenyl)tetrazolium chloride (DSTC)[56].

STC and ATC require lower alkali concentrations and their formazans are deep blue with absorption maxima at 580–615 nm.

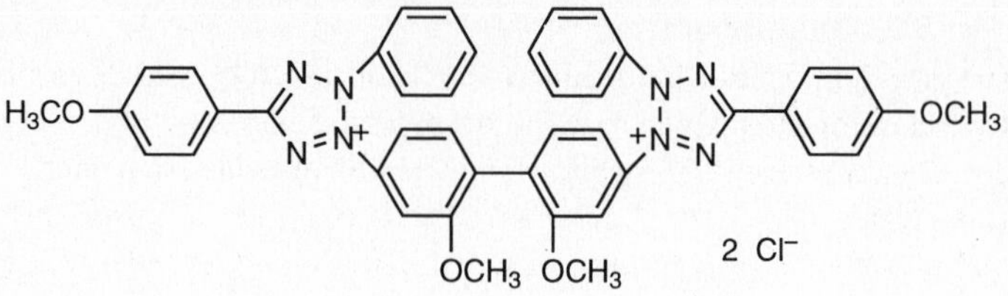

Figure 8.   Chemical Structure of the Tetrazolium Compounds ATC and STC

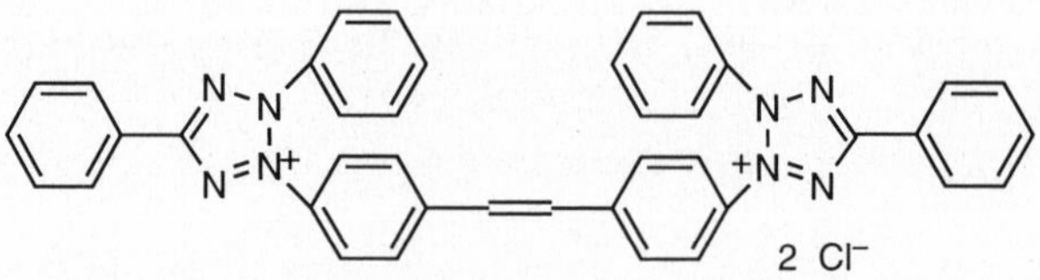

The reaction is only applicable to those sugars that yield enediols on treatment with alkali. The rate of the reaction is higher for ketoses than for aldoses[56] and is strongly dependent on the pH of the medium, for example, for TTC the reaction rate at pH 11.4 is only one fiftieth of that at pH 12.5.

Determination
Reaction with TTC and DSTC[56]
Reagent A: 1% aq solution of TTC or a 0.2% aq solution of DSTC.
Reagent B: methanol or 2-propanol with 10% acetic acid.
Reagent C: NaOH (0.5 M).
To 0.7 mL of the neutral sample in a test tube, pipet reagent A (0.1 mL) and reagent C (0.2 mL; pH 12.4) and keep this mixture at 37 °C for 12 min (ketoses) and 60 min (aldoses). The reaction is then terminated by addition of reagent B (methanol for TTC or 2-propanol for DSTC). The absorbencies are measured at 485 nm (TTC) and 510 nm (DSTC).
The procedures with STC and ATC are similar to those described above, except that they are carried out at higher temperatures and have shorter reaction times; the formazans are dissolved in either dioxane or pyridine.

1.1.7. Reaction with Cu(I) and Sodium 2,2'-Bicinchoninate

Reducing sugars react with the Cu(II)–sodium 2,2'-bicinchoninate system in a weak alkaline medium to give the deep red Cu(I)–bicinchoninate complex. This reaction is also suitable for quantitative sugar determination[57–60].

Determination

Reagent A: 160.9 g of sec. potassium phosphate, 10.4 g of prim. potassium phosphate and 870 mg of sodium 2,2'-bicinchoninate are dissolved in distilled water and made up to 1 L (pH 8.5). Addition of ethylene glycol (333 mL) before making up the solution enhances the sensitivity of the reaction.

Reagent B: 25 g aspartic acid, 33.4 g sodium carbonate (water free) are dissolved in distilled water (500 mL), and 6.7 g of $CuSO_4 \cdot 5H_2O$ (dissolved in 250 mL water) is added. Then this Cu(II) solution is also made up to 1 L.

Reagent C: 1 part reagent B is mixed with 23 parts of reagent A; this solution is stored overnight before use.

Pipet 3 mL of reagent C to 1 mL of the aq sample and shake intensively. Then heat the solution for 10 min in a boiling water bath and, after cooling to room temperature, read the absorbance at 560–562 nm. By calibration with standard solutions, the amounts of reducing sugars in the samples are estimated[58].

The sensitivity of this reaction is comparable to that of the method of Nelson–Somogyi. The intensity of the color is strongly dependent on the pH of the solution and extensive buffering is, therefore, necessary.

This reagent is used in a autoanalyzer and for postcolumn detection after sugar separations by borate ion exchange chromatography[59,60].

1.1.8. Reaction with 4-Hydroxybenzoic Acid Hydrazide

Reducing sugars react with hydrazides of benzoic acids in an alkaline medium to give the bis-benzoylhydrazones of glyoxal and methylglyoxal which have both intense yellow colors[61–66]. 4-Hydroxybenzoic acid was found to be the most suitable compound from this group of substances for the photometric determination of reducing sugars. The reaction is accelerated by the catalytic influence of bismuth ions, enabling the use of lower temperatures and the increasing sensitivity and yield of the final colored compound[64,65].

Determination

Reagent A: bismuth nitrate (1 mol), potassium sodium tartrate (1 mol) and NaOH (3 mol) are suspended in distilled water and heated until the Bi ions are dissolved. The solution in then made up to 1 L.

Reagent B: solution of 4-hydroxybenzoic acid hydrazide (0.05 M) in NaOH (0.5 M).

Reagent C: to the solution of reagent B, reagent A is added until the Bi ion concentration reaches 1–2 mM.

Pipet a small amount of the sample (5–50 µg sugar) to 5 mL of reagent C and heat for 10 min at 70 °C. After cooling read the absorbance at 410 nm[64].

The optimum temperature is 70 °C; a further increase leads to a decrease in the absorbance. The reaction has a good selectivity, however, Ca ions, proteins and the presence of chloroform interfere with this reaction[66]. This reagent has also been used in a sugar autoanalyzer[65] and for postcolumn detection[67].

1.1.9. Reaction with 2-Cyanoacetamide

With this reagent reducing sugars yield compounds with a strong fluorescence and UV absorption enabling fluorimetric, as well as photometric, determinations[68–71].

Determination
Reagent A: 1% aq solution of 2-cyanoacetamide.
Reagent B: borate buffer (0.1.M) pH 9.
Pipet reagent A (1 mL) and reagent B (2 mL) to 1 mL of a neutral aq solution of the sample (0.2–200 μg) and heat the mixture in a boiling water bath for 10 min. Then cool immediately and read the absorbance at 276 nm or the fluorescence at 382 nm with an excitation at 331 nm[71].
The reaction is strongly dependent on the pH of the medium with a maximum occurring at pH 9 where the blank has no absorption, however, the absorption increases with increasing pH. It is therefore necessary to stabilize the pH at 9 by buffering.
The presence of aliphatic and aromatic aldehydes interferes with the determination as they yield strong UV absorbing compounds.
This reaction is suitable for post column photometry or fluorimetry with high precision after HPLC separation of the sugars[71].

References

1   Wilt, H. G.; Kuster, B. M. F. *Carbohydr. Res.* **1971**, *19*, 5.
2   Degani, Ch. *Carbohydr. Res.* **1971**, *18*, 328.
3   Isbell, H.; Frush, S.; Wade, C.; Hunter, C. *Carbohydr. Res.* **1960**, *9*, 163.
4   Scherz, H. *Z. Lebensmittel. Unters. Forsch.* **1979**, *168*, 91.
5   Weygand, F. *Ark. Kemi* **1950/51**, *3*, 34.
6   Feather, M.; Harris, F. *Adv. Carbohydr. Chem.* **1973**, *28*, 161.
7   *Official Methods of Analysis of the Association of the Official Analytical Chemists (AOAC)* 14ᵗʰ ed. Nr. 31.037; AOAC: USA, 1984; p 580.
8   Acker, L. In *Handbuch der Lebensmittelchemie*; Schormüller, J., Ed.; Bd. II/2, *Analytik der Lebensmittel*; Acker, L. *Kohlenhydrate*; Springer: Berlin, 1967; p 338.
9   Schoorl, N. *Z. Unters. Lebensmittel* **1929**, *57*, 566.
10   Amtliche Sammlung von Untersuchungsverfahren nach § 35 LMBG L 31.00-11 Beuth: Hamburg, 1984.
11   Potterat, M.; Eschmann, H. *Mitt. Lebensmitteluntersuch. Hyg.* **1954**, *45*, 312.
12   Sichert, K.; Bleyer, B. *Z. Anal. Chem.* **1936**, *107*, 328.
13   Schormüller, J.; Andräss, W.; Lange, H. J.; Müller, K. H. *Z. Lebensm. Unters. Forsch.* **1962**, *117*, 379.
14   *Official Methods of Analysis of the Association of the Official Analytical Chemists (AOAC)*, 14ᵗʰ ed. Nr. 31.042; AOAC: USA, 1984; p 581.
15   Hadorn, H.; Fellenberg, Th. *Mitt. Lebensmittelunters. Hyg.* **1945**, *36*, 359.
16   Papastathopoulos, D. S.; Nikolelis, D. P.; Hadjiioaannou, T. P. *Analyst* **1977**, *102*, 852.
17   Amtliche Sammlung von Untersuchungsverfahren nach LMBG L 39.00 - E Beuth: Hamburg, 1981.
18   *Official Methods of Analysis of the Association of the Official Analytical Chemists (AOAC)*, 15ᵗʰ ed., Nr. 923.09, AOAC: USA, 1990; p 1016.
19   Barfoed, C. *Z. Anal. Chem.* **1873**, *12*, 27.
20   Myrbäck, K.; Leissner, C. *Ber. Dtsch. Chem. Ges.* **1942**, *75*, 1739.
21   Rotsch, A. *Getreide, Mehl Brot* **1947**, *1*, 10.
22   Hagedorn, H. C.; Jensen, B. *Biochem. Z.* **1923**, *135*, 46.
23   Hanes, C. S. *Biochem. J.* **1929**, *23*, 99.
24   Friedemann, T. E.; Weber, Ch. V.; Witt, N. F. *Anal. Biochem.* **1962**, *4*, 358.

25   Willstädter, R.; Schudel, G. *Ber. Dtsch. Chem. Ges.* **1918**, *51*, 780.
26   Auerbach, F.; Bodländer, E. *Angew. Chem.* **1923**, *36*, 602.
27   Isbell, H. S.; Frush, H. *Carbohydr. Res.* **1981**, *92*, 131.
28   Mazzuchin, A.; Thibert, R. J.; Walton, A. J.; Pedley, E. C. *Mikrochim. Acta* **1971**, 285.
29   Folin, O.; Wu, H. *J. Biol. Chem.* **1919**, *38*, 81; **1920**, *41*, 367.
30   Kakác, B.; Vejdelek, Z. In *Handbuch der photometrischen Analyse organischer Verbindungen*, Vol. 2; VCH: Weinheim, 1974; p 840.
31   Nelson, N. *J. Biol. Chem.* **1944**, *153*, 375.
32   Somogyi, M. *J. Biol. Chem.* **1945**, *160*, 61, 69; **1952**, *195*, 19.
33   Fratzke, R. F. *Proc. Ann. Biochem. Eng. Symp.* **1981**, *11*, 21.
34   Chien-Pen Lou; Lucy Ju-Yung *Ind. Eng. Chem., Anal. Ed.* **1944**, *16*, 637e.
35   Benham, G. H.; Petzing, V. E. *Anal. Chem.* **1949**, *21*, 991.
36   Kakác, B.; Vejdelek, Z. In *Handbuch der photometrischen Analyse organischer Verbindungen*, Vol. 2; VCH: Weinheim, 1974; p 852.
37   Hofstettler, F.; Borel, E.; Deuel, M. *Helv. Chim. Acta* **1951**, *34*, 2132.
38   Borel, E.; Hofstettler, F.; Deuel, H. *Helv. Chim. Acta* **1952**, *35*, 115.
39   Whistler, L.; Hickson, J. L. *Anal. Chem.* **1955**, *27*, 1514.
40   Bottle, R. T.; Gilbert, G. A. *Analyst* **1958**, *83*, 403.
41   Borel, E.; Deuel, H. *Helv. Chim. Acta* **1953**, *36*, 801.
42   Momose, T.; Inaba, A. *Chem. Pharm. Bull. (Tokyo)* **1961**, *9*, 263.
43   Momose, T.; Mukai, Y.; Watanabe, M. *Talanta* **1960**, *5*, 275.
44   Kakác, B.; Vejdelek, Z. In *Handbuch der photometrischen Analyse organischer Verbindungen*, Vol. 2; VCH: Weinheim, 1974; p 852.
45   Dehn, M.; Hartmann, F. A. *J. Am. Chem. Soc.* **1914**, *36*, 403.
46   Gardell, S. *Acta Chem. Scand.* **1951**, *5*, 1013.
47   Chavassieu, M. *C. R. Acad. Sci.* **1906**, *143*, 966.
48   Péronnet, M.; Hugonnet, J. *C. R. Acad. Sci.* **1951**, *232*, 2150.
49   Mattson, A. M.; Jensen, C. O. *Anal. Chem.* **1950**, *22*, 182.
50   Wallenfels, K. *Naturwissenschaften* **1950**, *37*, 491.
51   Corbett, W. M.; Kenner, J. *J. Chem. Soc.* **1953**, 2245; **1954**, 1789.
52   Mark, H.B.; Backes, L.M.; Pinkel, D. *Talanta* **1965**, *12*, 27.
53   Herb, W.; Venner, H. *Hoppe-Seyler's Z. Physiol. Chem.* **1957**, *308*, 36.
54   Cheronis, N.; Zymaris, M. C. *Mikrochim. Acta* **1957**, 775.
55   Livingston, E. M. *Microchem. J.* **1957**, *1*, 265.
56   Avigad, G.; Zelikson, R.; Hestrin, S. *Biochem. J.* **1960**, *80*, 57.
57   Gindler, M. *Clin. Chem.* **1970**, *16*, 519.
58   McFeeters, R. F. *Anal. Biochem.* **1980**, *103*, 302.
59   Mopper, K.; Gindler, M. *Anal. Biochem.* **1973**, *56*, 440.
60   Sinner, M.; Puls, J. *J. Chromatogr.* **1978**, *156*, 197.
61   Pinkus, G. *Ber. Dtsch. Chem. Ges.* **1898**, *31*, 31.
62   Russel, C. S.; Lyons, R. *Carbohydr. Res.* **1969**, *9*, 347.
63   Lever, M. *Anal. Biochem.* **1972**, *47*, 273.
64   Lever, M. *Anal. Biochem.* **1977**, *81*, 21.
65   Lever, M.; Waemsley, T.; Visser, R.; Ryde, S. J. *Anal. Biochem.* **1984**, *139*, 205.
66   Koziol, M. J. *Anal. Chim. Acta* **1981**, *128*, 195.
67   Mundie, C. M.; Cheshire, M. V.; Anderson, H. A.; Inkson, R. H. E. *Anal. Biochem.* **1976**, *71*, 604.
68   Honda, S.; Matsuda, Y.; Takahashi, M.; Kakehi, K.; Ganno, S. *Anal. Chem.* **1980**, *52*, 1079.
69   Honda, S.; Takahashi, M.; Kakehi, K.; Ganno, S. *Anal. Biochem.* **1981**, *113*, 130.
70   Honda, S.; Takahashi, M.; Nishimura, Y.; Kakehi, K.; Ganno, S. *Anal. Biochem.* **1981**, *118*, 162.
71   Honda, S.; Nishimura, Y.; Takahashi, M.; Chiba, H.; Kakehi, K. *Anal. Biochem.* **1982**, *119*, 194.

General References

Official Methods of Analysis of the Association of Official Analytical Chemists (AOAC); Arlington, USA.

Amtliche Sammlung von Untersuchungsverfahren nach §35LMBG; Bundesinstitut für gesundheitlichen Verbraucherschutz und Veterinärmedizin, Berlin, Bundesrepublik Deutschland, Ed.; Beuth: Hamburg.

Schweizer Lebensmittelbuch: Herausgabe im Auftrag des Bundesrates; Bearbeitet von der schweizerischen Lebensmittelbuchkommision; und dem Eidgenössischen Gesundheitsamt; Eidgenössische Drucksachen und Materialzentrale.

1.2. Reaction with Phenylhydrazine and Related Compounds

1.2.1. Reaction with Phenylhydrazine

Reducing sugars react with phenylhydrazine in weak acidic medium at a molar ratio of 1:1 to the corresponding phenylhydrazones[1-3]. With an excess of phenylhydrazine the reaction continues at pH 4–6 on formation of the deep yellow osazones which are scarcely soluble in water and whose use in the characterization of individual sugars has been well-established[4]. The reaction rates vary and the reaction itself is enhanced by an absence of oxygen. For aldoses the yields are between 50–80% and for ketoses, like fructoses, they are almost quantitative[5]. N-Glycoside and 1-amino-1-deoxyfructose compounds are also able to form osazones while for O-glycosides this is not possible[6,7]. In the case of glycoproteins, it is possible by using this method to distinguish between an O- and an N-glycosidic linkage. For the formation of the osazones see the following reaction scheme; an Amadori compound as intermediate is supposed followed by a redox reaction[8] (see Figure 1).

The presence of a sulfite prevents side reactions of the phenylhydrazine to colored compounds.

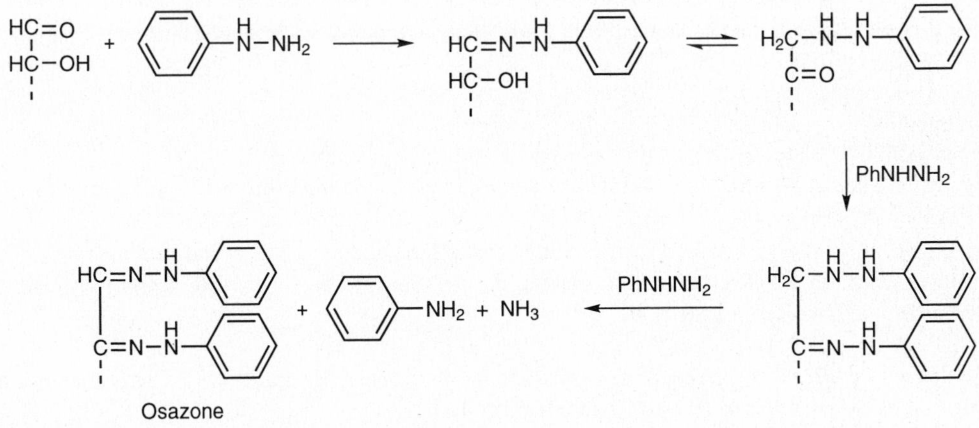

Figure 1.   Reaction between Reducing Sugars and Phenylhydrazine Forming Osazones (Mester)

Determination[6]

Reagent A: acetate buffer: 50 g of NaOAc and 50 mL of AcOH (pure) are dissolved in distilled water and the solution is made up to 200 mL.

Reagent B: 1 g of phenylhydrazine and 10 g of sodium bisulfite are dissolved in 100 mL distilled water. This reagent is stable for three days by storing in a refrigerator.

Pipet 1 mL of reagent A and 1 mL of reagent B into the sample (max 160 µg sugar) and dilute to 5 mL. Heat for 2 h in a boiling water bath and, after making the solution up to 10 mL, read the absorbance of the yellow osazones in solution at 405 nm against the blank solution. The calibration curve can be used to estimate each individual sugar since they differ in position from one another.

1.2.2. Reaction with 4-Nitrophenylhydrazine and with 2,4-Dinitrophenylhydrazine

With 4-nitrophenylhydrazine aldoses and ketoses yield, in acidic medium, the corresponding 4-nitrophenylhydrazones which are converted into chinoide structures under the influence of alkali resulting in a deepening of their color[9-11]. The alkaline medium can be formed either with sodium hydroxide[9], benzyltrimethylammonium hydroxide[10], or tetraethylammonium hydroxide[11] (see Figure 2).

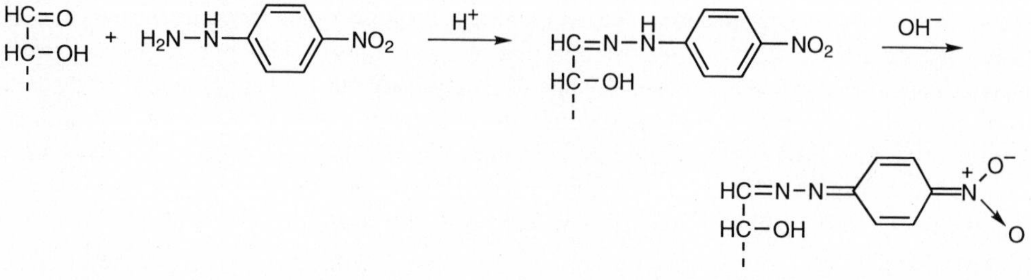

Figure 2.   Reaction between Reducing Sugars and 4-Nitrophenylhydrazine (Pesez and Bartos)

Determination[10]

Reagent A: 0.04% solution of 4-nitrophenylhydrazine in 0.1% ethanolic HCl.
Reagent B: 1 mL of 40% benzyltrimethylammonium hydroxide is made up to 100 mL with dimethyl-formamide.
Add 0.5 mL of reagent A to 0.5 mL of the sample (50–200 µg reducing sugar) and heat the mixture for 10 min at 70 °C in a water bath. After making the solution up to 10 mL with reagent B, read the absorbance at 510 nm against the blank.

In similar way the reaction can be carried out with 2,4-dinitrophenylhydrazine[12,13]. These reactions are obstructed by the presence of other aldehydes and ketones.

1.2.3. Reaction with 4-Amino-3-hydrazino-5-mercapto-1,2,4-triazole (AHMT)

This compound reacts with aldehydes and also, with a reduction in reaction rate, with ketones to produce deep red tetrazine compounds according to the following scheme[14] (Figure 3).
The reaction takes place in alkaline medium and has also been used for the determination of sugars[15,16].

Determination[15]

Reagent: 365 mg AHMT are dissolved in 100 mL aq NaOH (0.1 M).
Add 2 mL of the reagent to 1 mL a the neutral solution of the sample and heat for 5 min at 80 °C. Cool immediately and shake the solution for 5–10 min; then read the absorbance against the blank at 530 nm. Under these conditions, the absorbance over a concentration range of 0–200 µg sugar is linear.

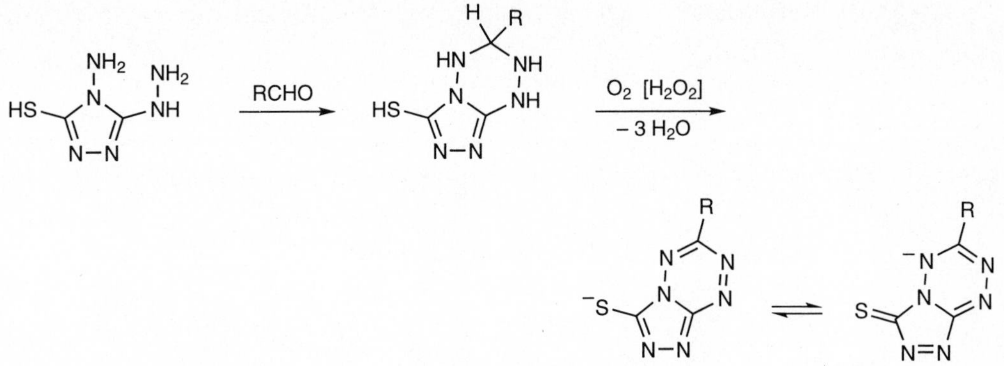

Figure 3. Reaction between Reducing Sugars and 4-Amino-3-hydrazino-5-mercapto-1,2,4-triazole (Dickinson and Jacobson)

Modified Procedure[16]

Reagent: 1.25 g AHMT are dissolved in 25 mL aq NaOH (0.1 M). (This reagent is stable only for two days).

Into 2.8 mL of the reagent pipet 0.2 mL of the sample and 0.2 mL of 30% hydrogen peroxide, and store the mixture at 25 °C for 2 h; then read the absorbance at 535–540 nm against a blank. The millimolar absorbance coefficients are different and vary in the range of 32 (glyceraldehyde), 22.5 (erythrose), 12.2 (ribose), 7.3 (fructose) and 0.48 (glucose).

1.2.4. Reaction of 3-Methyl-2-benzothiazolinone Hydrazone (MBTH)

This reagent which has been primarily developed for the identification of nonsubstituted aldehydes[17] can be also used for the determination of aldoses and ketoses[18,19].

The reaction mechanism can be explained by assuming that two equivalents of MBTH react with the reducing sugar to produce the corresponding osazone which further reacts with another molecule of MBTH, under the influence of an oxidant, to give a deep blue compound[19] with absorption maxima in the range of 630–670 nm (see Figure 4).

Other compounds like phenols and arylamines also react with this reagent to yield colored compounds.

Determination[19,20]

Reagent A: 0.5% aq solution of MBTH.

Reagent B: 2 g iron(III) perchlorate are dissolved in 8.3 mL concd HCl and the solution made up with distilled water to 100 mL; alternatively 0.25% aq solution of $FeCl_3$ can be used.

Pipet 1.0 mL of 0.1 M NaOH and 1.0 mL reagent A into 2 mL of the sugar sample and heat for 10 min in a boiling water bath. After cooling in cold water, add 1 mL of reagent B and stand for 30 min; then read the absorbance against the blank at 620 nm.

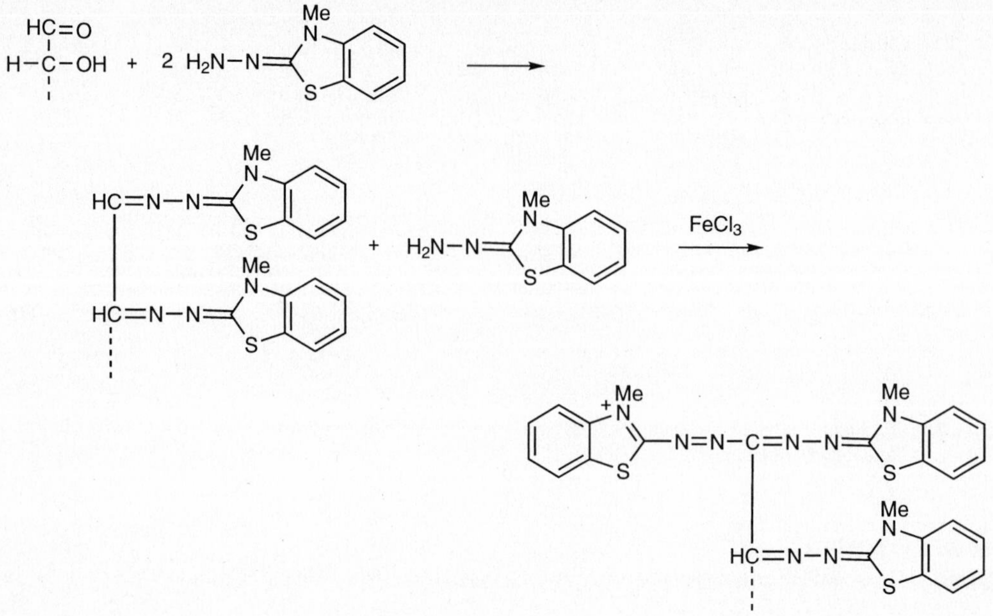

Figure 4. Reaction between Reducing Sugars and 3-Methyl-2-benzthiazolinone Hydrazone (Kakác and Vejdelek)

Positive results are obtained with monosaccharides, reducing disaccharides, uronic acids, and deoxy- and amino sugars. After reaction with periodic acid sugar alcohols and nonreducing disaccharides can also be determined with this reagent.

References

1   Ornig, A.; Stempel, G. *J. Org. Chem.* **1939**, *4*, 410
2   Kenner, J.; Knight, E.C. *Ber. Dtsch. Chem. Ges.* **1936**, *69*, 341
3   Kakác, B.; Vejdelek, Z. In *Handbuch der photometrischen Analyse organischer Verbindungen. Vol. 2*; VCH: Weinheim, 1974; p 858.
4   Fischer, E. *Ber. Dtsch. Chem. Ges.* **1884**, *17*, 579
5   Ashmore, J.; Renold A. E. *J. Am. Chem. Soc.* **1954**, *76*, 6189
6   Grassmann, W.; Hörmann, H.; Hafter, R. *Hoppe-Seyler´s Z. Physiol. Chem.* **1952**, *307*, 87.
7   Borsook, H.; Abrams, A.; Lowy, P. H. *J. Biol. Chem.* **1955**, *215*, 111
8   Mester, L. *Angew. Chem.* **1965**, *77*, 580; *Angew. Chem., Int. Ed. Engl.* **1965**, *4*, 574.
9   Webb, J. M.; Levy, H. B. *J. Biol. Chem.* **1955**, *213*, 107
10   Pesez, M.; Bartos, J. *Talanta* **1960**, *5*, 216
11   Sawicki, E.; Hauser, T.R.; Wilson, R. *Anal. Chem.* **1962**, *34*, 505
12   Pesez, M. *J. Pharm. Pharmacol.* **1959**, *11*, 475
13   Neuberg, C.; Strauss, E. *Arch. Biochem.* **1945**, *7*, 211
14   Dickinson, R. G.; Jacobson, N. W. *J. Chem. Soc.* **1970**, 1719
15   Reinefeld, E.; Bliesener, K. M.; van Malland, H.; Reichel, C. *Zucker* **1976**, *29*, 308.
16   Humeres, E.; Nome, F.; Aguirre, R. *Carbohydr. Res.* **1976**, *46*, 284

17   Sawicki, E.; Hauser, T. R.; Stanley, T. W.; Elbert, W. *Anal. Chem.* **1961**, *33*, 93.

18   Bartos, J. *Ann. Pharm. Fr.* **1962**, *20*, 650

19   Sawicki, E.; Schumacher, R.; Engel, C. R. *Microchem. J.* **1967**, *12*, 377

20   Kakác, B.; Vejdelek,. Z. In *Handbuch der photometrischen Analyse organischer Verbindungen. Vol. 2*; VCH: Weinheim, 1974; p 864.

1.3. Reagents Which Split α-Glycolic Linkages

1.3.1. Reaction with Periodic Acid

According to Malaprade[1-3] the C–C bonds of α-glycols are split with periodic acid yielding well-defined carbonyl and carboxyl compounds. Since this reaction is quantitative in many cases, it is applied as a general method for the quantitative determination of vicinal hydroxy compounds. Furthermore, the reaction is also widely used for elucidating the chemical structures of polyhydroxy compounds.

By splitting vicinal polyhydroxy compounds, the periodic acid converts the -CH_2OH group into 1 mole of formaldehyde ($HCH=O$) and the -$CH(OH)$ group into 1 mole of formic acid ($HCOOH$) as is demonstrated below using a polyalcohol. The reaction with the *cis*-hydroxyl groups is faster than with those in *trans* position[4,5].

$$
\begin{array}{l}
CH_2OH \\
| \\
CH(OH)_n \\
| \\
CH_2OH
\end{array}
\xrightarrow{HIO_4}
\quad 2\ CH_2{=}O \quad + \quad n\ HCOOH
$$

In the case of carbohydrates, two steps of this reaction occur which are named "selective oxidation" and "overoxidation".

In the first step, selective oxidation leads to the formation of formic acid and unique aldehydes, which themselves are not stable since they are hydrolyzed by the formic acid to lower polyhydroxy aldehydes. These are oxidized further by the periodic acid yielding the formic acid and formaldehyde. As an example, the individual steps of the reaction are demonstrated for D-(+)-glucose (see Figure 1)[6].

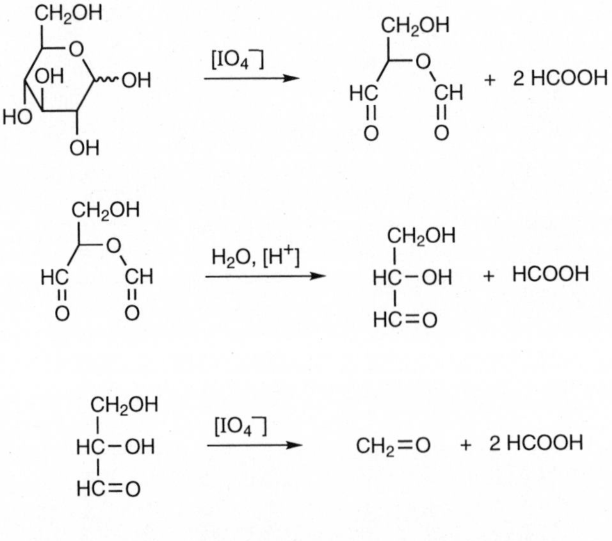

Figure 1.   Oxidation of D-Glucose with Periodic Acid (Franzke, Grunert and Obrikat)

The first step involving selective oxidation occurs at high speed. One mol D-(+)-glucose react with 3 mol periodic acid to give cleavage of the C-1–C-2, C-2–C-3 and C-3–C-4 bonds affording 2 moles of formic acid and the formyl derivative of glyceraldehyde. The latter compound would be stable against iodate ions but is hydrolyzed to glyceraldehyde and formic acid; the glyceraldehyde is then again oxidized to a further 2 moles of formic acid and to 1 mole formaldehyde. This second step of overoxidation requires a longer period of time. To summarize, the reaction between D-(+)-glucose and periodic acid yields 5 moles formic acid and 1 mole formaldehyde.

Substitution of the hydrogen of the hydroxyl groups with alkyl residues hinders the attack of periodic acid on the molecules. The same impedance occurs when these hydrogens are substituted by other monosaccharide molecules. The splitting products of the resulting oligosaccharides depend on the location of the substitution.

In some cases overoxidation occurs causing the intermediates of selective oxidation with periodic acid to react further in a so called "non-Malaprade way", e.g. the hydroxylation of active methylene groups and the splitting of enol and of enediol groups[5,7–9] as is demonstrated in the case of the periodic acid oxidation of 1,4-anhydro-D-allitol[10] (see Figure 2).

In order to minimize the overoxidation the following conditions should be observed when the periodic acid oxidation is carried out in practice[11]:

(a)   Exclusion of light.

(b)   Working within pH 3–5; at this range the optimum splitting of vicinal OH-groups coupled with a minimum of overoxidations has been found.

(c)   Performance of the reaction at room temperature.

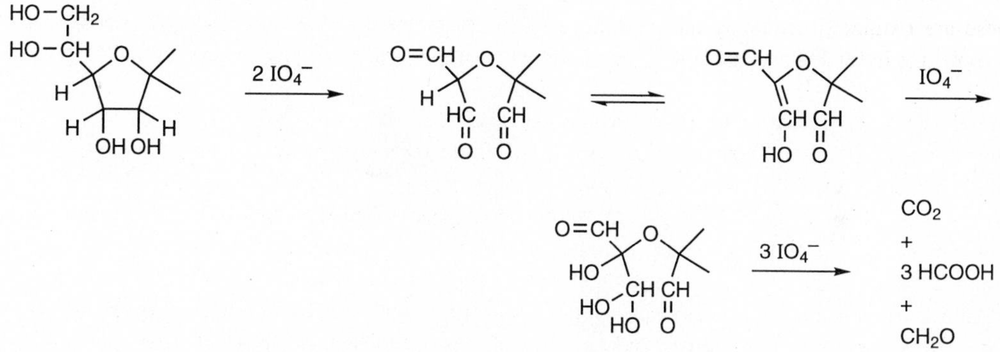

Figure 2.   Reaction between 1,4-Dehydro-D-allitol and Periodic Acid (Example of a Non-Malaprade Overoxidation) (Hudson and Barker)

1.3.1.a. Procedure I: Determination of the Yield of Formic Acid and the Consumption of Periodic Acid[11]

Determination

Reagent A: periodic acid (sodium periodate 0.3 M): 45 g sodium periodate are dissolved in 5% aq H_2SO_4. The solution is adjusted to pH 4, initially with 20% NaOH and then with 0.1 mol NaOH (titration against methyl red). Finally, it is made up to 500 mL.

Reagent B: pure ethylene glycol (spectroscopic grade).

Reagent C: sodium arsenite (0.4 M).

Determination of the Formic Acid

Pipet an aliquot part of reagent A to 10 mL of the sample and store the mixture for approximately 4 h; then destroy the excess periodic acid by adding of an excess of ethylene glycol (reagent B; this compound gives only formaldehyde). After a reaction time of 5 minutes titrate the generated formic acid with 0.1 M NaOH against methyl red.

Determination of Periodic Acid Consumption

Pipet an aliquot part of reagent A to 10 mL of the sample and after a distinct reaction time, adjust the mixture with NaOH to pH 8 (titration against phenolphthalein). Then, add 2 g of sodium bicarbonate, an exact volume of reagent C and solid potassium iodide. After standing for 15 min, titrate the excess arsenite with 0.1 M iodine solution.

1.3.1.b. Procedure II: Determination of Microamounts of Monosaccharides[12,13]

The periodate oxidation is carried out in acidic medium and the solution is buffered to a pH of 4.5–7.0. In this medium, the excess of periodic acid reacts with potassium iodide according to the following scheme.

$$IO_4^- + 2I^- + 2H^+ \longrightarrow IO_3^- + I_2 + H_2O$$

The liberated iodine is titrated with sodium thiosulfate. Under these conditions, the further reduction of iodate is kept negligible.

Determination

Reagent A: 1 g of potassium periodate is dissolved in warm water and after cooling made up to 1 L.

Reagent B: 5% H_2SO_4: 2.9 mL of concd H_2SO_4 ($\sigma = 1.84$) is diluted with water and made up to 100 mL.

Reagent C: 12% aq solution of *sec*-potassium phosphate.

Pipet to 1 mL of the sample, 8 mL of reagent A and 2 mL of reagent B and then heat the mixture for 20 min in a boiling water bath. After cooling add 4 mL of reagent C and solid potassium iodide and titrate the liberated iodine with 0.005 M sodium thiosulfate against starch. The same procedure is carried out with a blank and different standard solutions. The difference between the volumes of the blank and the standard solutions correlates to the amount of the polyhydroxy compound. By establishing a calibration curve, the amount of the polyhydroxy compound in the sample can be estimated. Complications are observed in the presence of ketoses and oligosaccharides[12]. Aldoses and sugar alcohols consume (n-1) mol periodate and ketoses (n-2) mol, where n is the number of carbon atoms in the sugar molecule.

1.3.1.c. Photometric Procedure for the Determination of Periodate[14]

A photometric method has been developed for the microdetermination of polyhydroxy compounds which works by determining periodic acid consumption.

The periodic acid oxidizes 4-(4-nitrophenoxy)-1,2-butanediol (PND) or 4-(2,4-dinitrophenoxy)-1,2-butanediol (PDB) to the corresponding aldehyde which undergo is β-elimination in alkaline medium giving nitrophenolate with an absorbance at 400 nm (Figure 3).

Determination

Reagent A: aq sodium periodate (0.0518 M).

Reagent B: acetate buffer pH 4.3 (1 M).

Reagent C: aq solution PNB or PDB both (0.013 M).
Reagent D: sodium carbonate–sodium bicarbonate buffer pH 9.5 (0.1 M).

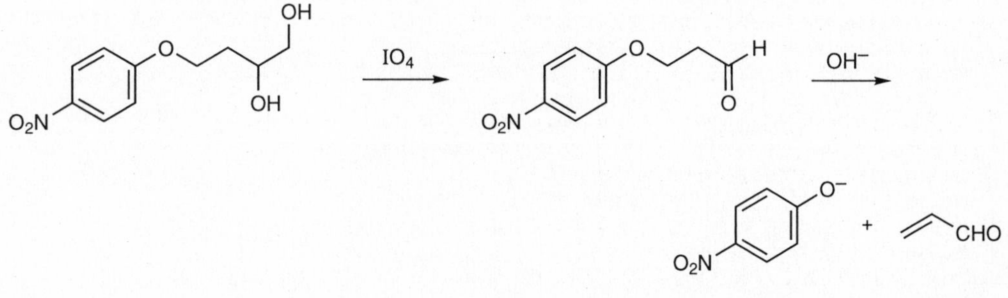

Figure 3. Reaction between Periodate and 4-(4-nitrophenoxy)-1,2-butanediol (Rammler, Bilton, Hangland and Parkinson)

Pipet together 0.1 mL of the sample, 0.2 mL of reagent A and 7.7 mL of reagent B and store the mixture for a distinct time, e.g. 4 h for monosaccharides like D-(+)glucose. Then, add to 0.8 mL of this mixture 0.2 mL reagent C (the amount of PNB or PDB must be in excess) to measure the nonreacted amount of periodic acid. After 15 min, pipet to 0.05 mL of this solution 0.95 mL of reagent D.

After 5 min the absorbance of this solution is read against a control which contains no periodate. All of these reactions are carried out at 50 °C. Parallel to this, the whole reaction is carried out with a blank and with standards of known content of the compound to be determined.

The application of this procedure is described not only for sugars and sugar alcohols but also for nucleosides, nucleotides and ribonucleic acids.

1.3.2. Other Reagents

In addition to periodic acid, a few other reagents split C–C bonds to vicinal hydroxyl groups. To this collection of compounds belongs:

lead(IV) acetate: this reagent can only be applied in nonaqueous medium due to its sensitivity against hydrolysis. The reaction rate is much higher for the *cis*-hydroxyl groups, than for those in the *trans* position[5,15];

active manganese dioxide[16];

potassium peroxodisulfate in presence of silver ions[17];

nickel peroxide[18];

sodium bismuthate(V) [19].

These reagents are widely used in the field of organic synthesis, but not for the determination of polyhydroxy compounds.

References

1   Malaprade, L. *C. R. Acad. Sci.* **1928**, *186*, 382.
2   Malaprade, L. *Bull. Soc. Chim. Fr.* **1928**, *43*, 683.
3   Fleury, P.; Lange, J. *C. R. Acad. Sci.* **1932**, *195*, 1395.
4   Buist, G. J.; Bunton, C. A. *J. Chem. Soc. B* **1971**, 2117, 2128.
5   Perlin, A. S. In Pigman, Horton, Wander (Eds.) *The Carbohydrates I B*; Academic: New York, 1980; p 1167.
6   Barbor, K.; Kalác, V.; Tihlárik, K. *Chem. Zvesti* **1973**, *27*, 676.
7   Schwarz, J. C. F.; McDougall, M. *J. Chem. Soc.* **1956**, 3065.
8   Bose, J. L.; Forster, A. B.; Stephens, R. W. *J. Chem. Soc.* **1959**, 3314.
9   Hesse, G.; Mix, K. *Chem. Ber.* **1959**, *92*, 2427.
10   Hudson, B. G.; Barker, R. *J. Org. Chem.* **1967**, *32*, 2101.
11   Franzke, Cl.; Grunert, K. S.; Obrikat, H. *Z. Lebensm. Unters. Forsch.* **1968**, *136*, 267, 324.
12   Rappaport, E.; Reifer, I.; Weinmann, H. *Mikrochim. Acta* **1937**, (1) 290, (2) 273.
13   Flood, A. E.; Priestley, C. A. *J. Sci. Food Agric.* **1973**, *24*, 945.
14   Rammler, D. H.; Bilton, R.; Haugland, R.; Parkinson, C. *Anal. Biochem.* **1973**, *52*, 198.
15   Criegee, R.; Höger, E.; Huber, G.; Kruck, P.; Marktscheffel, F.; Schellenberger, H. *Liebigs Ann. Chem.* **1956**, *599*, 81.
16   Ohloff, G.; Giersch, W. *Angew. Chem.* **1973**, *85*, 401; *Angew. Chem., Int. ed. Engl.* **1973**, *12*, 401.
17   Greenspan, F. P.; Woodburn, H. M. *J. Am. Chem. Soc.* **1954**, *76*, 6345.
18   Nakagawa, K.; Igano, K.; Sugita, J. *Chem. Pharm. Bull. (Tokyo)* **1964**, *12*, 403.
19   Rigby, W. *J. Chem. Soc.* **1950**, 1907.

1.4. Reaction with Acids

The action of concentrated acids on sugars leads to effective rearrangements due to multiple dehydration steps forming heterocyclic compounds like furfurals and ω-hydroxymethylfurfurals. These compounds react with phenols or aromatic amines giving deep colored substances. In a lot of cases such reactions enable photometric determinations of sugars. Here precise experimental conditions and careful control of the reaction time and acid strength are necessary to obtain reproducible results[1].

1.4.1. General Reactions for Aldoses and Ketoses

The reactions of sugars with phenols and related compounds in the presence of sulfuric acid are universal and positive for the whole group (e.g., aldoses, ketoses, mono-, oligo- and polysaccharides).

1.4.1.1. Reaction with Phenol

Mono-, oligo- and polysaccharides react with phenol and concentrated sulfuric acid at elevated temperatures resulting in the formation of colored substances with absorption maxima at 480–490 nm[1,2]. For the reaction mechanism, it is assumed that, in the case of oligo- and polysaccharides, the external ether bridges are split. Parallel dehydration reactions take place yielding furfural derivatives which condense with phenol to triarylmethane dyes according to the following scheme (see Figure 1).

Pentoses, methylpentoses and hexuronic acids yield orange colored compounds with absorption maxima (λ_{max}) at 490 nm, hexoses and their oligomers such ones with λ_{max} at 490 nm; 2- and 4-ketocarbonic acids interfere with the reaction[3].

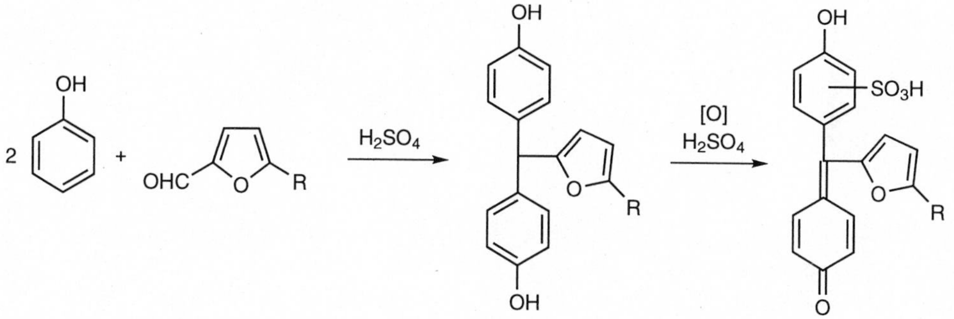

Figure 1.   Reaction between Phenol and Carbohydrates in the Presence of Concentrated Sulfuric Acid (Kakác and Vejdelek)

Determination[2,4]

Reagent: mixture of 80 g of cryst phenol with 20 mL of distilled water.

Pipet to 2.0 mL of the aq sample (10–200 µg sugar) 0.05–0.15 mL of the reagent and then carefully add, with intense shaking, 5 mL of concd sulfuric acid. Intensively shake the mixture again and let it stand for 10 min at room temperature and then for 10–20 min in a water bath at 25–30 °C. After this time, read the absorbance at 480–490 nm. This reaction is universal for all sugars and sugar-containing compounds. It is applied for the quantitative determination of sugar fractions after their separation by column, paper, or thin-layer chromatography.

A modification of this method enables the selective determination of ketoses in presence of aldoses[5].

1.4.1.2. Reaction with Anthrone

This compound yields, by condensation with carbohydrates in sulfuric acid, medium colored compounds with λ_{max} between 520–630 nm. Methylpentoses and hexoses give blue colored substances (λ_{max} = 620–630 nm), those obtained with pentoses are blue green (λ_{max} = 590–625 nm) and those with hexuronic acid are red (λ_{max} = 540–550 nm) after prolonged storage[6,7]. Glycerol, formaldehyde, acrolein and ascorbic acid can interfere by also yielding red colored compounds.

The reaction between anthrone and a sugar is complex; for a few reaction products, the following structures have been proposed[8,9] (Figure 2).

A lot of procedures exist which differ by applied concentrations of anthrone and sulfuric acid.

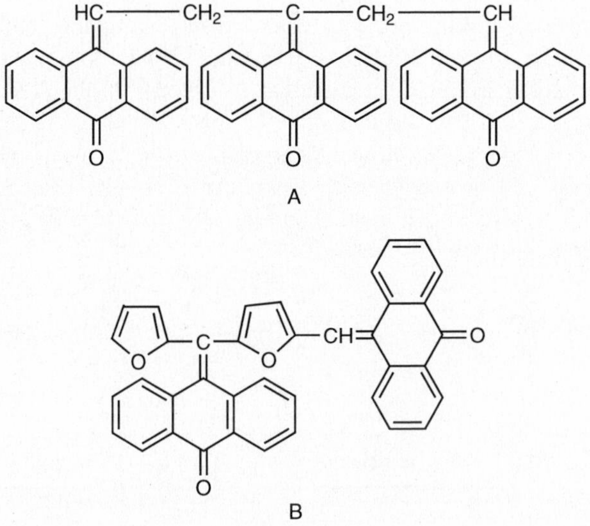

Figure 2. One Reaction Product Proposed for the Reaction between Anthrone and Hexoses (A) or Pentoses (B) (Momose et al.)

Determination of the Total Sugars[10]

Reagent: 0.2% solution of anthrone in 95% sulfuric acid.

To 1 mL of the solution, while continuously cooling with ice, pipet 2 mL of the reagent and then shake the solution intensively. Heat the mixture for a definite time in a boiling water bath and cooling, read the absorbance against a blank. The heating time for aldo- and ketohexoses is 10 min, while for pentoses the color reaches the maximum absorption after two minutes and then a strong decrease occurs. This reaction can be applied for the determination of mono-, oligo- and polysaccharides.

Determination of Hexoses[11–13]

Reagent: 0.2% solution of anthrone in 72% sulfuric acid.

Pipet 1 mL of the aqueous sugar sample to 5 mL of the reagent without mixing; cool intensively and then mix both phases by vigorously shaking. Heat the mixture for 10 min in a boiling water bath and after cooling, read the absorbance at 620 nm against a blank. When the reaction is carried out in presence of formic acid and hydrochloric acid the color intensity is doubled. Pentoses under this condition yield only a very weak colored compound.

Determination[14]

Reagent: solution of 20 mg of anthrone in 100 mL 80% sulfuric acid.

Mix 1 mL of the aqueous sample with 1 mL concd HCl and 0.1 mL of formic acid and then add 8.0 mL of the reagent; heat for 10–12 min in a boiling water bath and, after cooling, read the absorbance at 630 nm against a blank.

Other phenolic reagents which react with sugars in presence of sulfuric acid are listed in Table 1.

1.4.2. Specific Reactions for Ketoses

The reactions of sugars in a hydrochloric acid medium are more selective than those in sulfuric acid and enable differentiation within a distinct group of compounds.

1.4.2.1. Reaction with Resorcinol

When ketoses are heated in an ethanolic solution of resorcinol in the presence of hydrochloric acid a deep red color (λ_{max} = 515 nm) develops (Selivanoff reaction). The reaction with aldoses under these conditions is much weaker; therefore, it is possible to determine ketoses quantitatively in the presence of a 100-fold excess of aldoses[30,31].

It is assumed that the resorcinol reacts with furfural derivatives, which are formed from the ketoses under the influence of acid, to produce dyes with chinoide structures[32] (Figure 3).

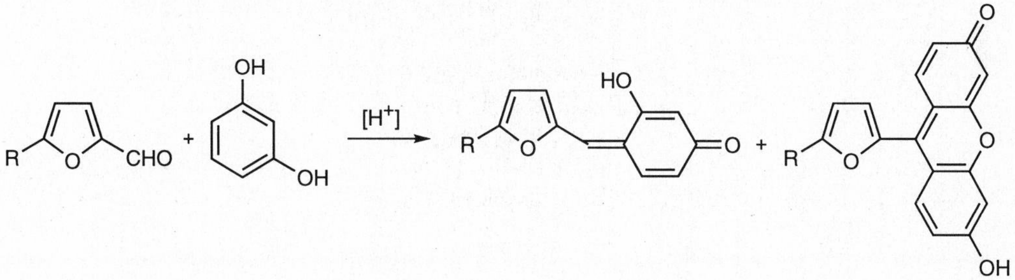

Figure 3.   Reaction between Ketoses and resorcinol in the Presence of Acetic Acid–Concentrated Hydrochloric Acid (Berner and Sandlie)

Table 1

Compound	Carbohydrates	Description of the Procedure	Absorption Maxima of the Colors
1. 1-Naphthol	Pentoses, hexoses and their oligo- and polysaccharides[15,16], and other carbohydrate compounds like glycoproteins[17] (Molisch reaction)	*Reagent A*: 2% ethanolic solution of 1-naphthol *Reagent B*: concd sulfuric acid Pipet 0.5 mL of reagent A to 2.0 mL of the sample, shake and underlay with 6 mL of reagent B then cool in an ice bath and mix. After 45 min read the absorbance at 550–570 nm	red-violet Pentoses: 550 nm Hexoses: 570 nm
2. Thymol	Pentoses, hexoses and their oligo- and polysaccharides[18,19]	*Reagent A*: a mixture of 23 mL of distilled water with 77mL concd sulfuric acid *Reagent B*: 10% solution of thymol in ethanol Underlay 1.0 mL of the aqueous sample with 7.0 mL of reagent A and leave to stand for 10 min; then cool the solution in an ice bath for 15 min, add 0.1 mL of reagent B and 0.9 mL of water, shake intensively, and heat for 20 min in a boiling water bath. Cool the mixture and after 25 min read the absorbance against a blank	red-orange Pentoses, heptoses: 490–495 nm Hexoses: 500–515 nm
3. Orcinol	Pentoses, hexoses, hexuronic acids and the corresponding oligo- and polysaccharides[20–22]	*Reagent A*: 2 g of orcinol are dissolved in a mixture of 75 mL of water and 25 mL of concd sulfuric acid *Reagent B*: a mixture of 40 mL of water with 60 mL of concd sulfuric acid Mix 1 mL of the aqueous sample with 2.0 mL of reagent A and 15.0 mL of reagent B; shake intensively and heat the mixture in a water bath at 80 °C for 20 min. Cool and read the absorbance against a blank	red-red orange Pentoses: 560 nm Hexoses: 530 nm
4. Carbazole	Hexuronic acid, pentoses, hexoses[23–25]	*Reagent A*: a mixture of water and concd sulfuric acid (1:8) v/v *Reagent B*: 0.5% solution of carbazole in ethanol (water free) To 9 mL of cooled reagent A pipet 1 mL of the aqueous sample and add 0.3 mL of reagent B. Heat for 10 min in a boiling water bath, cool and read the absorbance	red-brown Hexuronic acid, hexoses and pentoses: 520–540 nm
5. Tryptophan	Pentoses, hexoses, hexuronic acids[26–29]	*Reagent A*: 5.0 g of boric acid are dissolved in a mixture of 23 mL water and 77 mL of concd sulfuric acid *Reagent B*: 1% aq solution of tryptophan To 2.0 mL of the aqueous sample add 7.0 mL of reagent A whilst cooling with ice water; then pipet 1 mL of reagent B, heat for 20 min in a boiling water bath, and read the absorbance after cooling over 30 min	red Pentoses, hexoses, hexuronic acid: 520–540 nm

Determination[33]

Reagent: 0.1 g of resorcinol and 0.25 g of thiourea are dissolved in 100 mL of pure acetic acid; this solution is stored in the dark.

Mix 2.0 mL of the aqueous sample (10–50 µg ketose) with 2.0 mL of the reagent and 6.0 mL of concd HCl (d = 1.17). Heat for exactly 8 min at 80 °C in a water bath and after cooling, read the absorbance at 515 nm against a blank. This reaction is also positive for 3,6-anhydro sugars and is a specific assay for agar and carrageenan[34]. Addition of perchloric acid to the system enhances the sensitivity and specificity of the reaction[35].

In another modification of the procedure, 30% concd HCl (H_2O/concd HCl 1:5 v/v) is used instead of pure concd HCl (38%)[36]. Addition of acetaldehyde also increases the sensitivity of the reaction and shifts the absorption maximum toward 555 nm. The following procedure is based on this modification[37].

Determination[37]

Reagent A: 150 of mg resorcinol are dissolved in 100 mL of distilled water.

Reagent B: 820 mg of 1,1-diethoxyethane are dissolved in distilled water or ethanol and the solution is made up to 100 mL.

Reagent C: 100 mL of concd HCl is mixed with 9 mL of reagent A and 1 mL of reagent B.

Pipet in to a graduated test tube, which contains glass marbles and has been placed in an ice bath, 2 mL of the aqueous sample and 10 mL of reagent C. Cool for 3 min, then store for 4 min at 20 °C. After heating the solution for 10 min at 80 °C; cool the tube contents and read the absorbance at 550 nm.

The relationship between the absorbance and concentration is linear for a sample containing up to 50 µg of the ketose under these conditions. In Table 2 the specificity of this improved resorcinol reagent is listed.

Table 2.   Specificity of the Resorcinol Reagent[37]

Sugar	Color Ratio
Fructose	100
3,6-Anhydrogalactose	92
Glucose	2
Galactose	1
Mannose	2.5
Fucose	< 1
Rhamnose	1
Xylose	4
Arabinose	1
Ribose	5
Sorbose	69

1.4.2.2. Reaction with Skatole or β-Indolylacetic Acid

On reaction with ketoses in hydrochloric acid solution, skatole yield compounds with violet coloring (λ_{max} = 510–520 nm); the color development with aldoses is much weaker[38,39].

Determination[39]
Reagent A: 0.04% solution of skatole in ethanol.
Reagent B: a mixture of concd HCl and distilled water (5:1 v/v).
Pipet to 0.5 mL of the sample (~ 20 µg ketose) in a test tube, 0.5 mL of ice cold reagent A and 4 mL of reagent B. Shake thoroughly and heat at 60 °C for 60 min; then cool and add 5 mL of chloroform. Shake intensively again to transfer the colored compound into the organic phase. Read the absorbance immediately at 510 nm since the color is unstable. The concentration–absorbance correlation for ketoses is linear for up to 50 µg.

1.4.3. Specific Reactions for Aldoses

Aromatic amines only react with aldoses in acidic solutions to form colored condensation products; ketoses remain negative. The following compounds have been used for the determination of aldoses: aniline, *o*-toluidine, *p*-anisidine, 2-aminodiphenyl, benzidine, 4-aminobenzoic acid and 4-aminosalicylic acid[40].

1.4.3.1. Reaction with Aniline
The salts of aniline with phthalic acid, oxalic acid or trichloroacetic acid react with aldopentoses to form intensely colored compounds with absorption maxima at 520–540 nm; methylated aldopentoses yield such substances with λ_{max} at 460 nm, while those of aldohexoses have their maxima around 390 nm. Hexuronic acids and deoxysugars also give weakly colored compounds but ketoses do not when the following conditions with trichloroacetic acid are used.

Determination[41]
Reagent: 32 mL of aq trichloroacetic acid (8.5 M) is mixed with 50 mL of ethanol; then, while cooling with ice, 2.0 mL of freshly distilled aniline is added and the mixture is made up with cooled ethanol to 100 mL.
In a calibrated test tube, pipet to 1 mL of the sample containing 5–300 µg aldose 1 mL of the reagent stored in an ice bath; shake the contents of the tube thoroughly and then heat for 15 min in a boiling water bath. After cooling add ethanol (95% vol) to give a volume of 4 mL of the solution; read the absorbance at 370 nm.
Replacing the trichloroacetic acid with a mixture of phosphoric and acetic acid enables the photometric determination of ketoses[42].

1.4.3.2. Reaction with *o*-Toluidine
Aldopentoses react with this reagent to yield colored compounds with λ_{max} at 480 nm and aldohexoses with λ_{max} at 630 nm (there are additional weak peaks at 365 and 465 nm). This reaction has often been used for microdetermination of glucose in biological material (e.g., blood). The reaction conditions have been studied in detail. Besides acetic acid, solutions of *o*-toluidine in 1-propanol containing thiourea, boric acid or citric acid have been discovered to be effective catalysts for this reaction[43]. A mixture of 1,3-propylene glycol and acetic acid has been found to be the best solvent[44].

Determination[44]
Reagent: 1 g of thiourea is dissolved in 250 mL of pure acetic acid, then 50 mL of *o*-toluidine and 500 mg of sulfamic acid is added; this mixture is made up to 1 L with propylene glycol.

Pipet 3 mL of the reagent to 20 μL of the sample and heat the mixture for 10 min in a boiling water bath, then cool, and measure the absorbance at 590 nm. The same procedure is carried out with a blank and with standard solutions of the sugar to establish a calibration curve.

1.4.3.3. Reaction with Benzidine–Acetic Acid

Aldoses and alduronic acids react with this reagent to give compounds which have strong absorption maxima in the range of 320–420 nm. This reaction is more sensitive than the reaction with aniline[44,45].

Determination[45,46]

Reagent: 0.2 g of benzidine and 0.1 g of $SnCl_2$ are dissolved in 100 mL of pure acetic acid; this solution is stored in the dark.

Heat 1 mL of the aqueous sample (20–600 μg) with 5 mL of the reagent for 15–60 min in a boiling water bath, then cool the mixture and read the absorbance at 425 nm against a blank. Benzidine is a strong carcinogenic compound, therefore, this reaction must be carried out with particular precaution.

1.4.4. Specific Reactions for Pentoses

1.4.4.1. Reaction with Orcinol

Pentoses react with orcinol in presence of an iron(III) salt in a hydrochloric acid medium to form deep green compounds (Bial reaction). This reaction specifically relates to aldopentoses and aldopentose-containing compounds like polysaccharides, purine and pyrimidine nucleotides, and ribonucleic acids. The absorption maxima λ_{max} for the color is at 670 nm; ketopentoses result in a second λ_{max} at 540 nm[47–50]. At higher concentrations hexuronic acids, methylpentoses and trioses may interfere in this reaction as they also produce compounds with greenish colors.

The mechanism of the reaction has not yet been elucidated in detail. The following structure is proposed for the colored compound (see Figure 4)[48].

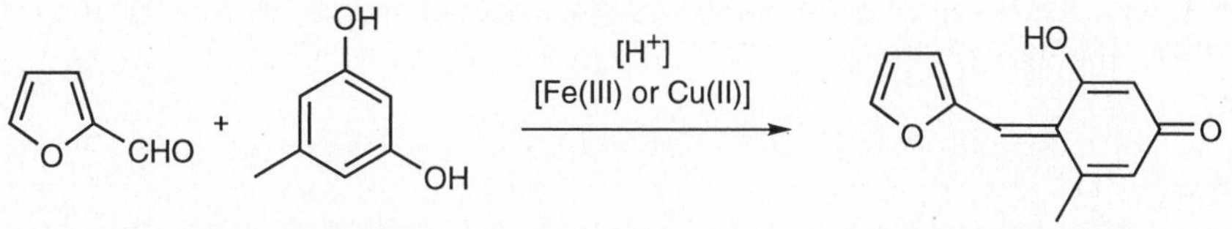

Figure 4.   Reaction between Pentoses and Orcinol in the Presence of Hydrochloric Acid and Iron(III) or Copper(II) ions (Kakác and Vejdelek)

Determination[49]

Reagent: 100 mg of $FeCl_3 \cdot 6H_2O$ are dissolved in 100 mL of concd HCl; just before use 100 mg of orcinol are dissolved in 10 mL of this solution.

Mix 1 mL of the aqueous sample (1–10 μg) with 1 mL of the reagent and heat for 20 min in a boiling water bath. Then, add 2 mL of distilled water and cool in an ice bath. Read the absorbance at 665–670 nm and in the presence of hexoses, then measure the absorbance difference E_{665}–E_{590}.

A variation of this procedure has been established for the determination of the pentose fragment in purine and pyrimidine nucleotides[51].

Another variation of this reaction is where the Fe(III) ions are replaced by Cu(II) ions. In this case, the interference by the color of the reagent is minimized[52–54].

Determination[53]

Reagent: 200 mg of orcinol are dissolved in 90 mL of concd HCl, and 10 mL of a 0.004 M CuCl$_2$ solution in concd HCl is then added. This reagent is stable only for a few hours.

To 5 mL of the sample in a test tube (2–20 µg pentose; 10–50 µg ribonucleic acid per mL) add 5 mL of the reagent and heat the mixture in a boiling water bath for 40 min. Cool and then extract the dye with 5 mL of isoamyl alcohol; read the absorbance of the organic phase at 665 nm against a blank which has been prepared in the same manner.

1.4.4.2. Reaction with Phloroglucinol

By warming an aqueous solution of pentoses with hydrochloric acid and phloroglucinol a deep red color appears (λ_{max} = 552 nm). This color is identical with that obtained by the reaction of furfural with phloroglucinol[55]. In terms of structure, that of a xanthene derivative is assumed which is formed by the following pathway (see Figure 5)[56]. This reaction is specific to aldopentoses, their glycosides, ribonucleosides, and ribonucleic acid[57,58]. Aldo- and ketohexoses reach only 2–10% of the color intensity at the same concentration.

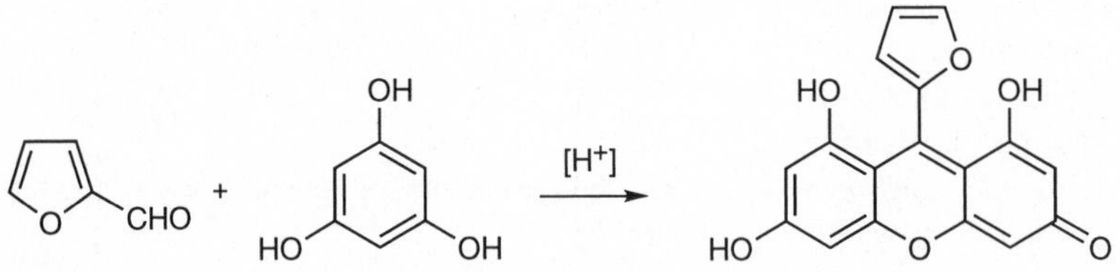

Figure 5.   Reaction between Pentoses and Phloroglucinol in the Presence of Concentrated Hydrochloric Acid (Sen and Sinha)

Determination[58]

Reagent: 2 mL of concd hydrochloric acid, 110 mL of pure acetic acid, 1 mL of 0.8% aq glucose solution, and 5 mL of a 5% ethanolic solution of phloroglucinol are mixed together just before use.

To 0.5 mL of the aqueous sample (5–25 µg pentose) add 0.5 mL of concd HCl and 4.5 mL of the reagent. Then shake intensively and heat in a boiling water bath for 5 min. After cooling in an ice bath read the absorbance at 552 and 510 nm against a blank. The difference E_{552}–E_{510} obtained with the test solution establishes calibration curves which are used for the estimation of the pentose concentrations.

1.4.4.3. Reaction with Pyrogallol

This compound reacts with pentoses, under similar conditions to those described in Section 1.4.4.2., in the presence of Fe(III) ions to give deep red dyes (λ_{max} = 500 nm). Hexoses and hexuronic acids do not interfere since they do not form colored compounds with this reaction[59,60].

1.4.5. Specific Reactions for Hexoses

These reactions are based, in part, on the formation of ω-hydroxymethylfurfural which is obtained by the treatment of hexoses with acids. This compound is then condensed with phenols or aromatic amines to give colored compounds.

1.4.5.1. Reaction with Diphenylamine

Warming hexoses with diphenylamine in a mixture of hydrochloric acid and acetic acid yields a deep blue solution[61,62] with two absorption maxima at 630 and 520 nm. Tetroses and pentoses yield only faint yellow-green colors and do not interfere with the determination; hexuronic acids, 2-deoxysugars and heptoses are, however, compounds which can interfere this reaction. The reaction with ketoses is faster than with aldoses.

Regarding the chemical mechanism, it has been found that two molecules of ω-hydroxymethylfurfural dehydrate to 5,5'-diformylfurfuryl ether, which reacts further with four molecules of diphenylamine to give bis(triarylmethane) dyes[63,64] (Figure 6).

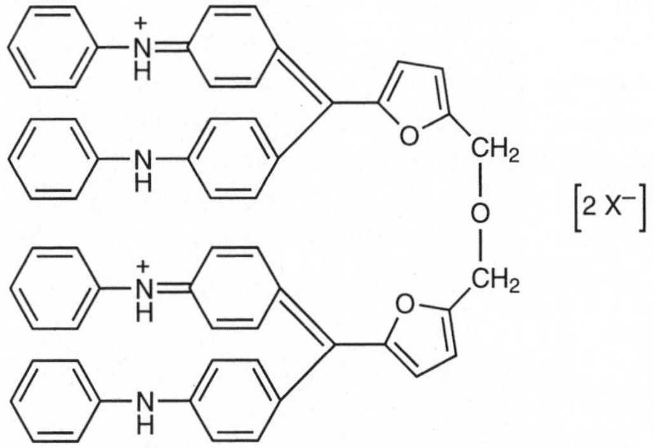

Figure 6.   Reaction between Hexoses and Diphenylamine at Presence of Acetic Acid Hydrochloric Acid (Momose et al.)

Determination[64]
Reagent: 10 mL of a 10% ethanolic solution of diphenylamine is mixed with 90 mL of pure acetic acid and 100 mL of concd hydrochloric acid.

Mix 1.5 mL of the aqueous sample with 3 mL of the reagent and heat this solution in a boiling water bath for 10 min (ketoses) or 30–40 min (aldoses). Then cool, make the solution up to 5 mL with ethanol and read the absorbance against a blank at 635–640 nm. This reaction has been applied for the determination of sugars after separation by thin-layer chromatography[65] and for the determination of polysaccharides[66].

1.4.5.2. Reaction with Chromotropic Acid

Hydroxymethylfurfural, formed by the treatment of hexoses with acid, loses formaldehyde quantitatively on reaction with concentrated sulfuric acid. This compound then reacts with chromotropic acid to give a deep red-violet dye ($\lambda_{max} = 570$ nm)[67–69].

Pentoses and methylpentoses yield yellow and orange colors, respectively. Substances which also give formaldehyde on treatment with sulfuric acid interfere with the reaction; to this group of compounds belong, for example, carboxymethyl compounds like carboxymethyl cellulose.

Determination[68]
Reagent: 0.2% solution of chromotropic acid (1,8-dihydroxynaphthalene-3,6-disulfonic acid) in 15 M sulfuric acid.
Mix 1.0 mL of the aqueous solution of the sample with 5 mL of the reagent. Heat at 100 °C for 30 min, cool, and then make up to 10 mL with 9 M sulfuric acid. Read the absorbance against water at 570 nm.

1.4.6. Specific Reactions for Uronic Acids

The treatment of uronic acids with concentrated mineral acids leads to the formation of carboxymethylfurfural, which forms deep colors with phenols and aromatic amines[70].

1.4.6.1. Reaction with 1,3-Dihydroxynaphthalene (Naphthoresorcinol)
Hexuronic acids in an acidic solution with this compound form blue-violet colors (λ_{max} = 570–620 nm)[71,72]. These colored compounds (for their proposed structures see Figure 7)[73] are extractable with organic solvents like diethyl ether, ethyl acetate, amyl acetate, benzene or toluene. Predominantly, the reaction is carried out in a hydrochloric acid medium[74,75]. Mono-, oligo- and polysaccharides form similar colors which are not extractable with organic solvents. Aldohexoses and hexosamines do not interfere in the reaction when used to a hundred fold excess. In contrast, those compounds which interfere with the reaction do so by forming keto compounds under the influence of acids (e.g., hydroxyketo acids)[76,77].

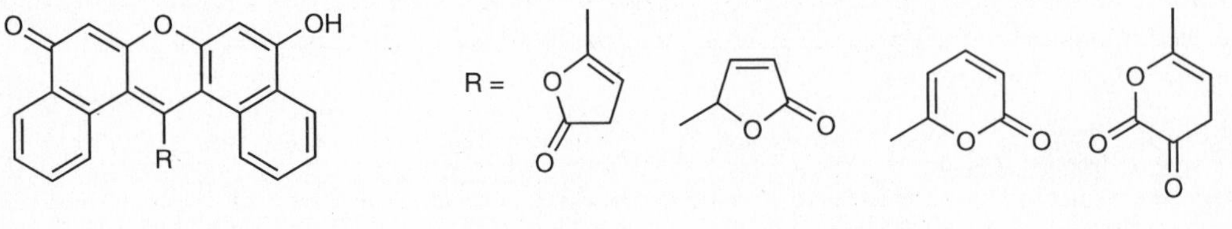

Figure 7.   Products Formed from the Reaction between Uronic Acids and Naphthoresorcinol in the Presence of Hydrochloric Acid (Momose, Ueda and Iwasaki)

Determination[77,78]
Procedure A[78]
Reagent: 200 mg of naphthoresorcinol are dissolved in 80 mL of water, then NaOH (0.5 M) is added till a pH of 8–8.5 is reached. This solution is stored for 15 min, then acidified with 10% phosphoric acid to pH 2–2.5. At this point, 100 mg of sodium bisulfite is added and the mixture is made up with distilled water to 100 mL. By storing in the dark at 5 °C, this reagent is stable for 5 months.
Mix 2 mL of the aq uronic acid containing sample with 2 mL of the reagent and 2 mL of concd hydrochloric acid. Heat the mixture for 30 min in a boiling water bath, then cool, extract the colored

compound with 10 mL of ethyl acetate, and read the absorbance of the organic phase at 580 nm against a blank, which has been prepared in the same manner.

Procedure B[77]

According to another procedure, naphthoresorcinol can be dissolved in pure acetic acid since it is stable in this medium. The reaction itself is carried out in a phosphoric acid medium.

Reagent: 0.2% solution of naphthoresorcinol in acetic acid.

Add, dropwise, 2 mL of concd phosphoric acid to 1 mL of the sample and mix immediately. Add a layer of 4 mL of the reagent, then mix and heat the homogeneous phase for 75 min at 70 °C. Cool, extract the colored compound with 2 mL of toluene, and read the absorbance of the organic phase at 570 nm against a blank, which has been prepared in the same manner.

1.4.6.2. Reaction with Carbazole

Hexuronic acids with carbazole in presence of concentrated sulfuric acid yield red colored compounds with absorption maxima at 520–530 nm[79].The colors are stabilized by borate ions, which also diminish the influence of neutral sugars on the absorption[79–81].The same effect is observed with sulfamate ions[82].

The compounds formed on reaction with hexoses are a red-brown color and those formed using pentoses are yellow. In the case of hexuronic acids, the absorbances[83] are dependent on individual chemical structure but not on the degree of polymerization.

Determination[82]

Reagent A: aqueous solution of potassium sulfamate (4 M).

Reagent B: aqueous solution of potassium tetraborate (1 M).

Reagent C: 0.2% solution of carbazole in ethanol (water free).

Pipet 0.1 mL of reagent A and B to 0.8 mL of the sample (10–100 µg hexuronic acid), cool this solution in an ice bath and then add 5.0 mL of concd sulfuric acid. Heat the mixture for 7 min in a boiling water bath, cool immediately and add 0.2 mL of reagent C. Heat the mixture again for 10 min in the boiling water bath, then cool again and read the absorbance at 525–530 nm against a blank prepared in the same manner.

1.4.6.3. Reaction with 3-Hydroxydiphenyl (*m*-Phenylphenol)

This compound reacts with a hexuronic acid in a concd sulfuric acid medium resulting in the formation of a deep red color (λ_{max} = 520 nm); this reaction is specific to hexuronic acids[84,85].

Determination[84]

Reagent A: 0.15% solution of 3-hydroxydiphenyl in 0.5% NaOH.

Reagent B: solution of sodium tetraborate in concd sulfuric acid (0.0125 M).

Cool 0.2 mL of the sample in ice water and immediately add 1.2 mL of cooled reagent B. Heat the mixture in a boiling water bath for 5 min, cool again and then pipet 20 µL of reagent A to the solution. Shake thoroughly and, within 5 min, read the absorbance at 520 nm against a blank which has been run without reagent A.

The colored compound from the reaction is stable for at least 12 h. Under these conditions, 10 µg of uronic acids have absorbances of 0.6–0.8 relative to the blank depending on the individual compounds. The reaction is more sensitive and selective than the carbazole method and amounts of up to 50 µg of neutral sugar in the solution can be used without interfering in the determination[85].

1.4.7. Specific Reactions for Deoxysugars

For this group of sugars, the distinction between such compounds with the methyl group at the end of the molecule (e.g., methylpentose = 6-deoxyhexose) and those where the deoxy group is located in the middle of the molecule (e.g., 2-deoxyribose) has to be made.

1.4.7.1. Specific Reactions for Methyl Sugars (e.g., Methylpentoses such as Rhamnose or Fucose)

1.4.7.1.a. Reaction with Thioglycolic Acid and Sulfuric Acid

Methylpentoses are dehydrated with concentrated sulfuric acid yielding 5-methylfurfural which reacts with thioglycolic acid to give a yellow-green colored compound (λ_{max} = 400 nm). By measuring the difference in absorbance E_{400}–E_{430}, pentoses, hexoses, hexuronic acids and hexosamines do not interfere[86–89].

Determination[89]
Reagent A: mixture of water–concd sulfuric acid (1:6 v/v).
Reagent B: 1 mL of thioglycolic acid is diluted in 30 mL of distilled water.
Mix 1 mL of the sample with 4.5 mL of reagent A whilst cooling with ice. Then, heat for 10 min in a boiling water bath, cool to room temperature and add 0.1 mL of reagent B. After storing for 3 h in the dark, read the absorption difference E_{400}–E_{430}.

1.4.7.1.b. Reaction with Periodic Acid, Sodium Nitroprusside and Piperazine

6-Deoxysugars are oxidized by periodic acid to acetaldehyde and formic acid. The acetaldehyde reacts with a secondary amine and sodium nitroprusside to form a red-violet color with λ_{max} at 600 nm (Rimini reaction)[90–92].

This reaction is not applicable in the presence of compounds which have a -CH_2CO- group in their structure or to those which give such compounds on splitting with periodic acid (e.g., 2-deoxy compounds giving malondialdehyde, threonine giving acetaldehyde, etc.). Pentoses and hexoses themselves do not interfere and formaldehyde, isobutyraldehyde, benzaldehyde and furfural do not give colored compounds with this reagent.

Determination[91]
Reagent A: 2.7 g of cryst periodic acid are dissolved in distilled water and the solution is adjusted to pH 6.8 with NaOH (1 M) and then made up to 100 mL.
Reagent B: immediately before use, 40 mL of a 5% ethanolic solution of ethylene glycol are mixed with 30 mL of a 20% ethanolic solution of piperazine and 10 mL of a 7% aqueous solution of sodium nitroprusside.
To 3 mL of the sample (0.1–0.6 mg methyl sugar) pipet 1 mL of reagent A and store for 10 min. Then, add 1 mL of reagent B and store the mixture for 60 min in the dark in a refrigerator at 4 °C. Read the absorbance at 600 nm against a blank. A linear correlation between absorbance and concentration is observed.

1.4.7.2. Specific Reactions for 2- and 3-Deoxysugars

1.4.7.2.a. Reaction with Trichloroacetic Acid and 4-Nitrophenylhydrazine

Heating an aqueous solution of 2-deoxysugars with trichloroacetic acid and 4-nitrophenylhydrazine, extraction of the solution with butyl acetate, and addition of alkali to the remaining aqueous layer yields a deep red-violet color ($\lambda_{max} = 560$ nm)[93–95].

α-Hydroxylevulinic aldehyde is assumed to be the reactive component derived from furan derivatives (see Figure 8).

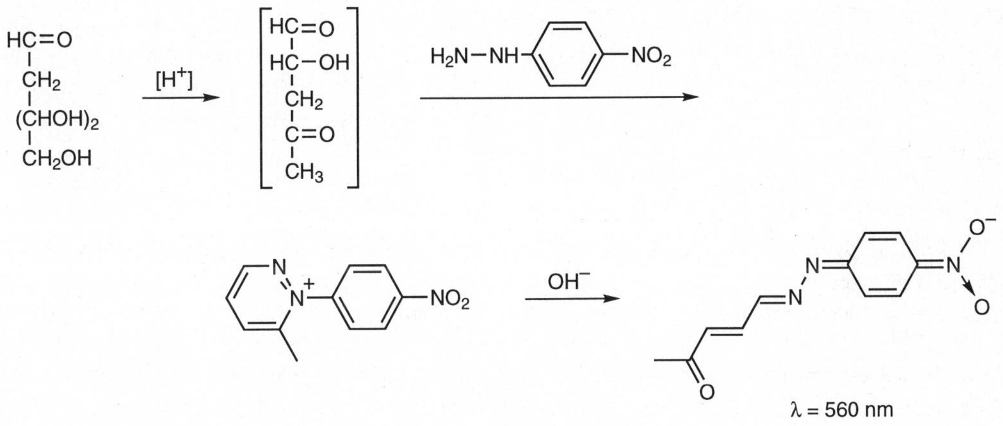

Figure 8. Reaction of 2-Deoxypentose and 4-Nitrophenylhydrazine in Alkaline Medium (Himmelspach and Westphal)

Determination[93,94]

Reagent A: 5% aqueous solution of trichloroacetic acid.

Reagent B: 0.5% solution of 4-nitrophenylhydrazine in ethanol.

Pipet to 2.0 mL of the sample (2–60 µg 2-deoxyaldose or 10–300 µg deoxyribonucleic acid) 2 mL of reagent A and 0.2 mL of reagent B. Heat for 20 min in a boiling water bath, then cool and add 10 mL of butyl acetate and shake for 5 min. After separation of the organic layer, add to 3 mL of the aqueous phase 1 mL of NaOH (2 M) and make the solution up to 5.0 mL with distilled water; then immediately read the absorbance at 560 nm against a blank. The reaction is positive with 2-deoxypentoses, 2-deoxyhexoses, 2,5-dideoxypentoses and 2,6-dideoxyhexoses but negative with 3-deoxysugars. Pentoses (e.g., in ribonucleic acids), hexoses and their oligo- and polysaccharides do not interfere under these conditions.

1.4.7.2.b. Reaction with Periodic Acid and Thiobarbituric Acid

Oxidation of 2- and 3-deoxysugars and related compounds (e.g., sugar alcohols) with periodic acid leads to the formation of malondialdehyde. This compound with thiobarbituric acid as well as 2-methylindole yields deep colors with λ_{max} at 532 and 555 nm[96,97], respectively.

2-Deoxysugars are oxidized at room temperature and 3-deoxysugars at higher temperatures[98]. Pentoses, hexoses and their oligo- and polysaccharides do not interfere since they yield only weak yellow colors with this reaction. Interference only occurs in the presence of such compounds which on oxidation with periodic acid yield malondialdehyde[98], or with compounds that perform similar color reactions such as maltol[99].

Determination[96]
Reagent A: solution of periodic acid (0.025 M) in sulfuric acid (0.063 M).
Reagent B: solution of 2% sodium arsenite in diluted hydrochloric acid (0.5 M).
Reagent C: 0.7 g of thiobarbituric acid are suspended in 90 mL of distilled water and 0.7 mL of aq NaOH (1 M) is then added. The solution is made up to 100 mL with distilled water and is filtered.
Pipet 0.5 mL of reagent A to 3.5 mL of the sample (1–15 µg deoxysugar) and store the mixture at room temperature for 20 min (splitting of *cis*-OH groups) or 40 min (splitting of *trans*-OH groups). Then add 1 mL of reagent B to reduce the excess periodic and iodic acid to iodide ions (the brown color which shortly appears must disappear again). After 2 min, add to 1 mL of this solution 2 mL of reagent C, heat for 20 min in a boiling water bath, cool to room temperature, and read the absorbance against a blank at 532 nm.
If the color is too faint for spectrophotometric measurement an extraction procedure can enhance the sensitivity. In this case, shake the aqueous solution with 1 mL of a mixture (1:1 v/v) of isoamyl alcohol–HCl (12 M) and separate the organic phase by low-speed centrifugation; measure the absorbance at 532 nm.
An excess of hexoses and pentoses do not interfere but it should be recognized that their compounds are also split by periodic acid. In the presence of such substances, the concentration of periodic acid has to be high enough to ensure that the complete splitting of all these compounds takes place.

1.4.8. Specific Reactions for Amino Sugars

These compounds react with carbohydrate specific as well as amino acid specific reagents (e.g., with ninhydrin).

1.4.8.1. Reaction with 4-Nitrobenzaldehyde and Pyridine
Hexosamines and other amino sugars react with 4-nitrobenzaldehyde in pyridine to form Schiff bases. Addition of tetraethylammonium hydroxide results in the formation of an intense red colored compound (λ_{max} = 500 nm). The chemical mechanism shown in Figure 9 occurs in this reaction. Hexoses, pentoses and their oligo- and polysaccharide derivatives do not react; it is a selective reaction for 2-amino-2-deoxy sugars. The presence of compounds with free amino groups does interfere with the reaction[100–102].

Determination[102]
Reagent A: 1% solution of 4-nitrobenzaldehyde in pyridine.
Reagent B: 3.0 mL of a 10% aqueous solution of tetraethylammonium hydroxide is diluted with ethanol to 100 mL.
To 1 mL of the sample (25–150 µg) pipet 0.5 mL of reagent A and let the mixture stand for 20 min at 27–28 °C. Then, cool the solution in an ice bath and make up the mixture with reagent B to 10 mL. After 25 min, measure the absorbance at 504 nm against a blank.

1.4.8.2. Reaction with Acetic Anhydride and 4-Dimethylaminobenzaldehyde

The reaction between hexosamines and acetic anhydride yields *N*-acetyl derivatives. By warming these compounds in diluted solutions of sodium carbonate or sodium borate, acidifying and then adding 4-dimethylaminobenzaldehyde, violet colors are formed which have two absorption maxima (544 and 585 nm; Morgan–Elson reaction)[103,104].

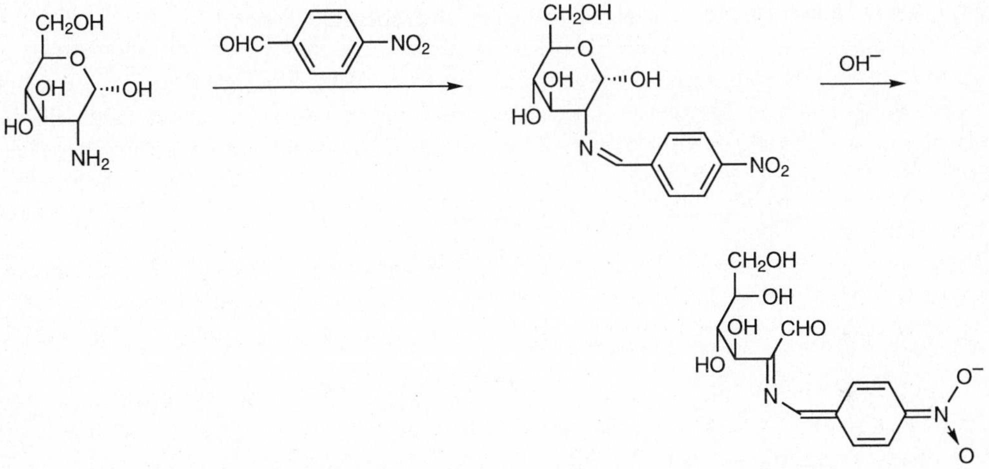

Figure 9.   Reaction between 4-Nitrobenzaldehyde and D-Glucosamine (Nakamura et al.)

In terms of the chemical reaction, it is assumed that furan derivatives are formed from the *N*-acetylhexosamines on treatment with alkali. These compounds undergo condensation reactions with the 4-dimethylaminobenzaldehyde in acidic medium to give triarylmethane dyes. *N*-Acetyl-3-aminofuran is one of such compounds which has been isolated. It is formed by dehydration of the furan derivative of *N*-acetyl-2-amino-2-deoxyhexose. When the C-1 or the C-4 position is substituted, the Morgan–Elson reaction is negative (e.g., *N*-acetylglucosamine 1-phosphate). Such compounds have to be hydrolyzed to obtain a positive reaction[105–107]; sugar and amino acids do not interfere with the reaction.

Determination[108]

Reagent A: 10 g of 4-dimethylaminobenzaldehyde are dissolved in an mixture of 87.5 mL pure acetic acid and 12.5 mL concd hydrochloric acid; shortly before use, 1 mL of this solution is diluted with 9 mL of pure acetic acid.

Reagent B: a solution of potassium tetraborate (0.8 M) is adjusted with diluted KOH to pH 9.0–9.2.

Reagent C: 5.0 mL of acetic anhydride is made up to 100 mL with ice water just before use; alternatively, a 5% solution of acetic anhydride in acetone can be used[109].

Procedure A: Reaction with *N*-Acetylhexosamines

Heat 0.5 mL of the sample (10–100 µg) with 0.1 mL of reagent B in a boiling water bath for 3 min; then cool and add 3 mL of reagent A to the mixture. After warming this solution for 20 min at 37 °C, read the absorbance at 544 or 585 nm against a blank.

Procedure B: Reaction with Hexosamines[108,109]
Pipet 0.1 mL of reagent C to 1 mL of the aqueous hexosamine solution. Add 0.1 mL sat. aq sodium bicarbonate and store the mixture at room temperature for 10 min. After heating this mixture for 3 min in a boiling water bath, cool and add aq sodium carbonate (0.5 M). Heat this mixture again at 100 °C for 3 min. Then cool the solution, add reagent A to an aliquot part and continue with Procedure A. Neutral sugars do not interfere with the reaction.

1.4.8.3. Reaction with Acetylacetone and 4-Dimethylaminobenzaldehyde
The condensation of hexosamines with acetylacetone in alkaline medium at pH 9.5 and temperatures around 90 °C leads to the formation of pyrrole derivatives (e.g., volatile methylpyrrole). These pyrrole compounds react with 4-dimethylaminobenzaldehyde to give deep colored compounds with two absorption maxima (450 and 510–545 nm). The following chemical mechanism is assumed (Figure 10)[108,110–112].

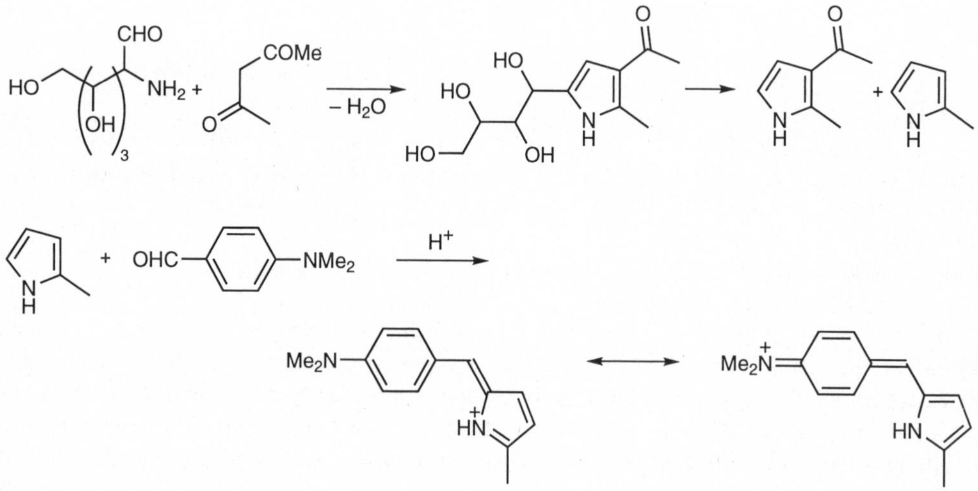

Figure 10. Reaction of 2-Amino Sugars with Acetylacetone and 4-Dimethylaminobenzaldehyde (Muhs and Weiss)

The reaction is positive only for free hexosamines; substituted compounds have to be split beforehand. Interference occurs when monosaccharides are heated with amino acids of primary amines. The compounds which are formed yield red colors with acetylacetone and 4-aminobenzaldehyde. Therefore, modifications are preferred which separate the volatile pyrrole compounds by distillation; in the distillate, the color reaction can be carried out without interference[113].

Determination
Modification A[22,108]
Reagent A: 4.0 mL of acetylacetone are diluted with aq sodium carbonate (0.63 M) to 100 mL.
Reagent B: 1.6 g of 4-dimethylaminobenzaldehyde are dissolved in a mixture of 30 mL of ethanol and 30 mL of concd hydrochloric acid.

Heat 2 mL of the sample (10–100 µg hexosamine) with 2 mL of reagent A for 60 min at 96 °C. After cooling, add 16 mL of ethanol and 2 mL of reagent B; read the absorbance of the red color at 450 and 530 nm against a blank.

Modification B[113]

Reagent A: 23 g of sodium carbonate (water free), 2.8 g of sodium bicarbonate and 5.8 g of sodium chloride are dissolved in 1 L of distilled water; in 100 mL of this solution, 1 mL of acetylacetone is dissolved and the pH is adjusted to 9.8 with sodium tetraborate. This reagent should be prepared just before use and remains stable for 1 day in a refrigerator.

Reagent B: 80 mg of 4-dimethylaminobenzaldehyde are dissolved in a mixture of 96.5 mL of ethanol and 3.5 mL of concd hydrochloric acid.

Pipet 2.0 mL of the sample (5–50 µg hexosamine) into a distillation flask, add 5.5 mL of reagent A and heat the mixture for 20 min in a boiling water bath. After this period, dilute the mixture with 5 mL of distilled water and distill 2 mL into a receiving vessel which contains 8 mL of reagent B. After 30 min, read the absorbance at 545 nm against reagent B.

The reaction between amino sugars and acetylacetone is strongly dependent on the pH of the medium (maximum pH 9.5–10). The amino acids do not interfere, only hydroxyproline and methylamine have faint reactions.

1.4.8.4. Reaction with Acetylacetone and Xanthydrol

The pyrrole compounds, formed by the reaction of hexosamines with acetylacetone in alkaline medium, react in acidic medium with xanthydrol to form a red-violet dye[114,115]. Compounds like 2-deoxysugars and even carbazole or indole compounds, interfere with the reaction.

Determination[114,115]

Reagent A: 1% solution of acetylacetone in 0.5 M sodium carbonate.

Reagent B: 100 mg of xanthydrol are dissolved in 100 mL of acetic acid and 1.0 mL of concd hydrochloric acid.

To 1 mL of the neutral aqueous sample, add 1 mL of reagent A and heat the mixture for 10 min in a boiling water bath. After cooling in ice water, add 10 mL of reagent B and heat again for 15 min in a boiling water bath; then cool and read the absorbance at 530 nm against a blank.

1.4.9. Specific Reactions for Neuraminic Acids (Sialic Acids)

These compounds are widely distributed in biological systems. The free neuraminic acids themselves do not exist, but do occur in the form of oxygen- and nitrogen-substituted compounds called sialic acids, e.g. β-*N*-acetylneuraminic acid (see Figure 11). This compound is formed by condensation of *N*-acetylmannosamine with pyruvic acid.

1.4.9.1. Reaction with Orcinol, Metal Salts and Hydrochloric Acid

Heating an aqueous hydrochloric acid solution of sialic acid with orcinol in presence of iron(III) or copper(II) ions yields red-violet colored compounds (λ_{max} = 540–580 nm), which are extractable with amyl or isoamyl alcohol[22,116,117]. The reaction is not specific since pentoses, deoxysugars, hexuronic acid and heptoses also yield blue and green colors. By extraction, the selectivity of the reaction is enhanced.

Determination[22]

Reagent: 100 mg of orcinol are dissolved in 41 mL of hydrochloric acid (12.5 M). Then, 1 mL of 1% aq $FeCl_3$ is added and the solution is made up with water to 50 mL.

To 2.0 mL of the sample (20–200 µg sialic acid) add 2 mL of the reagent and heat in a boiling water bath for 15 min. Then, cool in ice water and extract the colored solution with ice-cold amyl alcohol; read the absorbance of the organic phase at 540–580 nm against a blank.

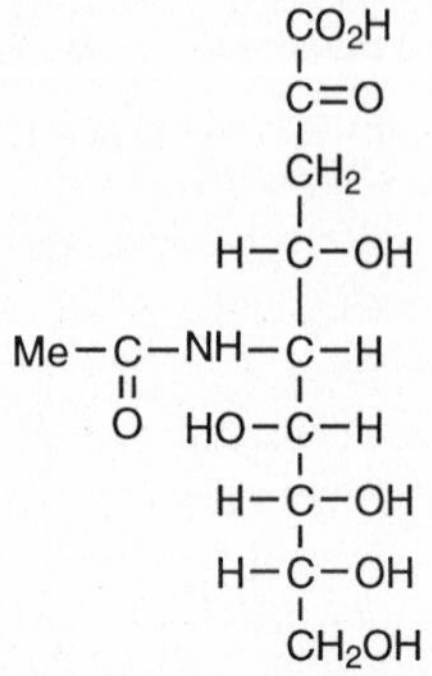

Figure 11.   Structure of a Sialic Acid (Acetylated Neuraminic Acid)

1.4.9.2. Reaction with 4-Dimethylaminobenzaldehyde and Hydrochloric Acid

Sialic acids react with 4-dimethylaminobenzaldehyde in hydrochloric acid solution to form a red-purple dye (λ_{max} = 565 nm). It is possible to distinguish between the direct reaction, where sugars and amino acids interfere, and the indirect reaction[118–120], where sialic acids are converted by alkali to pyrrolecarbonic acids which react further with the reagent to yield red colored compounds of the following chemical compositions (see Figure 12).

Determination[118]

Reagent A: aq boric acid solution (0.2 M) is adjusted with NaOH to a pH of 8.5.

Reagent B: 1.33 g of 4-dimethylaminobenzaldehyde are dissolved in a mixture of 50 mL of ethanol and 50 mL of concd hydrochloric acid.

To 0.5 mL of the aqueous sample (10–100 µg sialic acid) add 0.5 mL of reagent A and heat the mixture in a boiling water bath for 45 min. Then cool and add 3 mL of ethanol and 1 mL of reagent B. Heat for 20 min at 70 °C and, after cooling, read the absorption against a blank at 558 nm.

1.4.9.3. Reaction with Periodic Acid and Thiobarbituric Acid

The splitting of a sialic acid with periodic acid yields formylpyruvic acid which reacts with thiobarbituric acid to form a deep red dye (λ_{max} = 549 nm); this can be extracted by organic solvents[118,121–123]. The chemical structure of this compound is shown in Figure 13.

This reaction is positive for all 2-keto-3-deoxyaldonic acids as well as for 3-hydroxy-1,4-pyrones[99]. Interference occurs with 2-, 3- and 4-deoxysugars since they yield malondialdehyde, which also reacts with thiobarbituric acid to form deep red colored compounds (see Section 1.4.7.2.b.).

Determination[118]

The procedure is similar to that given in Section 1.4.7.2.b.

Reagent A: solution of periodic acid (0.025 M) in diluted sulfuric acid (0.063 M) (pH 1.2).

Reagent B: 2% solution of sodium arsenite in dilute hydrochloric acid (0.5 M).

Reagent C: aq thiobarbituric acid (0.1 M) adjusted to pH 9.0 with NaOH.

Reagent D: 1-butanol containing 5% hydrochloric acid (12 M).

Pipet 0.25 mL of reagent A to 0.5 mL of the aqueous sample and heat the mixture for 30 min in a water bath at 37 °C. Add 0.2 mL of reagent B and wait till the appearing brown color disappears again, then add 2 mL of reagent C, heat the solution for 8 min in a boiling water bath and cool in an ice bath. Extract the solution with reagent D and read the absorbance of the organic phase against a blank at 549 nm.

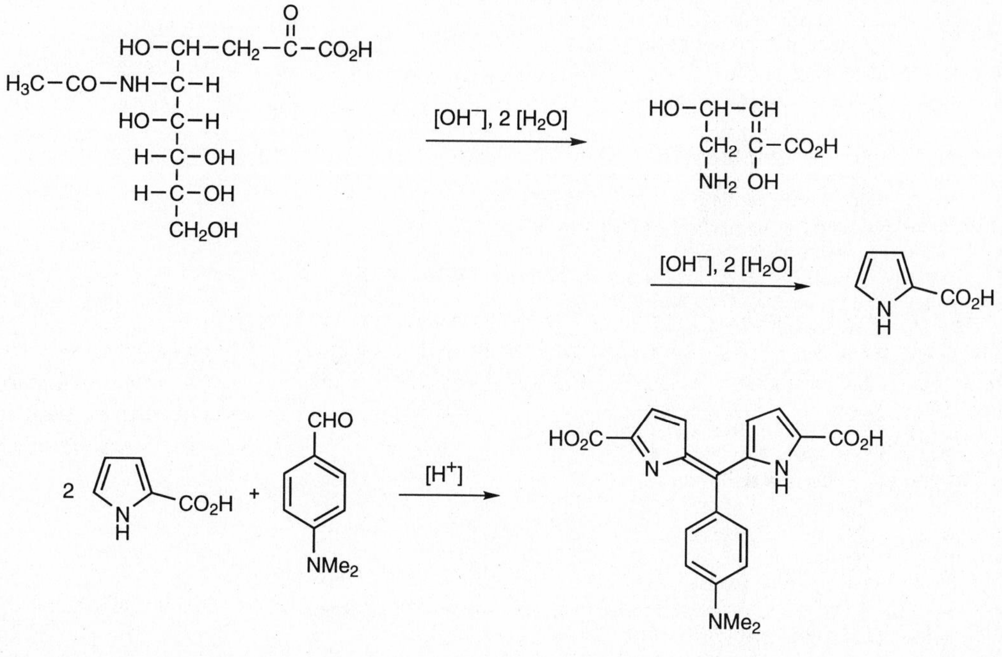

Figure 12.   Reaction between Sialic Acids and Alkali, and Subsequently with 4-Dimethylaminobenzaldehyde (Morgan and Schunior)

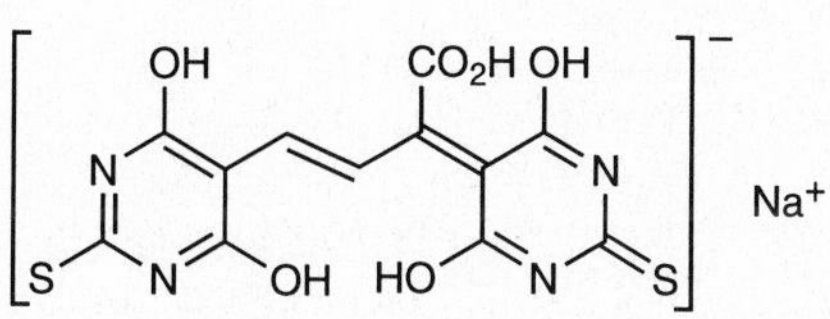

Figure 13.   Product Formed from the Reaction between Formylpyruvic Acid and Thiobarbituric Acid (Kuhn and Lutz)

References

1　　Kakác, B.; Vejdelek, Z. In *Handbuch der photometrischen Analyse organischer Verbindungen*, Vol. 2; VCH: Weinheim, 1974; p 871.
2　　Dubois, M.; Gilles, K. A.; Hamilton, J. K.; Rebers, P. A.; Smith, F. *Nature* **1951**, 168,167; *Anal. Chem.* **1956**, *28*, 350.
3　　Montgomery, R. *Biochim. Biophys. Acta* **1961**, *48*, 591.
4　　Whistler, R. L.; Hickson, J. L. *Anal. Chem.* **1955**, *27*, 1514.
5　　Livingston, E.; Maurmeyer, R. K.; Worthman, A. *Microchem. J.* **1957**, *1*, 261.
6　　Sattler, L.; Zerban, F. *Science* **1948**, *108*, 207.
7　　Dreywood, R. *Ind. Eng. Chem. Anal. Ed.* **1946**, *18*, 499.
8　　Momose, T.; Ueda, Y.; Sawade, K.; Sugi, A. *Chem. Pharm. Bull. (Tokyo)* **1957**, *5*, 31.
9　　Momose, T.; Okura, Y.; Hirauchi, K. *Chem. Pharm. Bull. (Tokyo)* **1963**, *11*, 1364.
10　　Koehler, L. H. *Anal. Chem.* **1952**, *24*, 1576.
11　　Yemm, E. W.; Willis, A. J. *Biochem. J.* **1954**, *57*, 509.
12　　Helbert, J. R.; Brown, K. D. *Anal. Chem.* **1955**, *27*, 1791.
13　　Jones, N. R.; Burt, J.-R. *Analyst* **1960**, *85*, 810.
14　　Jermyn, M. A. *Anal. Biochem.* **1975**, *68*, 332.
15　　Dische, Z. *Mikrochemie* **1929**, *7*, 33.
16　　Ujsaghy, P. *Biochem. Z.* **1938**, *298*, 141.
17　　Krainick, H. G. *Mikrochem. Mikrochim. Acta* **1941**, *29*, 45.
18　　Bollinger, A.; McDonald, N. D. *Aust. J. Sci.* **1947**, *9*, 189.
19　　Shetlar, M. R.; Masters, Y. R. *Anal. Chem.* **1957**, *29*, 402.
20　　Tillmans, J.; Philippi, K. *Biochem. Z.* **1929**, *215*, 36.
21　　Brückner, J. *Biochem. J.* **1955**, *60*, 200.
22　　Schultze, H. E.; Haupt, H.; Schmidtberger, R. *Biochem. Z.* **1958**, *329*, 490.
23　　Seibert, F. B.; Atno, J. *J. Biol. Chem.* **1946**, *163*, 511.
24　　Dische, Z. *Mikrochemie* **1930**, *8*, 4.
25　　Dische, Z.; Landberg, E. *Biochim. Biophys. Acta* **1957**, *24*, 193.
26　　Badin, J.; Jackson, C.; Schubert, M. *Proc. Soc. Exp. Biol. Med.* **1953**, *84*, 288.
27　　Solovjeva, T. P. *Lab. Delo* **1961**, *7 (11)*, 33.
28　　Sheppard, F.; Everett M. R. *J. Biol. Chem.* **1937**, *119*, 39.
29　　Cohen, S. S. *J. Biol. Chem.* **1944**, *156*, 691.
30　　Selivanoff, T. *Ber.* **1887**, *20*, 181.
31　　Roe, J. H. *J. Biol. Chem.* **1934**, *107*, 15.
32　　Berner, E.; Sandlie, S. *Chem. Ind. (London)* **1952**, 1221.
33　　Roe, J. H.; Papadopoulos N. M. *J. Biol. Chem.* **1954**, *210*, 703.
34　　Yaphe, W. *Anal. Chem.* **1960**, *32*, 1327.
35　　Foreman, D.; Gaylor, L.; Evans, E.; Trella, C. *Anal. Biochem.* **1973**, *56*, 584.
36　　Nakamura, M. *Agric. Biol. Chem.* **1968**, *32*, 696.
37　　Yaphe, W.; Arsenault, P. *Anal. Biochem.* **1965**, *13*, 143.
38　　Heyrovsky, A. *Collect. Czech. Chem. Commun.* **1957**, *22*, 43.
39　　Nakamura, M. *Agric. Biol. Chem.* **1968**, *32*, 689.
40　　Kakác. B.; Vejdelek, Z. In *Handbuch der photometrischen Analyse organischer Verbindungen*, Vol. 2; VCH: Weinheim, 1974; p 904.
41　　Gardell, S. *Acta Chem. Scand.* **1951**, *5*, 1011.
42　　Walborg, E.; Christensson, L. *Anal. Biochem.* **1965**, *13*, 186.
43　　Yee, H. Y.; Goodwin, J. *Anal. Chem.* **1973**, *45*, 2163.
44　　Abraham, C. V.; Gerarde, H. W. *Microchem. J.* **1976**, *21*, 14.
45　　Jones, J. K.; Pridham, J. B. *Nature* **1953**, *172*, 161.
46　　Jones, J. K.; Pridham, J. B. *Biochem. J.* **1954**, *58*, 288.
47　　Bial, H. *Dtsch. Med. Wochenschr.* **1902**, *29*, 253, 477.
48　　Kakác, B.; Vejdelek, Z. In *Handbuch der photometrischen Analyse organischer Verbindungen*, Vol. 2; VCH: Weinheim, 1974; p 927.
49　　Mejbaum, W. *Hoppe-Seyler's Z. Physiol. Chem.* **1939**, *258*, 117.
50　　McRary, W. L.; Slattery, M. C. *Arch. Biochem.* **1945**, *6*, 151.

51   Trim, A.; Parker, J. *Anal. Biochem.* **1970**, *35*, 475.
52   Barrenscheen, H.; Penham, A. *Hoppe-Seyler's Z. Physiol. Chem.* **1941**, *272*, 81.
53   Ceriotti, G. *J. Biol. Chem.* **1955**, *214*, 59.
54   Lin, R. I.; Schjeide, O. A. *Anal. Biochem.* **1969**, *27*, 473.
55   Widstoe, J. A.; Tollens, B. *Ber. Dtsch. Chem. Ges.* **1900**, *33*, 143.
56   Sen, R. S.; Sinha, N. N. *J. Am. Chem. Soc.* **1923**, *45*, 2984.
57   Dische, Z.; Borenfreund, E. *Biochim. Biophys. Acta* **1957**, *23*, 639.
58   Bolognani, L.; Coppi, G.; Zambotti, V. *Experientia* **1961**, *17*, 67.
59   Tomoda, M.; Kamyia, S. *J. Pharm. Soc. Jpn* **1962**, *82*, 1447.
60   Tomoda, M. *Chem. Pharm. Bull. (Tokyo)* **1963**, *11*, 806.
61   Vernon, L. P.; Avonoff, S. *Arch. Biochim. Biophys.* **1953**, *36*, 802d.
62   Stachey, M.; Deriaz, R. E.; Teece, E. G.; Wiggins, F. *Nature* **1946**, *157*, 740.
63   Momose, T.; Ueda, Y.; Nakamura, M. *Chem. Pharm. Bull. (Tokyo)* **1960**, *8*, 827.
64   Kakác, B.; Vejdelek, Z. In *Handbuch der photometrischen Analyse organischer Verbindungen*, Vol. 2; VCH: Weinheim, 1974; p 922.
65   Scherz, H.; Rücker, W.; Bancher, E. *Mikrochim. Acta* **1965**, 876.
66   Thiess, H.; Souci, S. W.; Kallinich, G. *Z. Lebensm. Unters.* **1952**, *94*, 240.
67   Eegrive, E. *Fresenius' Z. Anal. Chem.* **1937**, *110*, 22.
68   Klein, B.; Weissman, H. *Anal. Chem.* **1953**, *25*, 771.
69   Sawicki, E.; Hauser, R.; McPherson, S. *Anal. Chem.* **1962**, *34*, 1460.
70   Kakác, B.; Vejdelek, Z. In *Handbuch der photometrischen Analyse organischer Verbindungen*, Vol. 2; VCH: Weinheim, 1974; p 967.
71   Tollens, B.; Rorive, F. *Ber.* **1908**, *41*, 1783.
72   Tollens, B. *Ber.* **1908**, *41*, 1788.
73   Momose, T.; Ueda, Y.; Iwaski, M. *Chem. Pharm. Bull. (Tokyo)* **1955**, *3*, 321; **1956**, *4*, 49; **1962**, *10*, 546.
74   Kapp, E. M. *J. Biol. Chem.* **1940**, *134*, 143.
75   Hanson, S. W.; Hills, G. T.; Williams, R. T. *Biochem. J.* **1944**, *38*, 274.
76   Green, S.; Anstiss, C.; Fischman, W. *Biochem. Biophys. Acta* **1962**, *62*, 524.
77   Wagner, W. *Anal. Chim. Acta* **1963**, *29*, 182, 227.
78   Nir, I. *Anal. Biochem.* **1964**, *8*, 20.
79   Dische, Z. *J. Biol. Chem.* **1947**, *167*, 189.
80   Stutz, E.; Deuel, H. *Helv. Chim. Acta* **1956**, *39*, 2126.
81   Bitter, T.; Muir, H. *Anal. Biochem.* **1962**, *4*, 330.
82   Galambos, J. T. *Anal. Biochem.* **1967**, *19*, 119.
83   Knutson, C.; Jeans, A. *Anal. Biochem.* **1968**, *24*, 470.
84   Blumenkrantz, N.; Asboe-Hansen, G. *Anal. Biochem.* **1973**, *54*, 484.
85   Kinter, P. K.; van Buren, J. *J. Food Sci.* **1982**, *47*, 756.
86   Kakác, B.; Vejdelek, Z. In *Handbuch der photometrischen Analyse organischer Verbindungen*, Vol. 2; VCH: Weinheim, 1974; p 947.
87   Dische, Z. *Federation Proc.* **1947**, *6*, 248.
88   Dische, Z. *J. Biol. Chem.* **1947**, *171*, 725.
89   Gibbons, M. N. *Analyst* **1955**, *80*, 268e.
90   Edward, J. T.; Waldron, D. M. *J. Chem. Soc. (London)* **1952**, 3631.
91   Malpress, F. H.; Hytten, F. E. *Biochem. J.* **1958**, *68*, 708.
92   Waldron, D. M. *Nature* **1952**, *170*, 461.
93   Webb, J. M.; Levy, H. B. *J. Biol. Chem.* **1955**, *213*, 107.
94   Himmelspach, K.; Westphal, O. *Liebigs Ann. Chem.* **1963**, *668*, 165.
95   Fromme, T.; Himmelspach, K.; Lüderitz, O.; Westphal, O. *Angew. Chem.* **1957**, *69*, 643.
96   Waravdekar, V. S.; Saslaw, S. *J. Biol. Chem.* **1959**, *234*, 1945.
97   Scherz, H.; Stehlik, G.; Bancher, E.; Kaindl, K. *Mikrochim. Acta* **1967**, 915.
98   Cynkin, M. A.; Ashwell, G. *Nature* **1960**, *186*, 155.
99   Scherz, H. *Lebensm. Gerichtl. Chem.* **1974**, *28*, 337.
100   Nakamura, A.; Maeda, M.; Ikeguchi, K.; Kinoshita, T.; Tsuji, A. *Chem. Pharm. Bull. (Tokyo)* **1968**, *16*, 184.
101   Nakamura, A.; Maeda, M.; Kinoshita, T.; Tsuji, A. *Chem. Pharm. Bull. (Tokyo)* **1969**, *17*, 770.
102   Kakác, B.; Vejdelek, Z. In *Handbuch der photometrischen Analyse organischer Verbindungen*, Vol. 2; VCH: Weinheim, 1974; p 982.
103   Zuckerkandl, F.; Messiner-Klebermass, L. *Biochem. Z.* **1931**, *236*, 19.

104   Morgan, W. T.; Elson, L. A. *Biochem. J.* **1934**, *28*, 988.
105   Kuhn, R.; Krüger, G. *Chem. Ber.* **1956**, *89*, 1473.
106   Park, J. D. *J. Biol. Chem.* **1952**, *194*, 877, 885, 897.
107   Aminoff, D.; Morgan, W. T.; Watkins, W. M. *Biochem. J.* **1952**, *51*, 379.
108   Kakác, B.; Vejdelek, Z. In *Handbuch der photometrischen Analyse organischer Verbindungen*, Vol. 2; VCH: Weinheim, 1974; p 987.
109   Levy, C. A.; McAllen, A. *Biochem. J.* **1959**, *73*, 127.
110   Elson, L. A.; Morgan, W. T. J. *Biochem. J.* **1933**, *27*, 1824.
111   Boyer, R.; Fürth, O. *Biochem. Z.* **1935**, *282*, 242.
112   Muhs, M. A.; Weiss, F. T. *Anal. Chem.* **1958**, *30*, 259.
113   Cessi, C.; Piliego, F. *Biochem. J.* **1960**, *77*, 508.
114   Pesez, M.; Bartos, J.; Sezerat, A. *Bull. Soc. Chim. Fr.* **1961**, 567.
115   Kakác, B.; Vejdelek, Z. In *Handbuch der photometrischen Analyse organischer Verbindungen*, Vol. 2; VCH: Weinheim, 1974; p 990.
116   Klenk, E.; Langerbeins, H. *Hoppe Seyler's Z. Physiol Chem.* **1941**, *270*, 185.
117   Papadopoulos, N. M.; Hess, W. *Arch. Biochem. Biophys.* **1960**, *88*, 167.
118   Aminoff, D. *Biochem. J.* **1961**, *81*, 384.
119   Gottschalk, A. *Biochem. J.* **1955**, *61*, 298.
120   Morgan, L. R.; Schunior, R. *J. Org. Chem.* **1962**, *27*, 3696.
121   Warren, L. *J. Biol. Chem.* **1959**, *234*, 1971.
122   Weissbach, A.; Hurwitz, J. *J. Biol. Chem.* **1959**, *234*, 705.
123   Kuhn, R.; Lutz, P. *Biochem. Z.* **1963**, *338*, 554.

1.5. Enzymatic Methods in Aqueous Solutions

1.5.1. Introduction

The direct determination of individual substances, even in complicated mixtures, is possible with the use of specific enzymes. Such an approach can be applied in the field of carbohydrates since the necessary enzymes are commercially available.

For the determination of the compounds formed by the enzymatic reaction the following techniques have been applied: photometry, fluorimetry, electrochemistry (e.g., conductometric methods, measurement of redox potential, polarography), titrimetric analysis (e.g., stationary pH methods), polarimetry and manometric and radiometric measurements (isotope dilution analysis). For practical purposes, the most important techniques are those of photometry, fluorimetry and, more recently, electrochemical methods.

Enzymes are proteins with the properties of specific catalysts which accelerate the reaction of a substrate in a strictly defined direction.

Simple Enzyme Reactions

These can be used to distinguish between one- and two-substrate reactions.

The following is an example of a one-substrate reaction.

$$\text{oxalacetate} \xrightarrow{\text{enzyme}} \text{pyruvate} + CO_2$$

Pseudo-one-substrate reactions are those in which the concentration of one of the two reactants can be ignored as it is present in great excess, for example:

$$\text{sucrose} + H_2O \xrightarrow{\text{enzyme}} \text{glucose} + \text{fructose}$$

In the case of two-substrate reactions, the enzyme catalyzes the reaction between the two compounds. An example of this, shown below, is the phosphorylation of sugars with adenosine triphosphate (ATP) yielding phosphorylated sugars and adenosine diphosphate (ADP) which is a very important reaction in the analysis of carbohydrates.

$$\text{D-glucose} + \text{ATP} \xrightarrow{\text{enzyme}} \text{D-glucose-6-phosphate} + \text{ADP}$$

The one-substrate enzyme reaction takes place according to the following two steps:

$$S + E \underset{K_{-1}}{\overset{K_1}{\rightleftharpoons}} [ES] \xrightarrow{K_2} E + P$$

The enzyme reacts with the substrate S to form the complex ES in which the reaction occurs. This complex ES breaks down into the reaction product P and the unaltered enzyme. When the rate of the formation of [ES] is high and the enzyme–substrate complex is split irreversibly the Michaelis–Menten equation is valid.

$$v = K_2 \frac{[S]\,[E]_o}{[S] + K_m} = \frac{[S]\,V_{max}}{[S] + K_m}$$

v = reaction rate; $[S]$ = concentration of the substrate; $[E]_o$ = the total concentration of the enzyme; K_m = the Michaelis constant.

$$K_m = \frac{K_2 + K_{-1}}{K_1}$$

1. $[S] \gg K_m \quad v = K_2\,[E]_o = V_{max}$

2. $[S] = K_m \quad v = \dfrac{K_2}{2}\,[E]_o = \dfrac{V_{max}}{2}$

3. $[S] \ll K_m \quad v = \dfrac{K_2}{K_m}\,[E]_o\,[S] = \dfrac{V_{max}}{K_m}\,[S] = C\,[S]$

In the first case, the concentration of the substrate is much higher than K_m which leads to a zero order reaction rate. In the third case, the concentration of the substrate is significantly lower than K_m which results in a 1st order reaction where the reaction rate is proportional to the concentration of the substrate. K_m is not an absolute constant number but is dependent on temperature, buffer, pH, etc.
The two-substrate reaction is more mathematically complicated but this type of reaction can often be converted to a pseudo one-substrate reaction by careful choice of reaction conditions.

Coupled Enzyme Reactions
These are often used in those cases where either none of the enzymatic reaction products can be determined directly or if the substance to be determined has to be converted into a precursor compound with which the real enzymatic reaction for determination is carried out (e.g., phosphorylation of the sugar).

Possibilities of Measurement
Measurements are taken of either the rate of the reaction or, more usually, the concentration of one of the reaction products, determined by the methods previously described (e.g., UV photometry). Another possibility is to determine the change in redox potential against a standard electrode.

1.5.2. Determination of Monosaccharides, Oligosaccharides and Related Compounds

The enzymes used for these analyses are dehydrogenases or oxidases. They convert monosaccharides into oxidation products by transferring hydrogen to other substrates. The most important compounds are nicotinamide adenine dinucleotide (NAD) and its corresponding phosphoric acid ester (NADP), and elementary oxygen.

Enzymatic approaches have been used for the following monosaccharides and related compounds: glucose, fructose, galactose, mannose, arabinose, xylose, xylulose, ribulose, erythrose, 2-deoxyglucose, fucose, sorbitol, mannitol, *myo*-inositol, galacturonic acid, gluconic acid and its δ-lactone, and the corresponding phosphates of these compounds along with sedoheptulose 7-phosphate, and erythrose 4-phosphate[1].

1.5.2.1. Determination of D-Glucose

D-Glucose is the most important monosaccharide occurring in both the free and bound state and is one of the most widespread organic compounds on earth. The determination of this compound in many different fields (e.g., biochemistry, clinical chemistry, food chemistry, etc.) and matrices (e.g., blood for detecting and controlling diabetes) is carried out with great success using enzymatic methods.

1.5.2.1.a. Reaction with Hexokinase and Glucose-6-Phosphate Dehydrogenase

D-Glucose (Glu) is converted by hexokinase (D-hexose-6-phosphotransferase, EC 2.7.1.1) and adenosine triphosphate (ATP) to D-glucose-6-phosphate (D-Glu-6-P). This compound is dehydrogenated by D-glucose-6-phosphate dehydrogenase (G-6-P DH, EC 1.1.1.49) and NADP to D-glucono-δ-lactone-6-phosphate according to the following scheme.

$$\text{D-Glu + ATP} \xrightarrow{\text{hexokinase}} \text{D-Glu-6-P + ADP}$$

$$\text{D-Glu-6-P + NADP}^+ \xrightarrow{\text{G-6-P DH}} \text{D-glucono-δ-lactone-6-P + NADPH + H}^+$$

The formation of NADPH is determined by reading the enhancement of the absorbance at 346–365 nm. The difference $E–E_B$ (E_B: blank) is proportional to the concentration of glucose[2,3].

Hexokinase transfers a phosphate residue not only to D-glucose but also to D-fructose, D-mannose, D-glucosamine and 2-deoxy-D-glucose. Other sugars are not affected by this reaction. D-Glucose-6-phosphate dehydrogenase (G-6-P DH) is very specific and does not react with any other hexose or pentose phosphates. This property allows the determination of D-glucose in a mixture of other sugars. G-6-P DH reacts solely with NADP and not with NAD.

Determination[3]

Reagent A: 14.0 g of triethanolamine hydrochloride and 0.25 g of $MgSO_4 \cdot 7H_2O$ are dissolved in 80 mL of doubly distilled water. The solution is adjusted to a pH of 7.6 with 5 M NaOH (approximately 5 mL is required).

Reagent B: 60 mg of NADP–Na_2 are dissolved in 6 mL doubly distilled water; this solution is stable at 4 °C for approximately 4 weeks.

Reagent C: 300 mg ATP–Na_2H_2 (adenosine triphosphate–disodium salt) and 300 mg of $NaHCO_3$ are dissolved in 6 mL of doubly distilled water.

Reagent D: Hexokinase (2.0 mg/mL)/G-6-P DH (1.0 mg/mL) is commercially available and should not be diluted; it is stable at 4 °C for approximately 1 year.

Pipet the following quantities into glass, quartz or one-way cuvette.

	Sample (mL)	Blank (mL)
Reagent A (buffer)	1.0	1.0
Reagent B (NADP)	0.1	0.1
Reagent C (ATP)	0.1	0.1
Sample	0.1	-
Water	1.9	2.0

After 3–5 min, read the absorbance of the sample $(E_1)_S$ and the blank $(E_1)_B$ at 340, 334 (Hg lamp) or 365 nm (Hg lamp) against air or water. Start the reaction by adding 0.02 mL of reagent D, wait 10–15 min until the reaction comes to an end and read the absorbance of the sample $(E_2)_S$ and the blank $(E_2)_B$. Calculate the differences between the absorbances $E_2–E_1$.

$\Delta E_{Glucose} = \Delta E_S – \Delta E_B$

ΔE_S: Absorbance of the sample $(E_2)_S – (E_1)_S$

ΔE_B: Absorbance of the blank $(E_2)_B – (E_1)_B$

With an exactly known amount of D-glucose, determine the molar absorbance coefficient. This value is used for calculating the concentration of the glucose in the sample. The amount of glucose should range between 5–30 µg.

1.5.2.1.b. Reaction with D-Glucose Dehydrogenase

β-D-glucose dehydrogenase (GDH, EC 1.1.1.118) catalyzes the dehydrogenation of β-D-glucose in the presence of NAD to D-gluco-δ-lactone. To accelerate the conversion of α-D-glucose to β-D-glucose in solution, addition of mutarotase (aldose-1-epimerase, EC 5.1.3.3) is necessary[4-6].

$$\alpha\text{-D-glucose} \quad \underset{}{\overset{mutarotase}{\longleftrightarrow}} \quad \beta\text{-D-glucose}$$

$$\beta\text{-D-glucose} + NAD^+ \quad \overset{GDH}{\longrightarrow} \quad \text{D-gluco-}\delta\text{-lactone} + NADH + H^+$$

$$\text{D-gluco-}\delta\text{-lactone} + H_2O\ (OH^-) \quad \longrightarrow \quad \text{D-gluconate}$$

GDH reacts with a few other sugars with different rates of reaction, e.g. 2-deoxy-D-glucose (125), 2-amino-2-deoxy-D-glucose (31), D-xylose (15), D-mannose (8), cellobiose (1), D-ribose (0.8) and lactose (0.7) (where the rate of reaction with β-D-glucose = 100).

Determination[5]

Reagent A: 1.0 g $Na_2HPO_4 \cdot 2H_2O$, 55 mg of $NaH_2PO_4 \cdot H_2O$, 0.44 g of NaCl, 55.6 mg of EDTA–$Na_2H_2 \cdot 2H_2O$ and 112 mg of oxamic acid are dissolved in doubly distilled water, adjusted to pH 7.6 and made up to 50 mL. Addition of 50 mg NaN_3 prevents microbial growth.

Reagent B: 58.4 mg of NAD (free acid) and 1800 U mutarotase are dissolved in 40 mL of reagent A; the reagent should be stored at 2–5 °C.

Reagent C: 250 U glucose dehydrogenase is dissolved in 0.5 mL reagent A.

Pipet the following quantities into a glass, quartz or
one-way cuvette.

	Sample (mL)	Blank (mL)
Reagent B (NAD–mutarotase)	2.0	2.0
Sample	0.2	-
Water	-	0.2

Read the absorbance E_1 against the blank at 340 nm (Hg lamp with 334 or 365 nm filter). Start the reaction by adding of 0.02 mL of reagent C and read the absorbance E_2 against the blank after reaching a constant value (after approximately 15 min). Repeat the same procedure with a standard solution of D-glucose. Using the molar absorbance coefficient, calculate the concentration of the sample from the difference in the absorbances E_2–E_1. The standard deviations are between = 0.57 and 1.44% rel. The detection limit is approximately 15 µg/mL.

Souce of Errors

This method is almost insensitive to the presence of ascorbic acid, glutathione, hemoglobin, bilirubin, anticoagulants and inhibitors of glycolysis. At high concentrations, when the pH of the sample is below 7.0, the amount of glucose is underestimated.

1.5.2.1.c. Reaction with Glucose Oxidase

β-D-Glucose is oxidized, under the catalytic influence of glucose oxidase (GOD, EC 1.1.3.4), to D-glucono-δ-lactone by transferring two hydrogens to oxygen to form hydrogen peroxide according to the following scheme[7,8].

$$\alpha\text{-D-glucose} \longleftrightarrow \beta\text{-D-glucose}$$

$$\beta\text{-D-glucose} + O_2 \xrightarrow{\text{GOD}} \text{D-glucono-}\delta\text{-lactone} + H_2O_2$$

$$\text{D-glucono-}\delta\text{-lactone} \xrightarrow{\text{H}_2\text{O (OH}^-)} \text{D-gluconate}$$

Hydrogen peroxide is determined by photometric, fluorimetric or luminescence techniques, or by electrochemical methods like potentiometry. The enzyme is specific for β-D-glucose and a few other sugars react with the following relative reaction rates (where β-D-glucose: = 100): 2-deoxy-D-glucose: 25; D-mannose: 0.98; α-D-glucose: 0.64; D-xylose: >0.4; D-galactose: 0.14[8-10].

The most important techniques are based on photometric approaches. Peroxidase (EC 1.11.1.7) catalyzes the hydrogen peroxide oxidation of phenols, aromatic amines, etc., resulting in the formation of dyes.

The following compounds are used: o-dianisidine[11], 2,2'-azinobis(3-ethylbenzothiazoline-6-sulfonate) (ABTS)[12], o-toluidine[13] and 4-aminophenazone–phenol[14]. The equilibrium of the reaction lies completely to the right since D-glucono-δ-lactone spontaneously hydrolyzes to gluconate under the reaction conditions.

Determination[11,15]

The method described is based on the procedure of Hugett and Nixon.

Reagent A: 2.07 g of $Na_2HPO_4 \cdot 2H_2O$, 1.09 g of $NaH_2PO_4 \cdot 2H_2O$, 6 mg of peroxidase (POD, commercial product) and 38 mg of glucose oxidase (commercial product) are dissolved in doubly distilled water and the solution made up to 150 mL

Reagent B: 10 mg of *o*-dianisidine are dissolved in 2 mL distilled water.

Reagent C: 50 mL of reagent A is mixed with 0.5 mL of reagent B. This mixture must be freshly prepared or be stored in a dark bottle.

Pipet the following quantities into a test tube.

	Sample (mL)	Blank (mL)
Reagent C	5.0	5.0
Sample	0.2	-
Blank (distilled water)	-	0.2

Mix thoroughly and allow to stand for 30–40 min at 20–22 °C; then measure the absorbance at 436 nm against the blank. Carry out the same procedure with standard solutions of D-glucose and, from the calibration curve, estimate the concentration of the sample.

The reaction time can be reduced by addition of mutarotase (EC 5.1.3.3) as described in the procedure of Trinder[14].

This reaction is also the basis for the test strips used in the detection of glucose in urine (test for diabetes). *o*-Dianisidine is a critical substance from a medical point of view (also carcinogenic but weaker than benzidine) and can be replaced by 2,2'-azinobis(3-ethylbenzothiazoline-6-sulfonate) (ABTS). The reaction with H_2O_2 and peroxidase, yielding a blue-green color, is four times more sensitive than the reaction with *o*-dianisidine. The reagent is very water soluble and stable towards light and oxygen. The procedure for the determination of D-glucose with this reagent is similar to that described for *o*-dianisidine[12].

Sources of Error

The presence of large amounts of reducing substances like ascorbic acid, uric acid, glutathione, etc., encourage low glucose concentrations since they react with hydrogen peroxide. Fluoride and oxalate ions also interfere with the reaction resulting in an overestimation of the glucose concentration. Interference also occurs in the presence of disaccharides and if the GOD is contaminated with enzymes such as α-glucosidase, amylase, β-galactosidase, etc.

1.5.2.2. Determination of D-Fructose

This compound is phosphorylated by ATP using hexokinase (HK) as described in 1.5.2.1.a. to give fructose-6-phosphate

$$\text{D-fructose} + \text{ATP} \xrightarrow{\text{HK}} \text{D-fructose-6-P} + \text{ADP}$$

Fructose-6-phosphate is converted by glucose phosphate isomerase (PGI EC 5.3.1.9) into glucose-6-phosphate, which is then dehydrogenated with glucose-6-phosphate dehydrogenase (G-6-P DH) to phosphoglucono-δ-lactone (cf. Section 1.5.2.1.a.), which is further converted into 6-phospho-gluconate[16–18].

Determination

Reagent A: 14.0 g of triethanolamine hydrochloride and 0.25 g of $MgSO_4 \cdot 7H_2O$ are dissolved in 80 mL of water. The solution is adjusted to pH 7.6 with 5 M NaOH (approx 5 mL) and made up with distilled water to 100 mL.

Reagent B: 60 mg of NADP–Na_2 are dissolved in 6 mL of distilled water.

Reagent C: 300 mg of ATP–Na_2H_2 and 300 mg of $NaHCO_3$ are dissolved in 6 mL of water.

Reagent D: hexokinase/glucose-6-phosphate dehydrogenase (2 and 1 mg/mL, respectively); the enzyme mixture, which is commercially available for this specific purpose, is used without dilution.

Reagent E: glucose-6-phosphate isomerase (PGI), which is commercially available, is used without dilution.

Pipet into glass, quartz or one-way cuvettes the following reagents and mix thoroughly.

	Sample (mL)	Blank (mL)
Reagent A (buffer)	1.00	1.00
Reagent B (NADP)	0.10	0.10
Reagent C (ATP)	0.10	0.10
Reagent D	0.02	0.02
Sample	0.10	-
Water	1.90	2.00

After the end of the reaction (approx 15–20 min), read the absorbance of the sample $(E_1)_S$ and the blank $(E_1)_B$ at 340 nm (334 or 365 nm Hg lamps) against water or air. Add 0.02 mL of reagent E to both cuvettes, mix thoroughly and, after 15 min, read the absorbance of the sample $(E_2)_S$ against the blank $(E_2)_B$.

The difference in the absorbance $\Delta E = [(E_2)_S-(E_1)_S] - [(E_2)_B-(E_1)_B]$ is proportional to the fructose concentration. The proportional factor is determined with a standard solution of D-fructose.

This enzymatic assay for the determination of fructose is frequently applied in the fields of biochemistry, clinical chemistry and food chemistry.

Sources of Error

Since PGI is a very specific enzyme, this approach is disturbed by only a few substances. One of them is D-fructose-6-phosphate itself which is present in very low concentrations in blood or serum but not in food samples. Impurities in the enzyme are possible, e.g. contamination with gluconate 6-phosphate dehydrogenase. If the glucose concentrations are too high, the reaction rate is slow and, in this case, the addition of glucose oxidase and catalase is recommended[19].

1.5.2.3. Determination of D-Galactose

Two independent approaches are in existence, namely the reaction with galactose oxidase and the reaction with galactose dehydrogenase.

1.5.2.3.a. Reaction with Galactose Oxidase

The enzyme galactose oxidase (GaOD, EC 1.1.3.9) acts as catalyst with β-D-galactose to form D-galactose hexodialdose according to the following scheme.

$$\beta\text{-D-galactose} + O_2 \xrightarrow{\text{GaOD}} \text{D-galactose hexodialdose} + H_2O_2$$

$$H_2O_2 + DH_2 \xrightarrow{\text{POD}} D + 2\,H_2O$$

Hydrogen peroxide reacts with the aid of peroxidase (POD) with a chromogen DH_2 to give a colored compound, which is determined by photometry or fluorimetry[20-22]. The following compounds are used as chromogens: o-dianisidine, benzidine, o-cresol and o-toluidine[21-23].

Galactose oxidase is not a very specific enzyme; it also oxidizes D-galactose-containing oligo- and polysaccharides and the corresponding glycoproteins, and, to a lesser extent, 2-deoxy-D-galactose, D-galactosamine, D-talose, D-tagatose and D-gulose[24].

Determination[20]

Reagent A: 2.5 g of benzoic acid are dissolved in 100 mL of boiling water; at 0–4 °C this solution is saturated

Reagent B: phosphate buffer pH 7: 5.38 g of $NaH_2PO_4 \cdot H_2O$ and 10.86 g of $Na_2HPO_4 \cdot H_2O$ are dissolved in 1000 mL of distilled water.

Reagent C: lyophilized horse radish peroxidase (approx 10 mg) is dissolved in 1000 mL of phosphate buffer (reagent B); the solution should have an activity of 180 U/L.

Reagent D: 5 g of o-dianisidine are dissolved in 100 mL of acetone.

Reagent E: galactose oxidase; the amount which corresponds to 1200 U/L is dissolved in 1 mL of distilled water.

Reagent F: mixture of 190 mL of POD phosphate buffer (reagent C), 0.2 mL of Tween 20, and 0.2 mL of o-dianisidine (reagent D) which is made up to 200 mL.

Reagent G: 0.3 mL of galactose oxidase (reagent E) made up to 100 mL with reagent F.

Standard solution: 180 mg of D-(+)-galactose dissolved in 100 mL of reagent A; this stock solution is diluted to the concentration used for the determination (0.5–40 µg/mL).

Pipet the following into a glass cuvette or test tube.

	Sample (mL)	Standard (mL)
Reagent G (galactose oxidase)	2.0	2.0
Sample	1.0	-
Standard (diluted)	-	1.0

Shake thoroughly and store the mixtures for 60–90 min at room temperature. Measure the absorbance, at 440–460 nm, of the sample and standard against the sample and reagent F. From the absorbance of the standard, calculate the concentration of the sample (under these conditions this can range from 0.5–40 µg/mL). The procedure should be carried out in the absence of direct light.

Source of Errors

Substances which react with hydrogen peroxide, e.g. SH-containing peptides and proteins, ascorbic acid and uric acid, etc., must not be present.

1.5.2.3.b. Reaction with Galactose Dehydrogenase

The enzyme β-D-galactose dehydrogenase (GalDH, EC 1.1.1.48) transfers 2 moles of hydrogen to NAD to form D-galactono-δ-lactone and NADH according to the following scheme.

$$\text{β-D-galactose + NAD}^+ \xrightarrow{\text{GalDH}} \text{NADH + H}^+ \text{ + D-galactono-δ-lactone}$$

Under conditions of high pH, the D-galactono-δ-lactone hydrolyzes spontaneously and irreversibly to galactonate. NADH is determined by UV photometry or fluorimetry[25–27].

GalDH reacts not only with β-D-galactose but also with α-L-arabinose and β-D-fucose, 2-deoxy-D-galactose and 2-amino-2-deoxy-D-galactose. However, it does not react with β-L-fucose, galactitol, D-glucose, D-mannose, L-galactose, D-arabinose, D-ribose and D-xylose or galactosides.

Determination[25]

Reagent A: 0.1 M tris-buffer: 300 mg of tri(hydroxymethyl)aminoethane dissolved in distilled water, adjusted to pH 8.6 with 0.1 M hydrochloric acid and made up to 25 mL.

Reagent B: 0.017 M aqueous solution of NAD: 60 mg NAD dissolved in 5 mL distilled water.

Reagent C: D-galactose dehydrogenase (approximately 5 U/mg); the enzyme is commercially available and is used undiluted.

All solutions should be stored at 4 °C.

Pipet the following into a glass, quartz or one-way cuvette.

	Sample (mL)	Blank (mL)
Reagent A	2.68	2.68
Reagent B	0.10	0.10
Sample	0.20	-
Water	-	0.20

Read the absorbance E_1 at 340 nm in a spectrophotometer [or at 334 or 364 nm (Hg lamp)] against a blank. Add 0.02 mL of reagent C, mix the solutions in the cells and follow the change in the absorbance of E_2 until it remains constant.

The difference in the absorbance E_2–E_1 is proportional to the amount of D-galactose. A similar procedure exists for a microdetermination using 0.35 M phosphate buffer at pH 8.0 and a sample volume of 10 μL. The concentration of NADH is determined by fluorimetry using an excitation wave length of 340 nm and the emission is measured at 450 nm[26,27].

1.5.2.4. Determination of L-Arabinose by Reaction with β-D-Galactose Dehydrogenase

As remarked previously in Section 1.5.2.3.b., this enzyme also reacts with L-arabinose, which can be used for its determination according to the following scheme

$$\text{L-arabinose + NAD}^+ \underset{\longleftarrow}{\overset{\text{GalDH}}{\longrightarrow}} \text{L-arabono-1,4-lactone + NADH + H}^+$$

$$\text{L-arabono-1,4-lactone + H}_2\text{O (OH}^-) \longrightarrow \text{L-arabonate}$$

The procedures for this determination are the same as described in Section 1.5.2.3.b.[28,29].

1.5.2.5. Determination of Sorbitol

D-Sorbitol is dehydrogenated by the catalytic action of the enzyme sorbitol dehydrogenase (SDH, EC 1.1.1.14) and the aid of NAD^+ to give D-fructose and NADH.

$$\text{D-sorbitol} + NAD^+ \xrightleftharpoons{\text{SDH}} \text{D-fructose} + NADH + H^+$$

The equilibrium of the reaction is strongly dependent on pH. In alkaline medium (pH 9.5) and in the presence of excess NAD it lies completely on the side of fructose. This enzyme is not very specific and also reacts with glucose-6-phosphate and other hexose esters. For the exact quantitative determination of sorbitol it is necessary to measure the D-fructose formed before and after the reaction with SDH. Before the amount of D-fructose can be enzymatically determined, as described in Section 1.5.2.2., the NADH formed during the reaction must be converted to NAD by lactate dehydrogenase[30,31].

Determination[30]

Reagent A: glycine/$MgSO_4$ buffer (0.1 M glycin, 0.004 M $MgSO_4$): 750 mg of glycine and 100 mg of $MgSO_4 \cdot 7H_2O$ dissolved in 80 mL of distilled water, the pH adjusted to 9.5 with 1 M NaOH and the solution made up to 100 mL.

Reagent B: NAD/NADP–ATP solution: 60 mg of NAD (free acid), 29 mg of NADP–disodium salt, 150 mg of ATP–Na_2H_2 and 30 mg of sodium pyruvate salt dissolved in 3 mL of distilled water.

Reagent C: enzyme solution (this is prepared from commercial products): 0.12 mL of hexokinase (from yeast; suspension in 3.2 M ammonium sulfate, $\geq$ 140 U/mg protein), 0.12 mL glucose-6-phosphate dehydrogenase (from yeast; suspension in 3.2 M ammonium sulfate $\geq$ 140 U/mg protein), 0.08 mL glucose-6-phosphate isomerase (from yeast; suspension in 3.2 M ammonium sulfate $\geq$ 350 U/mg protein), 0.08 mL lactate dehydrogenase (from pig muscle; suspension in 3.2 M ammonium sulfate). The solutions of these four enzymes are mixed thoroughly and made up to 2.0 mL with 3.2 M ammonium sulfate

Reagent D: sorbitol dehydrogenase (SDH): 18 mg of commercial enzyme (6.7 U/mg lyophilisate) dissolved in 0.5 mL of distilled water.

All these reagents should be stored in the refrigerator between 0 and 4 °C. Reagent C is stable for a few months whilst reagent D is stable for only two weeks.

Pipet the following into a glass, quartz or one-way cuvette in the following range.

	Sample (mL)	Blank (mL)
Reagent A	2.00	2.00
Reagent B	0.10	0.10
Reagent C	0.10	0.10
Sample	0.10	-
Water	0.90	1.00

Mix thoroughly and after 5–10 min read the absorbance of the sample $(E_1)_S$ and the blank $(E_1)_B$ against water or air at 339 nm with a spectrophotometer (or at 334 or 365 nm with a Hg lamp). Add 0.02 mL of reagent D to the sample and the blank and read the absorbances $(E_2)_S$ and $(E_2)_B$ at the same wavelengths until they remain constant (~ 20 min).

The difference $\Delta E = [(E_2)_S-(E_1)_S] - [(E_2)_B-(E_1)_B]$ is proportional to the concentration of sorbitol. The same procedure should be carried out with a standard sorbitol solution which allows determination of the sorbitol content of the unknown sample. The absorbance/concentration ratio is linear between 1–300 µg/mL under these conditions.

Source of Errors

The presence of lead ions inhibits sorbitol dehydrogenase which leads to retardation of the reaction. In the presence of large amounts of glucose and fructose the concentrations of ATP and NADP (reagent B) must be increased.

1.5.2.6. Determination of *myo*-Inositol

This substance is widely distributed in living organisms (plant and animals), for example, as its phosphate esters known as phytic acid in cereals and legumes.

myo-Inositol is dehydrogenated by inositol dehydrogenase (IHD, EC 1.1.1.18) and NAD forming *scyllo*-inosose (2,3,4,5,6-pentahydroxycyclohexanone) and NADH according to the following scheme.

$$myo\text{-inositol} + NAD^+ \xrightarrow{\quad IHD \quad} scyllo\text{-inosose} + NADH + H^+$$

The reaction does not proceed to completion under these conditions; the maximum extent of reaction being proportional to the concentration. The NADH concentration is determined by spectrophotometry at 339 nm[32–34].

Determination[33]

Reagent A: 0.5 M carbonate buffer pH 9.5: aqueous 0.5 M $NaHCO_3$ solution (42 g/L) is added to 0.5 M Na_2CO_3 solution (53 g/L) until pH 9.5 is reached.

Reagent B: NAD^+ solution: 73.5 mg of NAD^+ (free acid) dissolved in 5 mL of distilled water.

Reagent C: inositol dehydrogenase, commercial product of the lyophilisate of enterobacteria aerogenes. Reagents B and C should be stored in a refrigerator. Both are stable at –20 °C for several months.

Pipet the following successively into a glass, quartz or one-way cuvette.

	Sample (mL)	Blank (mL)
Reagent A	0.20	0.20
Reagent B	0.10	0.10
Reagent C	0.05	0.05
Sample	0.65	-
Water	-	0.65

Mix the solution thoroughly and measure the absorbance at 339 nm (or at 334 or 365 nm with a Hg lamp) against the blank. Carry out the same procedure with standard solutions of *myo*-inositol to establish a calibration curve, which can be used to determine the concentrations of the samples.

The enzyme is not very specific for *myo*-inositol; D-glucose, dihydroxyacetone and glyceraldehyde also react but with a much slower reaction rate.

1.5.2.7. Determination of D-Gluconate and D-Glucono-δ-lactone

In the presence of ATP, D-gluconic acid is converted by the catalysis of gluconate kinase (GK; EC 2.7.1.12) to 6-phosphogluconic acid, which is dehydrogenated by 6-phosphogluconate dehydrogenase (6-PGDH, EC 1.1.1.44) in the presence of NADP to form ribulose-5-phosphate.

$$\text{1. D-gluconate + ATP} \xrightarrow{\text{GK}} \text{D-gluconate-6-P + ADP}$$

$$\text{2. D-gluconate-6-P + NADP}^+ \xrightarrow{\text{6-PGDH}} \text{D-ribulose-5-P + NADHP + CO}_2 + \text{H}^+$$

The equilibrium of reaction 2 at pH 8 is totally on the side of ribulose-5-phosphate and is quantitative under these conditions[3,35,36]. The reaction with gluconate kinase is specific as neither L-gluconic acid nor any other sugar acid react with this enzyme.

Determination[3]

Reagent A: glycylglycine buffer pH 8.0, 0.25 ml Mg^{2+} (17 mM): 1.98 g of glycylglycine and 210 mg $MgSO_4 \cdot 6H_2O$ are dissolved in 50 mL water, adjusted to pH 8.0 with 2 M KOH and the solution made up to 60 mL with distilled water.

Reagent B: NADP solution (11.5 mM): 50 mg of NADP–Na_2 are dissolved in 5 mL distilled water and stored at 4 °C.

Reagent C: ATP solution: 250 mg of ATP–Na_2H_2 and 250 mg of $NaHCO_3$ are dissolved in 5 mL of doubly distilled water.

Reagent D: 6-phosphogluconate dehydrogenase (6-PGDH); the commercial suspension in 3.2 M ammonium sulfate (3.2 M) is used without dilution (2 mg/mL).

Reagent E: gluconate kinase; the commercial suspension in 3.2 M ammonium sulfate (3.2 M) is used without dilution.

Reagents B and C are stable in a refrigerator for 14 days only, and reagents D and E for 1 year at 4 °C.

Pipet the following successively into glass, quartz or one-way cuvettes.

	Sample (mL)	Blank (mL)
Reagent A	1.00	1.00
Reagent B	0.10	0.10
Reagent C	0.10	0.10
Reagent D	0.05	0.05
Sample	0.10	-
Water	1.40	1.50

Table. Survey of Other Enzymatic Determinations of Mono- and Oligosaccharides

Number	Compound	Reaction Scheme	Buffer and Enzyme Systems	Ref
1.5.2.8.	D-Mannose	Mannose + ATP $\xrightarrow{\text{HK}}$ Mannose-6-P + ADP Mannose-6-P $\xrightarrow{\text{PMI}}$ Fructose-6-P Fructose-6-P $\xrightarrow{\text{PGI}}$ Glucose-6-P Glucose-6-P + NADP$^+$ $\xrightarrow{\text{G-6-P DH}}$ Gluconate-6-P + NADPH + H$^+$	0.05 M Triethanolamine buffer, pH 7.6; 0.007 M MgCl$_2$; HK: Hexokinase EC 2.7.1.1; PMI: Phosphomannose isomerase EC 5.3.1.8; PGI: Phosphoglucose isomerase EC 5.3.1.9; G-6-P DH: Glucose-6-phosphate dehydrogenase EC 1.1.1.49	25,37
1.5.2.9.	D-Xylose	Xylose + NADP$^+$ $\xrightarrow{\text{XDH}}$ Xylono-δ-lactone + NADPH + H$^+$ Xylono-δ-lactone + H$_2$O $\xrightarrow{\text{OH}^-}$ Xylonate	0.1 M Phosphate buffer, pH 8.0; XDH: Xylose dehydrogenase EC 1.1.1.175; AEP: Aldose-1-epimerase (Mutarotase) EC 5.1.3.3	38,39
1.5.2.10.	L-Fucose	Fucose + NAD$^+$ $\xrightarrow{\text{FDH}}$ Fucono-1,5-lactone + NADH - H$^+$	1 M Tris-buffer, pH 8.0; FDH: Fucose dehydrogenase EC 1.1.1.122	40,41
1.5.2.11.	D-Mannitol	Mannitol + NAD(P)$^+$ $\xrightarrow{\text{MDH}}$ Fructose + NAD(P)H + H$^+$	0.2 M Glycine buffer, pH 9.8; 1 M Hydrazine; MDH: Mannitol dehydrogenase EC 1.1.2.2 or EC 1.1.1.67	42
1.5.2.12.	Sucrose	Sucrose + H$_2$O $\xrightarrow{\text{Inv.}}$ D-Glucose + D-Fructose Glucose + Fructose + 2 ATP $\xrightarrow{\text{HK}}$ Glucose-6-P + Fructose-6-P + 2 ADP Fructose-6-P $\xrightarrow{\text{PGI}}$ Glucose-6-P 2 Glucose-6-P + 2 NADP$^+$ $\xrightarrow{\text{G-6-P DH}}$ Gluconate-6-P + NADHP + H$^+$	0.32 M Citrate buffer, pH 4.6; 0.75 M Triethanolamine buffer, pH 7.6; 10 mM Mg^{2+}; Inv.: Invertase (β-Fructosidase) EC 3.2.1.26; HK: Hexokinase EC 2.7.1.1; PGI: Phosphoglucose isomerase EC 5.3.1.9; G-6-P DH: Glucose-6-phosphate dehydrogenase EC 1.1.1.49	3,43

Table.   (Continued)

Number	Compound	Reaction Scheme	Buffer and Enzyme Systems	Ref
1.5.2.13.	Lactose	Lactose + H_2O $\xrightarrow{\beta\text{-Gals}}$ D-Galactose + D-Glucose β-D-Galactose + NAD^+ $\xrightarrow{\beta\text{-GalDH}}$ D-Galactono-δ-lactone + $NADH + H^+$	0.4 M Phosphate buffer, pH 7.5; 0.004 M Mg^{2+} β-Gals: β-Galactosidase EC 3.2.1.23; β-GalDH: β-Galactose dehydrogenase EC 1.1.1.48	3,44,45
1.5.2.14.	Maltose	Maltose + H_2O $\xrightarrow{\alpha\text{-GS}}$ 2 D-Glucose 2 Glucose + 2 ATP $\xrightarrow{HK}$ 2 Glucose-6-P + 2 ADP 2 Glucose-6-P + 2 $NADP^+$ $\xrightarrow{\text{G-6-P DH}}$ Gluconate-6-P + 2 $NADPH + 2H^+$	0.1 M Acetate buffer, pH 6.6; 0.75 M Triethanolamine buffer, pH 7.6; 0.01 M Mg^{2+}; α-GS: α-Glucosidase EC 3.2.1.20; HK: Hexokinase EC 2.7.1.1; G-6-P DH: Glucose-6-phosphate dehydrogenase EC 1.1.1.49	3,46,47
1.5.2.15.	Raffinose	Raffinose + H_2O $\xrightarrow{\alpha\text{-Gals}}$ D-Galactose + Sucrose D-Galactose + NAD^+ $\xrightarrow{\beta\text{-GalDH}}$ Galactono-δ-lactone + $NADH + H^+$	0.05 M Citrate buffer, pH 4.6 or 0.67 M acetate buffer pH 4.6; 0.51 M Phosphate buffer, pH 8.6 α-Gals: α-Galactosidase EC 3.2.1.22 β-GalDH: β-Galactose dehydrogenase EC 1.1.1.48	48-51

Mix the solution thoroughly and after 5 min read the absorbance of the sample $(E_1)_S$ and the blank $(E_1)_B$ against water or air at 340 nm (or at 335 or 365 nm with a Hg lamp). Add to the sample and the blank 0.01 mL of reagent E and when the reaction is completed (10–15 min) read the absorbance of the sample $(E_2)_S$ and the blank $(E_2)_B$. When the absorbance rises constantly, measure the absorbance every 2 min and extrapolate it from the time of the addition of reagent E.

$$\Delta E \quad = \quad [(E_2)_S - (E_1)_S] - [(E_2)_B - (E_1)_B]$$

ΔE is proportional to the concentration of D-gluconate in the solution (calibrated with a standard solution of D-gluconate).

The reaction rate with glucono-δ-lactone is slow and it must, therefore, be converted into gluconate by the addition of alkali (e.g., heating with 0.01 M KOH in a boiling water bath for 1 min). If proteins are present in high concentration as in meat products they must be removed prior to determination. Inorganic salts inhibit gluconate kinase and the reaction rate becomes slower such that more enzyme must be added to the assay mixture.

A survey of enzymatic determinations for additional mono- and oligosaccharides is shown (see Table).

References

1   Bergmeyer, H. U. In *Methods of Enzymatic Analysis*, 3rd ed., Vol. VI; Bergmeyer, H. U., Ed.; VCH: Weinheim, 1984.

2   Kunst, A.; Draeger, B.; Ziegenhorn, J. In *Methods of Enzymatic Analysis*, 3rd ed., Vol. VI; Bergmeyer, H.U., Ed.; VCH: Weinheim, 1984; p 163.

3   *Methoden der enzymatischen Lebensmittelanalytik*; Boehringer Mannheim GmbH, 1984.

4   Banauch, D.; Brümmer, W.; Ebeling, W.; Metz, H.; Rindfrey, H.; Lang H. *Z. Klin. Chem. Klin. Biochem.* **1975**, *13*, 101.

5   Vormbrock, R. In *Methods of Enzymatic Analysis*, 3rd ed., Vol. VI; Bergmeyer, H. U., Ed.; VCH: Weinheim, 1984; p 172.

6   Passey, R. B.; Gillum, R. L.; Fuller, R. B.; Urry, F. M.; Giles, M. L. *Clin. Chem.* **1977**, *23*, 131.

7   Kunst, A.; Draeger, B.; Ziegenhorn, J. In *Methods of Enzymatic Analysis*, 3rd ed., Vol. VI; Bergmeyer, H. U., Ed.; VCH: Weinheim, 1984; p 178.

8   Keilin, D. K.; Hartree, E. F. *Biochem. J.* **1952**, *50*, 331.

9   Sols, A.; de la Fuente, G. *Biochim. Biophys. Acta* **1957**, *24*, 206.

10   Mueller, H. *Stärke* **1967**, *19*, 55.

11   Huggett, A. St.; Nixon, A. D. *Biochem. J.* **1957**, *66*, 12P.

12   Werner, W.; Rey, H. G.; Wielinger, H. *Fresenius' Z. Anal. Chem.* **1970**, *252*, 224.

13   Salomon, L. L.; Johnson, J. E. *Anal. Chem.* **1959**, *31*, 453.

14   Trinder, P. *Ann. Clin. Biochem.* **1969**, *6*, 24.

15. Bergmeyer, H. U.; Bernt, E. In *Methoden der enzymatischen Analyse*, Vol 3.; VCH: Weinheim, 1974; p 1251.

16   Beutler, H. O. In *Methods of Enzymatic Analysis*, 3rd ed., Vol. VI; Bergmeyer, H. U., Ed.; VCH: Weinheim, 1984; p 321.

17   Schmidt, R. H. *Klin. Wochenschr.* **1961**, *39*, 1244.

18   Kubadinow, N. *Zucker* **1974**, *27*, 72.

19   Weichel, H. *Dtsch. Lebensm.-Rundsch.* **1965**, *61*, 53.

20   Hjelm, M.; de Verdier, C. H. In *Methods of Enzymatic Analysis*, 3rd ed., Vol. VI; Bergmeyer, H. U., Ed.; VCH: Weinheim, 1984; p 281.

21   Hjelm, M. *Clin. Chim. Acta* **1967**, *15*, 87.

22   Roth, H.; Segal, S.; Bertoli, D. *Anal. Biochem.* **1965**, *10*, 32.

23   Fischer, W.; Zapf, J. *Hoppe-Seyler's Z. Physiol. Chem.* **1964**, *337*, 186.

24   Avigdad, G.; Amaral, D.; Asensio, C.; Horecker, B. L. *J. Biol. Chem.* **1962**, *237*, 2736.

25 Sturgeon; R. J. In *Methods of Carbohydrate Chemistry*, Vol. VIII; Whistler, R. L., BeMiller, J., Eds.; Academic: New York, 1980; p 131.

26 Fujimura, Y. In *Methods of Enzymatic Analysis*, 3rd ed., Vol. VI; Bergmeyer, H. U., Ed.; VCH: Weinheim, 1984; p 288.

27 Fujimura, Y.; Ishii, S.; Kawamura, M.; Naruse, N. *Anal. Biochem.* **1981**, *117*, 187.

28 Sturgeon, R. J. In *Methods of Enzymatic Analysis*, 3rd ed., Vol. VI; Bergmeyer, H. U., Ed.; VCH: Weinheim, 1984; p 427.

29 Melrose, J.; Sturgeon, R. *Carbohydr. Res.* **1983**, *118*, 247.

30 Beutler, H. O. In *Methods of Enzymatic Analysis*, 3rd ed., Vol. VI; Bergmeyer, H. U., Ed.; VCH: Weinheim, 1984; p 356.

31 Beutler, H. O.; Becker, J. *Dtsch. Lebensm.-Rundsch.* **1977**, *73*, 182.

32 Weissbach, A. *Biochim. Biophys. Acta* **1958**, *27*, 608.

33 Weissbach, A. In *Methods of Enzymatic Analysis*, 3rd ed., Vol. VI; Bergmeyer, H. U., Ed.; VCH: Weinheim, 1984; p 366.

34 Berman, T.; Magasanik, B. *J. Biol. Chem.* **1966**, *241*, 800.

35 Moellering, H.; Bergmeyer, H. U. *Z. Lebensm. Unters. Forsch.* **1967**, *135*, 198.

36 Moellering, H.; Bergmeyer, H. U. In *Methods of Enzymatic Analysis*, 3rd ed., Vol. VI; Bergmeyer, H. U., Ed.; VCH: Weinheim, 1984; p 220.

37 Gawehn, K. In *Methods of Enzymatic Analysis*, 3rd ed., Vol. VI; Bergmeyer, H. U., Ed.; VCH: Weinheim, 1984; p 262.

38 Wissler, J.H. *Fresenius' Z. Anal. Chem.* **1978**, *290*, 179

39 Wissler, J.H.; Logemann, E. In *Methods of Enzymatic Analysis*, 3rd ed., Vol. VI; Bergmeyer, H. U., Ed.; VCH: Weinheim, 1984; p 449.

40 Maler, T. *Anal. Biochem.* **1980**, *102*, 340.

41 Morris, J. In *Methods of Enzymatic Analysis*, 3rd ed., Vol. VI; Bergmeyer, H. U., Ed.; VCH: Weinheim, 1984; p 386.

42 Horikoshi, K. In *Methods of Enzymatic Analysis*, 3rd ed., Vol. VI; Bergmeyer, H. U., Ed.; VCH: Weinheim, 1984; p 271.

43 Outlaw, W. H.; Tarczynski, M. C. In *Methods of Enzymatic Analysis*, 3rd ed., Vol. VI; Bergmeyer, H. U., Ed.; VCH: Weinheim, 1984; p 96.

44 Kulski, J. K.; Buehring, C. *Anal. Biochem.* **1982**, *119*, 341.

45 Kulski, J. In *Methods of Enzymatic Analysis*, 3rd ed., Vol. VI; Bergmeyer, H. U., Ed.; VCH: Weinheim, 1984; p 112.

46 Beutler, H. O. In *Methods of Enzymatic Analysis*, 3rd ed., Vol. VI; Bergmeyer, H. U., Ed.; VCH: Weinheim, 1984; p 119.

47 Schmidt, F. H.; Heidrich, P.; v. Dahl, K.; Kühnle, H. F. *Klin. Wochenschr.* **1977**, *55*, 965.

48 Schiweck, H.; Büsching, L. *Zucker* **1975**, *28*, 242.

49 Hollaus, F.; Wieninger, L.; Brauensteiner, W. *Zucker* **1977**, *30*, 653.

50 Wallenfels, K.; Kurz, G. *Biochem. Z.* **1962**, *335*, 559.

51 Beutler, H. O. In *Methods of Enzymatic Analysis*, 3rd ed., Vol. VI; Bergmeyer, H. U., Ed.; VCH: Weinheim, 1984; p 90.

1.6. Enzymatic Methods with Immobilized Enzyme Systems

1.6.1. Introduction

Commercial enzymes are expensive due to the cost of raw materials and the need for complicated production techniques and it is, therefore, preferable to recycle them. This can be achieved by immobilization on suitable carrier materials using one of the following procedures[1]:

1.6.1.1.   Physical entrapment.
1.6.1.2.   Adsorption on an inert carrier.
1.6.1.3.   Covalent cross-linking with a bifunctional reagent yielding macroscopic particles.
1.6.1.4.   Covalent binding to a water insoluble matrix.

1.6.1.1. Physical Entrapment

The enzyme is entrapped in a gel matrix which allows diffusion and the reaction of small molecules. Polyacrylamide is the most suitable gel material, since it is inert and the activity of the entrapped enzyme remains unaltered over long periods of time (e.g., 3 months)[1,2]. Enzyme-starch gels are not stable due to the retrogradation of the matrix and escape of the enzymes[3].

A special variation of this physical method of inclusion is microcapsulation. Two modifications exist:

(a)   Inclusion in spherical particles with thin walls.

The walls of these particles must be semipermeable to allow the passage of low molecular weight reactants; materials: polyacrylamide[4] or calcium alginate[5].

(b)   Hollow fibres.

The enzyme is included in thin fibres of a semipermeable material which are then sealed at both ends. This modification is particularly suitable for industrial enzyme reactors[6].

1.6.1.2. Adsorption on Insoluble Material

The adsorption is carried out either by ionic or hydrogen bonding, hydrophobic interactions or π-electron interactions using materials such as silica gel, active carbon, quartz and glass powders, bentonite, aluminum oxide and ion exchangers[7,8]. One advantage of this process lies in its simple production; although a disadvantage is its great dependence on external parameters such as pH, ionic strength, temperature and solvents.

1.6.1.3. Covalent Cross-Linking

The enzymes are coupled with bifunctional compounds to form macroscopic particles. The following compounds have been used:

Glutardialdehyde[9,10], bis-diazobenzidine[11], 4,4'-difluoro-3,3'-dinitrophenyl sulfone[12], diphenyl-4,4'-dithiocyanate-2,2'-disulfonic acid[13], 3-methoxydiphenylmethane-4,4'-diisocyanate[14], and trichloro-s-triazine[15].

The active groups of the enzyme molecules for attachment are the terminal amino groups such as arginine and lysine. The production of these materials is simple, although the loss of some of the activity is disadvantageous.

1.6.1.4. Covalent Binding to an Insoluble Matrix

This is the most frequent method for immobilizing enzymes. The following materials have been used: porous glass[16], polyacrylamide[17], polyacrylic acid derivatives[18], nylon[19], cellulose[20], Sephadex[21], ethylene–maleic acid anhydride copolymers[22], agarose[23], carboxymethyl cellulose[20] and collagen[11,24].

The enzymes are attached to the insoluble matrix via special reactions with, for example, diazo compounds, isocyanates, carbodiimides, bromocyanogen and azides[25,26]. The following groups are suitable for binding reactions:

α- and ε-amino groups;

the phenol group of tyrosine;

β- and γ-carboxyl groups;

the hydroxyl group of serine;

the sulfhydryl group of cysteine;

imidazole group of histidine.

The enzymes which are bonded to an insoluble matrix are no longer detachable by external influences such as pH, ionic strength and temperature. In a lot of cases their activities are more stable than those of the free enzyme in solution.

1.6.2. Enzyme Electrodes

1.6.2.1. Introduction

Ion-selective electrodes enable the quantitative determination of a substance alone or in a mixture without temporal delays. They are used usually in cases where the result of the analysis is required immediately or when the concentration of a substance must be measured continuously.

Ion selective electrodes exist mainly for inorganic ions such as Na^+, K^+, Ca^{2+}, Cl^-, or for H^+. For the determination of organic compounds enzymes are used which react at room temperature according to the following scheme.

$$A + B \xrightarrow{\text{enzyme}} C + D$$

One of the reactants is measured by a so called "sensor electrode" which can be a gas electrode (oxygen electrode), H^+-electrode (measuring changes in pH) or an ion selective electrode (e.g., CN, F, I). Two basic types of sensor electrodes[27] are described below (see Figure 1).

Type A: Dialysis Membrane Electrode

The sensor electrode is surrounded by a layer containing the appropriate enzyme which can be dissolved in an aqueous solution. The enzyme is attached to a high molecular weight soluble matrix or to an insoluble matrix and is covered with a sheet of semipermeable membrane such as cellophane which is fastened by special rings (mostly rubber rings, see Figure 1). These electrodes are stored in buffered solutions overnight to remove included air from the layer.

Electrodes with enzymes attached to a soluble matrix have life times of on average of one week. In contrast to those where the enzymes are attached to an insoluble matrix which are very stable and can be used to carry out up to 10 000 measurements.

Type B: Surface Membrane Electrode
The enzyme is attached directly to the surface of the sensor electrode and the surface covered by a thin
cellophane or nylon net.

Stability of Enzyme Electrodes
The life time of such electrodes is dependent on the following factors:
(a)   Type of the enzyme application.
(b)   Concentration of the enzyme in solution.
(c)   Reaction conditions.
(d)   Stability of the sensor electrode.

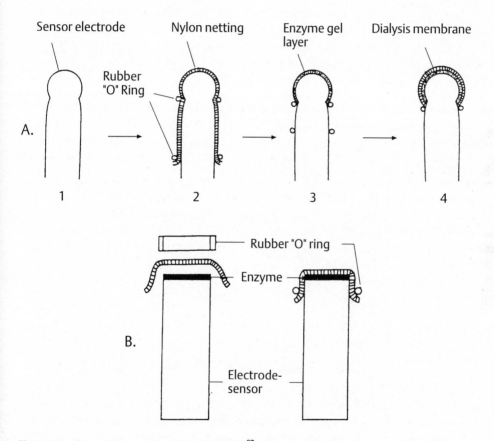

Figure 1.   Type of Enzyme Electrodes (Guilbault[27]; with permission)

Depending on the enzyme application, the following life times can be assumed:

Soluble enzyme-dialysis membrane:	1 week	10–50 determinations
Physically entrapped:	3–4 weeks	50–100 determinations
Chemically bound:	4–14 months	10 000 determinations

Response Time

This is the time required to reach the final state; it is dependent on the following factors:

(a)   Stirring rate of the solution.

(b)   Concentration of the substrate.

(c)   Concentration of the enzyme.

(d)   pH and temperature of the solution.

(e)   Quality of the semipermeable membrane, e.g. thickness, porosity, etc.

On average, the response time of an enzyme electrode is around 1–2 minutes.

1.6.2.2. Enzyme Electrodes for Determination of D-Glucose

The most important type of electrode uses the reaction between D-glucose and glucose oxidase and measures either the decrease of oxygen concentration in the surrounding solution of the electrode or the formed hydrogen peroxide. Such electrodes were developed, for example, for the determination of glucose in blood[28]. It consists of an oxygen electrode, which is covered with a layer of polyacrylamide gel in which glucose oxidase is physically entrapped. The decrease of the oxygen concentration is equivalent to the glucose concentration in the blood and the response time of the systems is less than 1 minute. This type of electrode has been modified several times[29,30], especially with regard to the matrix. In one case, an electrode is constructed with gelatine as the inert matrix for glucose oxidase, which is entrapped between a semipermeable membrane and a polyethylene sheet. The oxygen which diffuses through the polyethylene sheet is measured by an oxygen electrode (Figure 2)[30]. In another modification glucose oxidase is coupled by the carboxyl-azide method to a collagen membrane and the hydrogen peroxide determined by anodic oxidation[31].

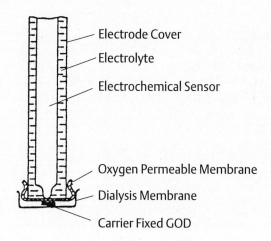

Electrode Cover

Electrolyte

Electrochemical Sensor

Oxygen Permeable Membrane

Dialysis Membrane

Carrier Fixed GOD

Figure 2.   Construction of a Glucose Oxidase (GOD) Enzyme Electrode (Weise, Scheller and Siegler[30]; with permission)

In yet another modification, the hydrogen peroxide is measured by an iodide specific electrode according to the following scheme[32].

$$\beta\text{-D-glucose} + O_2 + H_2O \xrightarrow{\text{glucose oxidase}} \text{D-gluconate} + H_2O_2$$

$$H_2O_2 + 2I^- + 2H^+ \xrightarrow{\text{peroxidase}} H_2O_2 + I_2$$

The decrease in I^- concentration at the surface of the electrode is proportional to the glucose concentration. Interfering reducing substances such as ascorbic acid, tyrosine and uric acid must be removed from the sample before the determination. Molybdate ions can be used instead of peroxidase for catalyzing the reaction between H_2O_2 and I^- ions[33].

Another method for establishing a glucose electrode is based on the following scheme[34].

$$\beta\text{-D-glucose} + H_2O + 2\,[Fe(CN)_6]^{3+} \xrightarrow{\text{glucose oxidase}} \text{D-gluconate} + 2\,H^+ + 2\,[Fe(CN)_6]^{4+}$$

In this case, the concentration of $[Fe(CN)_6]^{4+}$ is measured by reoxidation to $[Fe(CN)_6]^{3+}$ with a platinum electrode.

Glucose oxidase is much more reactive towards β-D-glucose than to α-D-glucose. To increase the conversion rate of α- to β-D-glucose, aldose-1-epimerase (mutarotase) can be added to the enzyme complex[35].

1.6.2.3. Electrodes for Determination of Mono- and Oligosaccharides Except D-Glucose

1.6.2.3.a. Maltose and Sucrose

For this purpose a double-layer electrode has been developed in which glucoamylase or invertase are fixed at the bulk side of the semipermeable collagen membrane. Maltose or sucrose is hydrolyzed and the glucose produced by this reaction migrates through the membrane and is oxidized on the inner face of the electrode by the immobilized glucose oxidase. The sensitivity and linearity of such electrodes is excellent[36,37,38].

1.6.2.3.b. Galactose and Lactose

The galactose sensitive electrodes utilize membranes with immobilized galactose oxidase. D-Galactose reacts with this enzyme according to the following scheme.

$$\text{D-galactose} + O_2 \xrightarrow{\text{galactose oxidase}} \text{D-galactose hexodialdose} + H_2O_2$$

The oxygen consumption is measured with an oxygen electrode[37,39]. For the determination of D-galactose in blood and plasma a microelectrode has been constructed consisting of an amperometric hydroperoxide electrode whose surface is covered by a membrane containing galactose oxidase[40].

The lactose sensitive electrode consists of a combination of β-galactosidase and glucose oxidase and is based on the same principle as that described in Section 1.6.2.3.a.[41].

1.6.2.4. Immobilized Cells

The alternative of utilizing specific enzymes is the entrapping of whole cells. Such systems can be applied as bioselective sensors and offer the following advantages:

(a)   The utilization of pure enzymes is not necessary; such products are often very expensive.

(b)   The regeneration of the electrodes can be carried out by immersion into a simple nutrient broth.

(c)   The possibilities of combined determinations are greater than with pure enzymes.

The disadvantages of these electrodes include:

(a)   Reduced selectivity; microorganisms contain a lot of enzyme systems which react with several different compounds simultaneously instead of the desired one.

(b)   Long response times; some of the enzymes are present in the microorganisms only in small quantities causing long reaction times. Increasing the quantity of the microorganism increases the time of diffusion.

Microbiological Carbohydrate Electrodes

For the determination of the amount of D-glucose, D-fructose and sucrose the cells of *Brevibacterium lactofermentum* in combination with an oxygen electrode has been described[42]. Other microorganisms, which are used for this purpose are: *Hansenula anomala* for the amount of D-glucose, sucrose, L-lactate and pyruvate[43], and *Pseudomonas fluorescens* for D-glucose alone[44].

1.6.3. Enzyme Reactors and Enzyme Membranes

1.6.3.1. Introduction

The support with the immobilized enzyme is poured into a column, the substrate which contains the compound to be determined passes through the column where the conversion occurs. The reaction product is determined by a detector after its emergence from the column. When the dead volume is kept small by completely filling the column a quantitative reaction is possible by means even of rapid elution. This requires mechanically stable carriers which are resistant to the high pressures of the eluting liquids. These types of systems were described for the first time by Guilbault and Kramer[45].

In an alternative technique the enzymes are fixed at the inner surface of nylon or polypropylene tubes. In the first case, nylon is treated with 3.7 M hydrochloric acid for 37 minutes at 35 °C. Under these conditions a few peptide bonds are cleaved to free carboxyl and amino groups. Treatment with glutardialdehyde binds the enzymes to the free amino groups of the polymer[46,47]. In the second case, the enzymes are bound to the free carboxyl groups of the polypropylene which are formed by treatment with chromic acid–sulfuric acid[48].

The advantage of these systems is the possibility to work under low pressure at high elution rates. Such systems are used for commercial current flow analyses and enable a high through put of samples. Disadvantageous for these systems are the great lengths of the enzyme tubes required to obtain quantitative conversions and for the instability of nylon against acids.

1.6.3.2. Systems for Determination of D-Glucose

The majority of these systems use columns filled with glucose oxidase as the specific enzyme which is fixed either on polyacrylamide[49] or on porous glass[50]. The decrease in the oxygen concentration of the liquid emerging from the column is measured by an oxygen electrode. This system was developed for the rapid determination of glucose in blood samples (Figure 3)[51]. By bubbling air into the buffer vessel a constant concentration of oxygen in the solution is maintained. This buffer is pumped in as a constant stream through the enzyme column and an aliquot of the sample introduced through the inlet. The oxygen concentration of the solution is measured after the column by an oxygen electrode. Hereafter the buffer returns to the buffer vessel to be saturated with oxygen again.

With this system up to 60 measurements per hour and up to 10 000 measurement for one enzyme column are possible. Instead of the decrease in oxygen in the solution the concentration of hydrogen peroxide at the end of the column is measured by a special Pt electrode[52].
An alternative to the electrochemical detection of hydrogen peroxide is photometeric determination after its emergence from the column. Here, the following methods have been described:
(a)   Reaction with iodide ions and determination of elementary iodine[53].
(b)   Reaction with peroxidase, 4-aminophenazone and 3,5-dichloro-2-hydroxybenzenesulfonate[54].
(c)   Reaction with hexacyanoferrate(IV) and peroxydase to hexacyanoferrate(III); the latter reacts further with phenol and 4-aminoantipyrine to form a colored compound (λ_{max}: 525 nm)[55].
(d)   A further alternative is the determination of hydrogen peroxide by chemiluminescence, where special molecules are converted into excited states which return to the ground state with emission of light.

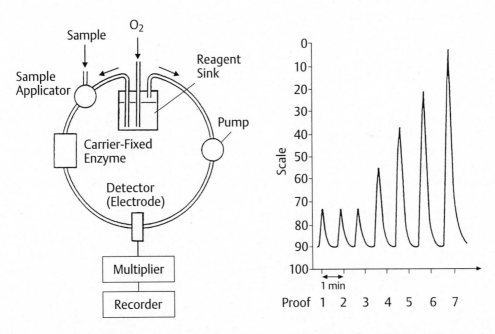

Figure 3.   System for Permanent Determination of D-Glucose with Carrier-Fixed Enzymes (Bergmeyer and Hagen[51]; with permission)

For the determination of D-glucose with glucose oxidase the formed hydrogen peroxide reacts after passage through the column with the immobilized enzyme with luminol (5-amino-2,3-dihydrophtalazine-1,4-dione) in the presence of hexacyanoferrate(III) which serves as a catalyst. The emitted light is measured by a photomultiplier and the chemiluminescence is proportional to the concentration of D-glucose in the range of 10^{-8} to 10^{-5} M[56,57]. For this purpose the apparatus shown in Figure 4 has been described.

Determination
Plastic syringes of (50 mL) containing the following reagents:
Reagent A: acetate buffer 0.004 M, pH 5.6.
Reagent B: luminol 2×10^{-4} M in borate buffer 0.1 M, pH 10.5.
Reagent C: $K_3[Fe(CN)_6]$ 10^{-2} M.
The syringes are driven by special equipment with a microslide. By means of a special ventile, the sample is introduced into the acetate buffer system (reagent A) and a reaction occurs in the column filled with immobilized glucose oxidase. The formed hydrogen peroxide reacts with reagents B and C and the chemoluminescence is measured by the photomultiplyer, which is connected to the reaction cell.

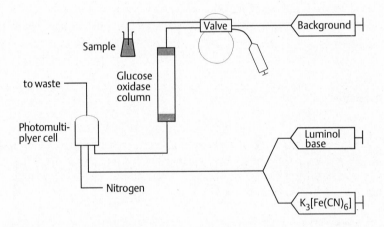

Figure 4.   Scheme for the Determination of D-Glucose by Chemiluminescence (Bostick, Hercules[56]; with permission)

Another detection method is based on measurement of the heat production of the enzymatic reactions. These enthalpies in the range of 10–100 kJ/mol can be measured by highly sensitive thermistors, e.g. H_2O_2 decay by catalase: 100 kJ/mol; oxidation of D-glucose by glucose oxidase: 80 kJ/mol; phosphorylation of glucose by hexokinase: 28 kJ/mol. In most cases the following measurement principle is applied[58].
A well-thermostated system (fluctuation ± 0.01 °C) is equipped with a highly sensitive thermistor at the end of the enzyme reactor. The buffer, in which the sample is introduced, is brought to the temperature of the system by passage through a heat exchanger. The thermistor measures the temperature continuously and records its change when the sample reacts with the immobilized enzyme. The enzyme is immobilized on porous glass. For improving the stability of the system the buffer stream is split after the inlet valve. One part is passed through the enzyme column and the other

through the reference column with the inert carrier (covered with deactivated enzyme). The thermistors are attached at the end of both columns. By means of compensation circuits for both thermistors (Wheatstone bridge), interference by unspecific heat formation can be avoided. Figure 5 shows a graphical representation of this equipment.

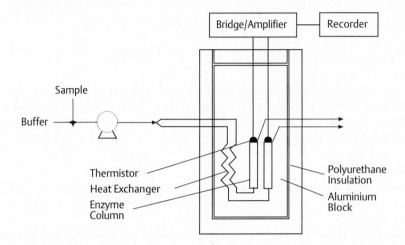

Figure 5.   Calorimetric Equipment for Enzymatic Analysis of an Immobilised Enzyme (Mosbach, Danielsson[58]; with permission)

This method has also been applied for the assay of D-glucose using glucose oxidase (range: 0.03–0.5 mmol)[59] or by a co-immobilization of glucose oxidase and catalase (range: 0.01–0.9 mmol)[60]. Another more unspecific system using the phosphorylation of immobilized hexokinase (range: 50–500 mg D-glucose/100 mL) has also been described[61].

1.6.3.3. Determination of Other Sugars

1.6.3.3.a. D-Galactose
β-D-galactose dehydrogenase is fixed to Sephadex G-200 by reaction with bromocyanogen. For the determination of D-galactose the immobilized enzyme is put into a small filter tube fitted with a sinter plate at the bottom and which rotates slowly around its own axis. The reaction itself is carried out at pH 8.6 (tris-buffer) in the presence of NAD. After 30 minutes the reaction is complete and the solution is filtered from the immobilized enzyme. The NADH concentration in the clear solution is measured by photometry at 360 nm, which is proportional to the concentration of D-galactose[62].

1.6.3.3.b. Lactose
β-Galactosidase and glucose oxidase are immobilized on a phenol–formaldehyde resin and the reaction takes place according to the following scheme.

$$\text{lactose} + H_2O \xrightarrow{\;\;\beta\text{-galactosidase}\;\;} \text{D-galactose} + \text{D-glucose}$$

$$\text{D-glucose} \xrightarrow{\;\;\text{glucose oxidase}\;\;} \text{D-gluconate} + H_2O_2$$

$$H_2O_2 + 2\,I^- + 2\,H^+ \xrightarrow{\;\;[MoO^4]\;\;} I_2 + H_2O$$

The elementary iodine is measured by photometry[63]. In a second system, which contains only glucose oxidase, the influence of the occasional presence of glucose on the results can be excluded. Another system uses the heat liberated from oxidation of D-galactose by immobilized D-galactose oxidase after cleavage of lactose by immobilized β-galactosidase[67].

1.6.3.3.c. Sucrose

Invertase and glucose oxidase are immobilized on the internal surface of a nylon tube. During passage of the sample the sucrose is cleaved to glucose and fructose; the first compound reacts with glucose oxidase forming hydrogen peroxide which is then determined[64].

Another system consists of a combination of four immobilized enzymes, invertase, hexokinase, phosphoglucose isomerase and D-glucose-6-phosphate dehydrogenase fixed to Sepharose by the bromocyan reaction. The concentration of NADPH is measured by photometry[65].

A direct determination of sucrose can be performed by measurement of the heat liberated from cleavage with invertase by a thermistor[66,67]. This system is used for continuous monitoring in fermentation processes.

References

1   Guilbault, G. In *Analytical Uses of Immobilized Enzymes*; M. Dekker, Inc.: New York, Basel, 1984; p 78.
2   Hicks, G. P.; Updike, S. J. *Anal. Chem.* **1966**, *38*, 726.
3   Guilbault, G. In *Handbook of Enzymatic Analysis*; M. Dekker, Inc: New York, 1977.
4   Chang, T. M. S. *Science* **1964**, *146*, 524.
5   Klein, J. In *Polysaccharide*; Burchard, W., Ed.; Springer: Berlin, 1985; p 221.
6   Rony, P. R. *Biotechnol. Bioeng.* **1971**, *13*, 431.
7   Silman, I. H.; Katchalski, E. *Ann. Rev. Biochem.* **1966**, *35*, 873.
8   Guilbault, G. In *Analytical Uses of Immobilzed Enzymes*; M. Dekker, Inc.: New York, Basel, 1984; p 83.
9   Jansen, E. F.; Olson, A. C. *Arch. Biochem. Biophys.* **1969**, *129*, 221.
10   Haynes, R.; Walsh, K. A. *Biochem. Biophys. Res. Commun.* **1969**, *36*, 235.
11   Silman, I. H.; Albu-Weissenberg, M.; Katschalski, E. *Biopolymers* **1966**, *4*, 441.
12   Wold, F. I. *J. Biol. Chem.* **1961**, *236*, 106.
13   Manecke, G.; Guenzel, G. *Naturwissenschaften* **1967**, *54*, 647.
14   Schick, H. F.; Singer, S. J. *J. Biol. Chem.* **1961**, *236*, 2447.
15   Kay, G.; Crook, E. M. *Nature* **1967**, *216*, 514.
16   Weetall, H. H. *Nature* **1969**, *233*, 959.
17   Barker, S. A.; Somers, P. J.; Epton, R.; McLaren, J. V. *Carbohydr. Res.* **1970**, *14*, 287.
18   Erlanger, B. F.; Isambert, M. F.; Michelson, A. M. *Biochem. Biophys. Res. Commun.* **1970**, *40*, 70.
19   Inman, D. J.; Hornby, W. E. *Biochem. J.* **1972**, *129*, 255.

20   Wheeler, K. P.; Edwards, B. A.; Whittam, R. *Biochim. Biophys. Acta* **1969**, *191*, 187.
21   Axen, R.; Porath, J. *Nature* **1966**, *210*, 367.
22   Goldstein, L. *Meth. Enzymol.* **1970**, *19*, 935.
23   Green, H. L.; Crutchfield, G. *Biochem. J.* **1969**, *115*, 183.
24   Coulet, P. R.; Juillard, J. H.; Gautheron, D. C. *Biotechnol Bioeng.* **1974**, *16*, 1055.
25   Guilbault, G. In *Analytical Uses of Immobilized Enzymes*; M. Dekker, Inc.: New York, Basel, 1984; p 86.
26   Vallee, B. L.; Riordam, F. *Ann. Rev. Biochem.* **1969**, *38*, 733.
27   Guilbault, G. In *Analytical Uses of Immobilized Enzymes*; M. Dekker, Inc.: New York, Basel, 1984; p 134.
28   Updike, S. J.; Hicks, P. G. *Nature (London)* **1971**, *214*, 986.
29   Scheller, F.; Jänchen, M.; Pfeiffer, D.; Seyer, J.; Müller, K. *Z. Med. Labor Diagnost.* **1977**, *18*, 312.
30   Weise, H.; Scheller, F.; Siegler, K. *Nahrung* **1981**, *25*, 127.
31   Thevenot, D. R.; Coulet, P. R.; Sternberg, R.; Laurent, J.; Gautheron, D.C. *Anal. Chem.* **1979**, *51*, 96.
32   Nagy, G.; von Storp, L. H.; Guilbault, G. *Anal. Chim. Acta* **1973**, *66*, 443.
33   Llenado, R. A.; Rechnitz, G. A. *Anal. Chem.* **1973**, *45*, 826.
34   Mor. J. R.; Guarnaccia, R. *Anal. Biochem.* **1977**, *79*, 319.
35   Bachner, J. *Sci. Pharm.* **1980**, *48*, 156.
36   Coulet, P. R.; Bertrand, C. *Anal. Lett.* **1979**, *12*, 581.
37   Bertrand, C.; Coulet, P. R.; Gautheron, D. C. *Anal. Chim. Acta* **1981**, *126*, 23.
38   Satoh, J.; Karube, I.; Suzuki, S. *Biotechnol. Bioeng.* **1976**, *18*, 269.
39   Cheng, F.; Christian, G. *Anal. Chim. Acta* **1979**, *104*, 47.
40   Taylor, P.; Kmetec, E.; Johnson, J. *Anal. Chem.* **1977**, *49*, 789.
41   Cheng, F.; Christian, G. *Analyst* **1977**, *102*, 124.
42   Hikuma, M.; Obana, H.; Yasuda, T.; Karube, I.; Suzuki, S. *Enzym. Microb. Technol.* **1980**, *2*, 234.
43   Kulys, J.; Kadziauskiene, K. *Biotechnol. Bioeng.* **1980**, *22*, 221.
44   Suzuki, S.; Satoh, I.; Karube I. *Biochem. Biotechnol.* **1982**, *7*, 147
45   Guilbault, G.; Kramer, *Anal. Chem.* **1965**, *37*, 1675.
46   Hornby, W.; Filipusson, H.; McDonald, A. *FEBS-Lett.* **1970**, *9*, 8.
47   Hornby, W.; Sundaram, P. V. *FEBS-Lett.* **1970**, *10*, 325.
48   Ngo, T. *Int. J. Biochem.* **1980**, *11*, 459.
49   Updike, S. J.; Hicks, G. P. *Science* **1967**, *158*, 270.
50   Weibel, M. K.; Dritschilo, W.; Bright, H. J.; Humphrey, A. *Anal. Biochem.* **1973**, *52*, 402.
51   Bergmeyer, H. U.; Hagen, A. *Fresenius' Z. Anal. Chem.* **1972**, *261*, 333.
52   Watson, B.; Stiffel, D. N.; Semersky, F. E. *Anal. Chim. Acta* **1979**, *106*, 233.
53   Campbell, J.; Hornby, W.; Morris, D. *Biochim. Biophys. Acta* **1975**, *384*, 307.
54   Leung, F. Y.; Ward, P.; Janik, B. *Clin. Biochem.* **1977**, *10*, 4.
55   Marconi, W.; Bartoli, F.; Gulinelli, S.; Morisi, F. *Proc. Biochem. May* **1974**, 22.
56   Bostick, D. T.; Hercules, D. M. *Anal. Chem.* **1975**, *47*, 447.
57   Auses, J. P.; Cook, S.; Maloy, J. *Anal. Chem.* **1975**, *47*, 244.
58   Mosbach, K.; Danielsson, B. *Anal. Chem.* **1981**, *53*, 83A.
59   Mattiasson, B.; Danielsson, B.; Mosbach, K. *Anal. Lett.* **1976**, *9*, 217.
60   Danielsson, B.; Gadd, K.; Mattiasson, B.; Mosbach, K. *Clin. Chim. Acta* **1977**, *81*, 163.
61   Bowers. K. D.; Carr, P. W.; Schifreen, R. S. *Clin. Chem.* **1976**, *22*, 1427.
62   Fleischmann, W. D.; Scherz, H. *Mikrochimica Acta* **1976**, *II*, 443.
63   Voleski, B.; Emond, C. *Biotechnol. Bioeng.* **1979**, *21*, 1251.
64   Inman, D. J.; Hornby, W. E. *Biochem. J.* **1974**, *137*, 25.
65   Fresenius, R. E.; Wönne, K. G.; Flemming, W. *Fresenius' Z. Anal. Chem.* **1974**, *271*, 194.
66   Mandenius, C. F.; Danielsson, B.; Mattiasson, B. *Acta Chem. Scand. B* **1980**, *34*, 463.
67   Mattiasson, B.; Danielsson, B. *Carbohydr. Res.* **1982**, *102*, 273.

1.7. Polarimetry

1.7.1. Introduction

Sugars are optically active as a result of their chiral centres and rotate the plane of the polarized light. According to the Biot's rule the optical rotation is proportional to the concentration of the solution, and the length of the sample tube. The specific rotation $[\alpha]$ of a compound is calculated by the following equation:

$$[\alpha] = \frac{100 \cdot \alpha}{L \cdot c} = \frac{100 \cdot \alpha}{L \cdot p \cdot d}$$

α = angle of the optical rotation of the solution of the substance with concentration c (g substance in 100 mL of the solution) and with length of sample tube L in dm. When the concentration c is replaced by the concentration p (g/100 g solution) it must be multiplied by the density d.

The specific rotation is dependent on the wavelength of the light. In general the rotation is determined with Na light of wavelength 589 nm (labelled D) and at 20 °C. This specific rotation is indicated as $[\alpha]_D^{20}$. An alternative modification [light electric polarimeter] uses green Hg light of wavelength 546 nm and is indicated as $[\alpha]_{546}^{20}$.

1.7.2. Polarimetric Determination of Sugars

The determination of sugars by polarimetry is carried out preferably with pure substances in higher concentrations. The specific rotation of sugars is dependent not only on the wavelength of the light and the temperature, but also to a small extent on the concentration as shown in Table 1[1].

Table 1.   Dependence of the Specific Rotation of Some Sugars on Concentration[1]

	Specific Rotation $[\alpha]_D^{20}$ at Given Concentrations (g/100 mL)			
	10	15	20	25
Glucose[a]	+ 52.7	+ 52.9	+ 53.1	+ 53.5
Fructose[a]	− 90.7	− 92.0	− 93.3	− 94.6
Maltose[a]	+ 138.2	+ 138.2	+ 138.1	+ 138.0
Sucrose	+ 66.5	+ 66.5	+ 66.5	+ 66.5

[a]Values at the mutarotational equilibrium.

Also some metal ions have an influence on the specific rotation, for example, in the case of D-(+)-glucose: in aqueous solution = +52.5°; in 4 M potassium iodide = +47.4°; in 4 M calcium chloride = +61.2°[1]; such influences can interfere with the polarimetric determination of sugars in food stuffs.

Mutarotation

When the anomeric centre of a sugar is unsubsituted, up to five tautomeric structures which are in equilibrium with each other can exist in solution. Each of these structures has its own chiral properties and its own specific rotation. In the solid state sugars exist in one of these structures. On solvation changes in the optical rotation occur as a result of the formation of other tautomeric structures, which is complete at equilibrium; this effect is called mutarotation (see page 8).

The following table lists the specific rotations of some common sugars before and after mutarotation[2].

Table 2.   Specific Rotation of Common Sugars Before and After Mutarotation

Compound	$[\alpha]_D^{20}$ Initial	$[\alpha]_D^{20}$ Final	Compound	$[\alpha]_D^{20}$ Initial	$[\alpha]_D^{20}$ Final
α-D-Glucose	+ 112.2	+ 52.7	α-D-Lactose·H_2O	+ 85.0	+ 52.6
β-D-Glucose	+ 18.7	+ 52.5	β-D-Lactose	+ 34.9	+ 54.5
α-D-Galactose	+ 150.7	+ 80.2	β-D-Maltose·H_2O	+ 117.7	+ 130.4
β-D-Galactose	+ 52.8	+ 80.2	Sucrose	-	+ 66.5
α-L-Arabinose	− 77.3	− 104.5	Raffinose·$5H_2O$	-	+ 105.0
β-L-Arabinose	− 190.6	− 104.5	Stachyose·$4H_2O$	-	+ 131.0
β-D-Fructose	− 132.2	− 92.4	Starch, glycogen, dextrin	-	+ 196.0

The dependence of the specific rotation on the temperature can be expressed by the following equation[4]:

$$[\alpha]_D^t = [\alpha]_D^{20} [1-0.000184(t-20)]$$

Taking into account these parameters allows the sugar concentration to be calculated from the rotation values according to the following equations[3,4]:

$$\text{glucose} \quad C = \frac{100 \times \alpha}{2 \cdot 52.8} = 0.9470 \cdot \alpha$$

$$\text{fructose} \quad C = \frac{100 \times \alpha}{2 \cdot (-92.5)} = 0.5405 \cdot \alpha$$

$$\text{sucrose} \quad C = \frac{100 \times \alpha}{2 \cdot 66.5} = 0.7519 \cdot \alpha$$

$$\text{lactose} \quad C = \frac{100 \times \alpha}{2 \cdot 52.5} = 0.9524 \cdot \alpha$$

$$\text{maltose} \quad C = \frac{100 \times \alpha}{2 \cdot 138} = 0.3623 \cdot \alpha$$

The length of the polarimeter tube is 2 dm; α is given in degrees.

Before recording the optical rotation it is essential that mutarotation is complete. For this purpose solutions should be stored overnight; alternatively, mutarotation can be accelerated by heating or by addition of a few drops of ammonium hydroxide to the solution.

1.7.2.1. Determination of Sucrose[4]

In practice the polarimetric method is most frequently used for the determination of sucrose. Special equipment (saccharimeter) exists which quotes the optical rotation in international sugar degrees (°S) according to a decision of the ICUMSA[5]. 100°S is defined as the polarization of a sucrose solution of 26.000 g/100mL in a 200 mm tube at 20 °C with white light filtered by potassium dichromate [calculation of the concentration of dichromate: % x length of the tube (cm) = 9]. Alternatively yellow sodium light (λ = 589.25 nm) can be used. 100°S is 34.620° (degrees).

For the determination of sucrose in food samples, high molecular weight compounds such as proteins and polysaccharides must be removed. This can be done by precipitation of these substances with basic lead acetate, neutral lead acetate, aluminum hydroxide or with a combination of Carrez [Carrez I: potassium hexacyanoferrate(IV); Carrez II: Zinc acetate].

Reducing sugars which interfere with the determination can be destroyed by heating with alkali.

Determination[4,6]

Pour 50 mL of the sample into a 100 mL volumetric flask and add 1 g of barium hydroxide; heat the mixture for 1 h at 70–80 °C in a water bath; neutralize the solution with 15% acetic acid measured against phenolphthalein; pour 25 mL of this solution into a 100 mL volumetric flask, add 25 mL of water and 2.315 g of sodium chloride; make up to 100 mL and read the optical rotation in a 200 mm tube, multiply the obtained value by 2.

Inversion

When it is necessary to determine sucrose in the presence of sugars such as common monosaccharides, the optical rotation must be recorded before and after hydrolysis. Hydrolysis is carried out with either dilute hydrochloric acid or with special enzymes such as invertase. From the difference in the optical rotation before and after hydrolysis the concentration of sucrose can be determined.

Polarimetric sugar determinations are one of the common methods used in the field of food analysis and have been used officially in the case of the analysis of sucrose in milk and dairy products[7] and in the analysis of starch in cereal products[8,9].

References

1   Kearsley, M. W. In *Analysis of Food Carbohydrates*; Birch, G. G., Ed.; Elsevier: London, New York, 1985; p 19.
2   Shallenberger, R. S. In *Analysis of Food Carbohydrates*; Birch, G. G., Ed.; Elsevier: London, New York, 1985; p 50.
3   Browne, C. A.; Zerban, F. *Chem. Methods of Sugar Analysis*, 3rd ed.; Wiley: New York, 1955.
4   Acker, L. In *Handbuch der Lebensmittelchemie*, Vol. II/2; Schormüller, J., Ed.; Springer: Berlin, Heidelberg, New York, 1967; p 363.
5   ICUMSA: International Commission for Uniform of Sugar Analysis (AOAC-number: 29.020).
6   AOAC-Methods No. 31.026, Off. *Methods of Association of Official Agricultural Chemists*, 14. Ed., 1984.
7   *Amtliche Sammlung von Untersuchungsverfahren* nach § 35 LMBG, L 02.06-E; Beuth: Berlin, Köln, 1981.
8   *Amtliche Sammlung von Untersuchungsverfahren* nach § 35 LMBG, L 17.00/5; Beuth: Berlin, Köln, 1982.
9   Hadorn, H.; Doevelar, F. *Mitt. geb. Lebensmitteluntersuch. Hyg.* **1960**, *51*, 1.

1.8. Electrochemical Methods

1.8.1. Polarography

The property of the electrochemical reduction of sugars was observed for the first time by Heyrovsky et al.[1,2] who obtained polarographic steps only for monosaccharides and not for disaccharides. In the first case, the steps observed only for ketoses were in the same magnitude for those of other reducing substances at the same concentration. The steps observed for aldoses were much lower than expected from their concentrations.

These results were confirmed later[3] and it was determined that the polarographic step results only from reduction of the free aldehyde form of the aldoses in solution at the surface of the mercury drop. The half-wave potential for sugars lies between −1.5 and −1.8 V. From the height of the polarographic steps concentrations of the free aldehyde in addition to the pyrano and the furano cyclic semiacetals of the monoaldoses can be estimated. These are usually very low except for D-allose and D-ribose (see Table).

Table. Half-Wave Potential and Concentration of Free Aldehydes (in %) in Aqueous Solutions for Common Monoaldoses (0.1 M; pH 7.0[3])

Compound	$E_{(1/2)}$ (V)	% Free Aldehyde
D-Xylose	− 1.54	0.13
D-Lyxose	− 1.54	0.18
L-Arabinose	− 1.58	0.22
D-Ribose	− 1.81	0.5
D-Glucose	− 1.58	0.022
D-Mannose	− 1.55	0.062
D-Galactose	− 1.59	0.085
L-Allose	− 1.71	1.10

The height of the waves are further dependent on temperature and pH; with increasing pH they become increasingly negative. Polarography has been used frequently for the selective determination of fructose, for example, in honey (electrolyte: 0.01 M LiCl; −1.7 V)[1], in wine[2], in raw sucrose (electrolyte: 1 M CaCl₂; −1.6 V)[4] and in fruits[5]. In the last case the following procedure was used.

Determination[5]
Extract the fruit samples with 80% ethanol until the tests for sugars are negative; then, evaporate the ethanol and filter the aqueous residue through Celite. Dilute the clear solution with distilled water to a concentration of 2–10 mg fructose/L and remove the cations and anions (fruit acids) by treatment with ion exchangers: pour an aliquot of this solution into a 50 mL volumetric flask and add 5 mL 1 M aqueous CaCl₂, and 5 mL 0.2% aqueous gelatine solution and make up with distilled water. In parallel to this, make up solutions of D-fructose with different concentrations: Measure the heights of the polarographic waves at −1.7 V (against saturated calomel electrodes) for the sample and the standard solutions and determine the fructose concentration of the sample by reference to a calibration curve drawn up from the standard solutions.

1.8.2. Amperometric Oxidation

Amperometric oxidation is carried out with carbon,[6] nickel[7] or precious metals[8] and occurs on the working electrode according to the following equation:

$$A = B + n \cdot e$$

A yields $n \cdot e$ electrons during formation of compound B. The potential of the working electrodes must be selected such that A is oxidized almost quantitatively to B. The current at this potential drops after a short time due to the absorbance of the oxidation product on the electrode surface. By means of the so-called triple-impulse-amperometry this effect can be prevented. This equipment consists of a working electrode (gold, platinum), a reference electrode (silver/silver chloride) and a counter electrode (precious metal)[9] or modified glassy carbon[10].

The compounds to be oxidized (carbohydrates, sugar alcohols, etc.) are oxidized at the measurement potential E_1 and the current is proportional to their concentrations. The adsorbed oxidation products at the surface of the electrodes are removed by the cleaning potential E_2, which is noticeably more positive than E_1, where metal oxides such as PtO, Au_2O are formed at the surface and oxidize the adsorbed products totally to CO_2 and H_2O, which are removed. In the third period the metal oxides are reduced by the negative potential E_3. Then the cycle begins again.

Today this technique is used for the determination of mono- and oligosaccharides by flow-injection analysis[8] after separation by HPLC[9,11,12]. The limit of detection for D-glucose and D-sorbitol is 0.2 µg/100µL (cf. refraction 50 µg/100 µL).

For the determination of sugars after separation by HPLC the following conditions are used: Material for the working electrode: gold; potentials E_1: 0.05 V [t_1 = 360 milliseconds (ms)], E_2: 0.8 V (t_2 = 120 ms) and E_3: –0.6 V (T_3 = 420 ms) (see also Section 2.1.3.4.)[13].

References

1   Heyrovsky, J.; Smoler, I. *Collect. Czechoslov. Chem. Commun.* **1932**, *4*, 521.
2   Heyrovsky, J.; Smoler, I.; Stasny, J. *Vest. Cs. Akad. Zemedel.* **1933**, *9*, 599.
3   Cantor, S. M.; Peniston, Q. P. *J. Am. Chem. Soc.* **1940**, 62, 2113.
4   Vavruch, I.; Rubeš, E. *Listy Cukrovar* **1947/48**, *64*, 185.
5   Williams, K. T.; McComb, E. A.; Potter, E. F. *Anal. Chem.* **1950**, *22*, 1031.
6   Brunt, K. *Analyst* **1982**, *107*, 1261.
7   Buchberger, W.; Winsauer, K.; Breitwieser, Ch. *Fresenius' Z. Anal. Chem.* **1983**, *315*, 518.
8   Hughes, S.; Meschi, P. L.; Johnson, D. *Anal. Chim. Acta* **1981**, *132*, 1, 11.
9   Hughes, S.; Johnson, D. C. *J. Agric. Food Chem.* **1982**, *30*, 712.
10   Weiss, J. *G. I. T. Supplement* **1986**, *3*, 65.
11   Hughes, S.; Johnson, D. C. *Anal. Chim. Acta* **1983**, *149*, 1.
12   Johnson, D. C.; Polta, T. Z. *Chromatogr. Forum* **1986**, *1*, 37.
13   Hardy, M.; Towsend, R. R.; Lee, Y. C. *Anal. Biochem.* **1988**, *170*, 54.

2. Analytical Methods Using Separation Procedures

In the most cases, mixtures of monosaccharides and oligosaccharides are encountered in sugar analysis (e.g., in biological materials, foods or drugs). Application of the methods described in Sections 1.1.–1.4. allows only determination of the total amount of sugars. This is no longer sufficient for the requirements of today. In most cases it is also important to know the concentrations of the individual components of such sugar mixtures. For this purpose, it is necessary to separate the mixtures before carrying out the definitive sugar determinations. These separation procedures are based on the following principles:

(a)   Chromatographic methods which are based on the different distribution properties between two phases in a flowing system. The most important of these are paper chromatography, thin-layer chromatography, column liquid chromatography, high pressure liquid chromatography (HPLC) and gas chromatography.

(b)   Separation based on different molecular masses, e.g. ultrafiltration, gel chromatography (column and thin-layer procedures).

(c)   Separation based on different mobilities in an electric field, e.g. paper electrophoresis, thin-layer electrophoresis, gel electrophoresis and capillary electrophoresis.

2.1. Chromatographic Methods

2.1.1. Paper Chromatography

The general procedure, which is valid today, was first described by Consden, Gordon and Martin[1] and was used for the separation of amino acids. This method was also a very great break through for the separation of mono- and oligosaccharides and is carried out according to the general procedure for this method[2,3,4].

The solid phases are special filter papers with high degrees of purity containing a highly homogenous felt. They consists of cotton lint, which should be free from soluble substances. Today many commercial varieties are available for various special purposes.

According to the capillary speed of the mobile phase, papers of high mobility rate (70–10 min for 30 cm height), middle mobility rate (140–220 min for 30 cm height) and low mobility rate (70–10 min for 30 cm height) can be differentiated. The average thickness of such papers is 0.16–0.19 mm (except those for micropreparative purposes) and the average weight/m^2 is 80–150 g[5].

For special purposes, (e.g., separation of sugar phosphates) commercial papers must be cleaned by washing with acids before use. Such papers are also commercially available.

As standard papers for the separation of sugars Whatman 1 and Schleicher-Schüll 2043b are recommended.

For practical purposes, papers are used either in their original dimensions or they are cut to suitable sizes (e.g., ascending procedure in small glass vessels). A solution of the samples and of the substances are spotted on starting points marked on the baseline which is drawn in pencil at a distance of 50 mm from the edge of the paper. A distance of 30–40 mm between the starting points is recommended for one dimensional chromatography. The paper is then equilibrated in the chromatographic chamber, containing an atmosphere of the mobile phase. The volume of the sample solutions applied to the paper should be between 10–200 μL.

The sugars samples are dissolved either in aqueous solutions, in alcoholic solutions (methanol or ethanol) or in a mixture of both solvents. The amounts of the single components in the sugar mixtures of the samples should range between 10–50 μg. The solutions should be free of accompanying substances especially electrolytes or high molecular compounds such as proteins or polysaccharides which interfere with the separation. They must be removed before the chromatographic procedures.

Desalting of the Samples

The presence of inorganic cations and anions in the sample solutions causes a distortion of the equilibrium between the aqueous or organic phase and therefore a distortion in the zones of the separated substances. Their removal can usually be effected by treating the samples with a mixture of cation and anion exchangers (e.g., desalting of urine: Dowex 50X8 H⁺-form 25–50 mesh; Dowex 1–8, COO⁻ form 25–50 mesh; ~300 g per 1 L urine[6] or Zerolite–DMF = Biodeminrolite in H⁺ and acetate form[7]). In another variation the salt-containing samples are evaporated to dryness and the sugars are extracted with pyridine[8].

Removal of Proteins

The following procedures are used:

(a)   Heating of the aqueous samples with trichloroacetic acid or addition of sulfosalicylic acid without heating. Precipitation of the protein then occurs. The reagents are then removed by extraction with organic solvents[9]. The disadvantage of this method is the possibility of cleaving oligosaccharides with the acids (especially heating with trichloroacetic acid).

(b)   Addition of 1 volume of ethanol to 1 volume of the extract; precipitation of the proteins with basic lead acetate (30% solution) and removing the excess of Pb ions as sulfate, phosphate, carbonate or sulfide[10].

(c)   Precipitation with barium hydroxide and zinc sulfate[11] or with a mixture of Carrez [zinc acetate and potassium cyanoferrate(IV)].

(d)   Separation of the high molecular part of the sample from the low molecular compounds by gel chromatography; passage through a column filled with Sephadex or biogel.

After these procedures the solutions of the sugar samples are spotted on the chromatographic paper which is developed by the mobile phases in the special chromatographic chamber. When working with the descending technique the trough with the mobile phase is fixed at the upper end of the chamber to which the paper is attached. The mobile phase flows down through the paper as a result of capillary action and gravity. When the mobile phase is not stopped before reaching the lower end of the paper but allowed to drop by cutting a serrated edge at this end of the paper, this prolonged chromatographic development (continuous flow) improves the separation of sugar mixtures. The advantage of the technique is limited by the diffusion of the separated compounds into the mobile phase causing spreading of the spots.

For ascending paper chromatography the mobile phase is at the bottom of the chromatographic vessel in which the chromatographic paper is placed. The ascending flow of the mobile phase is caused only by capillary action and it stops, when an equilibrium between the capillary action and gravity is reached. Prolonged passage of the mobile phase is obtained only if evaporation at the upper end of the paper occurs.

The chromatographic chambers are preferably closed glass vessels.

The chromatographic development is finished by opening the chamber and removing the paper. The mobile phase is removed by evaporation; for this purpose special drying equipment exists. Visualization of the separated substances is carried out using a suitable color reaction. The reagents are generally applied by spraying and in a few cases by dipping the paper into the reagent. The positions of the zones are characterized by their R_f values:

$$R_f \text{ value } = \frac{\text{Distance (Start - Middle of the spot)}}{\text{Distance (Start - Front of the mobile phase)}}$$

For the continuous development, the relative mobility R_{rel} of a substance is given by the following expression.

$$R_{rel} = \frac{\text{Distance (Start - Middle of the spot of the sample)}}{\text{Distance (Start - Middle of the spot of the reference substance)}}$$

2.1.1.1. Paper Chromatography of Mono- and Oligosaccharides

2.1.1.1.a. Monosaccharides

Several solvent combinations are in existence and those which are applied frequently are listed below:

1.   Phenol–water: 100 mL of distilled water is added to 900 g of freshly distilled phenol and the mixture is liquefied. Before use this solvent is saturated with water in a separating funnel. For the separation of acid compounds 1% ammonium hydroxide should be added to improve the shape of the spots[2,3,12].

2.   Collidine (S-trimethylpyridine)–water; application of the water saturated organic phase[2,3].

3.   Picoline (2-methylpyridine)–water 6:4 v/v[12].

4.   1-Butanol, saturated with 1% ammonium hydroxide[2,3].

5.   1-Butanol–ethanol–water 4:1:4 v/v or 4.5:0.5:4.9 (1% ammonium hydroxide); the organic phase is used as mobile phase[3].

6.   1-Butanol–acetic acid–water 4:1:5 v/v; the organic phase is used as mobile phase; the mixture has only a limited stability[3].

7.   1-Butanol–pyridine–water 3:2:1.5 v/v; to the separated upper layer addition of 1 part of pyridine[13].

8.   Ethyl acetate–pyridine–water 2:1:2 v/v; the phase has a low viscosity and therefore a high mobility. It is particularly suitable for the continuous descending procedure, but is not stable and must be prepared just before use[4].

9.   Ethyl acetate–acetic acid–water 3:1:3 v/v; this phase also has a high mobility and must be prepared just before use[4].

10.   Methyl ethyl ketone–aqueous 1% ammonium hydroxide; the organic phase is used[3].

11.   1-Propanol–ethyl acetate–water 7:1:2 v/v[14].

12.   2-Propanol–pyridine–water–acetic acid 8:8:4:1 v/v[15].

13.   Acetone–water 9:1 v/v[16].

14.   Benzyl alcohol–1-propanol–water–88% formic acid 7.2:5:2:2 v/v[17].

15.   Ethyl acetate–acetic acid–0.55% aqueous phenylboric acid 9:2:2 v/v[18].

The R_f values (average data) of a few of these mobile phases are listed in Table 1.

The R_f values are dependent on the hydrophilicity of the compound. The greater the number of the CH–OH groups the lower are the R_f values (R_f tetroses > R_f pentoses > R_f hexoses). Monosaccharides generally have higher R_f values than oligosaccharides.

Within a group containing equal numbers of C-atoms the differences in the R_f values of the individual monosaccharides are small. For aldohexoses, mannose has the highest and galactose the lowest value (exception: in water saturated phenol). The separation of these three hexoses, which can occur in the hydrolysis products of some polysaccharides is only possible by the continuous flow technique. Common mobile phases such as 1-butanol–acetic acid–water 4:1:5 v/v allows a good separation of mixtures of glucose, fructose and sucrose (see Table 1).

Table 1. R_f Values of Common Monosaccharides

Compound	A	B	C	D	E	F	G	H
D-Glucose	0.39	0.39	0.18	0.28	0.64	0.11	0.26	0.08
D-Galactose	0.44	0.34	0.16	0.24	0.62	0.09	-	0.08
D-Mannose	0.45	0.46	0.20	0.32	0.69	0.13	0.30	0.09
D-Fructose	0.51	0.42	0.23	0.32	0.68	0.14	0.32	0.12
L-Sorbose	0.42	0.40	0.20	-	0.68	0.12	-	0.16
D-Xylose	0.44	0.50	0.28	0.38	0.73	0.17	0.33	0.15
D-Arabinose	0.54	0.43	0.21	0.33	0.70	0.15	0.32	0.11
D-Ribose	0.59	0.56	0.31	-	0.76	0.21	0.41	0.50
D-Lyxose	-	-	-	-	0.73	-	-	0.18
D-Erythrose	-	-	-	-	-	-	-	0.84
L-Threose	-	-	-	-	-	-	-	0.53
L-Rhamnose	0.59	0.59	0.37	0.49	0.82	0.29	-	-
L-Fucose	0.63	0.44	0.27	-	0.75	-	-	-
D-Deoxyribose	0.73	0.60	-	-	-	-	-	0.41

A = Phenol, water saturated (descending) + 1% NH₄OH (Patridge).
B = S-Collidine, water saturated (descending) (Cramer).
C = 1-Butanol–acetic acid–water 4:1:5 v/v (descending) (Patridge).
D = Ethyl acetate–pyridine–water 2:1:2 v/v (Jermyn and Isherwood).
E = 2-Propanol–pyridine–water–acetic acid 8:8:4:1 v/v (Gordon et al.).
F = 1-Butanol–ethanol–water (NH₄OH) 4.5:0.5:4.9 v/v (Patridge).
G = Benzyl alcohol–1-propanol–water–85% formic acid 7.5:5:2:2 v/v (Giovannozzi-Sermani).
H = Ethyl acetate–acetic acid–0.55% phenylboric acid 9:2:2 v/v (Bourne, Lees and Weigel).

2.1.1.1.b. Oligosaccharides

The solvents used for the separation of monosaccharides are less suitable for oligosaccharides since either the R_f values are too low necessitating an extremely long continuous flow or the R_f values are almost identical and separations are not possible. An alteration in the proportions of the single components in the mixture leads to suitable mobile phases for the separation of oligosaccharides. The following systems have been proved to be suitable:

1.   Ethyl acetate–acetic acid–formic acid–water 9:1.5:0.5:2 v/v[19,20].

2. Benzyl alcohol–1-propanol–formic acid–water 7.2:5:2:2 v/v[17].
3. Ethyl acetate–1-propanol–water 5.7:3.2:1.3 v/v[21].
4. 1-Butanol–pyridine–water 5:3:2 v/v[22].
5. Ethyl acetate–pyridine–water 10:4:3 v/v[23].

For the most of these systems continuous development (duration of 1–4 days) is necessary. R_f values or the respective R_G values of some oligosaccharides are listed in Table 2.

Table 2. R_f Values of Oligosaccharides

Compounds	A	B	C	D[a]	E[a]
Sucrose	0.40	0.14	0.62	0.71	-
Maltose	0.32	0.11	0.57	0.45	0.62
Lactose	0.24	0.09	0.46	-	-
Melibiose	-	-	0.45	-	-
Gentiobiose	-	-	-	-	-
Cellobiose	-	-	0.53	-	-
Turanose	-	-	0.64	0.74	-
Isomaltose	-	-	0.47	0.41	0.47
Maltulose	-	-	-	0.58	-
Isomaltulose	-	-	-	0.53	-
Leucrose	-	-	-	0.50	-
Panose	-	-	-	-	0.28
Raffinose	0.20	-	0.44	-	-
Melezitose	-	-	0.54	-	-
Maltotriose	-	-	-	0.23	0.41
Isomaltotriose	-	-	-	0.18	-
Isomaltotriulose	-	-	-	0.26	-
Maltotriulose	-	-	-	0.34	-
Maltotetraose	-	-	-	-	0.26

[a]R_G = R_f relative to R_f Glucose = 1.00

A = S-Collidine–water saturated (Patridge).
B = 1-Butanol–acetic acid–water 4:5:1 v/v (Patridge).
C = 2-Propanol–pyridine–water–acetic acid 8:8:4:1 v/v (Gordon).
D = 1-Butanol–pyridine–water 5:3:2 v/v (Avigad).
E = Ethyl acetate–pyridine–water 10:4:3 v/v (Archibald and Manners).

Reducing oligosaccharides can be converted by reaction with benzylamine into *N*-benzylglycosides. With the mobile phase 1-butanol–ethanol–water–concd ammonium hydroxide 4.5:0.5:4.9:0.1 v/v good separations of the oligosaccharides can be obtained[24].
Oligomeres of a single sugar type can be separated by the following systems:
1. 1-Butanol–pyridine–water 6:4:3 v/v[25–27] or 6:5:5 v/v[28].
2. 2-Propanol–acetic acid–water 7:1:2 v/v[29].

With these solvents, separations of oligomers up to a degree of 12 can be obtained by continuous development.

Visualization of the Separated Sugars

Since paper is also a carbohydrate material (cellulose) the choice of the color reagents is limited to those which react only with mono- and oligosaccharides as otherwise they would also react with the chromatographic paper giving a strong background coloration.

The first group of color reactions are based on the reducing properties of the sugars. They are not specific and are also positive for other reducing compounds. A number of such reactions are described in Section 1.1. and special visualization procedures have been established.

Silver Nitrate–Ammonium Hydroxide; Original Procedure[2,3,30]

Reagent: 1 volume 0.1 M silver nitrate is mixed with 1 volume 5 M aqueous ammonium hydroxide. Solid NaOH is added until a concentration of 1 M is reached. After development the dry chromatographic paper is sprayed thoroughly with the reagent and dried for 5–10 minutes at 105 °C. The reducing sugars as well as other reducing compounds appear as brown-black spots on a light background. Nonreducing sugars such as sucrose only afford weak spots.

The dipping procedure is more sensitive than spraying[31,32].

Reagent A: 61 g silver nitrate is dissolved in 26 mL distilled water (stock solution); 2 mL of this solution is mixed with 200 mL acetone and 8 mL water.

Reagent B: 0.5 M ethanolic NaOH.

Reagent C: 10% aqueous sodium thiosulfate.

After development and drying, dip the chromatographic paper in reagent A then dry for 3–5 minutes at room temperature and dip in reagent B and dry by pressing between filter paper. After approximately 10 minutes when the paper should be almost dry, dip in reagent C to fix the precipitated elementary silver and complex the excess of silver ions; the latter are removed by intensive washing of the paper with running water.

Quantitation of the paper chromatograms is carried out either by direct densitometry of the spots on the paper[33,34], by estimation of the area of the spots[35], or by determination of the silver concentration by atom absorption techniques[32].

Triphenyltetrazolium Chloride (TTC)[36]

As it is in the liquid state, this compound also reacts on paper with reducing substances to form the deep red triphenylformazan.

Reagent: 1 volume 1 M methanolic NaOH is mixed with 1 volume 4% TTC solution in methanol just before use. The methanol must be free of traces of acetone.

Spray the developed dry paper chromatogram with the reagent, or draw it through a vessel containing the reagent, and dry at 65 °C for 60 minutes or at 70 °C for 30 minutes. The reducing sugars appear as deep red spots on a pink background. Nonreducing sugars react even after spraying with dilute hydrochloric acid before the TTC reaction.

3,5-Dinitrosalicylic Acid (3,5-DS)[26,37]

Reagent: 0.5% solution of 3,5-DS in 4% aqueous NaOH. The reagent is stable at room temperature.

Spray the chromatogram with the reagent and heat for 4–15 minutes at 100 °C. The reducing sugars appear as brown spots on a yellow background.

3,4-Dinitrobenzoic Acid[38]

Reagent: 1% solution of 3,4-dinitrobenzoic acid in 1 M aqueous sodium carbonate.

Spray the dry developed paper chromatogram with the reagent and dry for 5–10 minutes at 100 °C. The reducing sugars appear as blue spots which turn brown on prolonged heating. Ketoses react faster than aldoses; the reaction rates with oligosaccharides are low.

Malonamide[39]

Reagent: 1% solution of malonamide in 1 M sodium carbonate buffer (pH 9.2).

Spray the dry paper chromatogram with the reagent and heat for 5–10 minutes at 120 °C. The reducing sugars appear as intense bright fluorescences at an excitation wavelength of 360 nm; detection limit: 0.05 µg for reducing mono- and disaccharides. The fluorescence is highly selective; ordinary alcohols, phenols, ketones, aldehydes, carboxylic acids, amino acids do not show fluorescence at this wavelength.

Sodium Periodate[40]

This reaction is valid for all compounds containing vicinal OH groups, e.g. sugar alcohols, non-reducing sugars, aldonic acids, etc.

Reagent A: 6.4 g sodium periodate is dissolved in 750 ml distilled water and the solution made up to 1000 mL with *tert.*-butanol

Reagent B: 5.5 g *o*-dianisidine or benzidine (carcinogenic!) is dissolved in 500 mL of *tert.*-butanol and the solution made up to 1000 mL with distilled water containing 48 g of ammonium nitrate

Spray the dry paper chromatogram with reagent A, dry for 30 minutes at room temperature and then spray with reagent B. The separated substances appear as white spots on a blue background.

Aniline–Phthalic Acid[41,42]

Reagent: 0.93 g of aniline (freshly distilled) and 1.66 g of phthalic acid are dissolved in 100mL of water saturated 1-butanol[41] or in a mixture of acetone–water 95:5 v/v[42].

Spray the dry paper chromatogram intensively with the reagent and heat for 5–10 minutes at 105 °C. Aldohexoses, methyl aldopentoses and hexuronic acids show olive brown spots with a yellow fluorescence, aldopentoses show reddish brown spots with a red fluorescence. Ketoses yield only weak spots with a distinct yellow fluorescence. Sensitivity: approximately 5 µg.

Other modifications of this reagent are: aniline–phosphoric acid[43], aniline–oxalic acid[44].

m-Phenylenediamine[13]

Reagent: 0.2 M solution of *m*-phenylenediamine in 76% ethanol.

Spray the dry paper chromatogram with the reagent and heat for 5 minutes at 105 °C. Under the UV lamp aldopentoses show an intense orange fluorescence, aldo-and ketohexoses show an intense yellow fluorescence all with sensitivities around 10 µg. Disaccharides and hexuronic acids react weakly and sugar alcohols do not react with this reagent.

p-Aminohippuric Acid[45]

Reagent: 0.3% solution of *p*-aminohippuric acid in ethanol.

Spray the dry paper chromatogram with the reagent and heat for 8–10 minutes at 140 °C. Hexoses and pentoses appear as orange spots which have an intense orange fluorescence with sensitivities around 1 µg. Addition of 3% phthalic acid to the reagent makes it possible to detect reducing and nonreducing oligosaccharides, which are easily hydrolyzed (e.g., sucrose).

Naphthoresorcinol–Trichloroacetic Acid[3]
Reagent A: 0.2% naphtholresorcinol in ethanol.
Reagent B: 2% aqueous trichloroacetic acid

Mix both reagents just before use in a ratio of 1:1 v/v and spray the dry paper chromatogram; after briefly drying at room temperature heat for 5–10 minutes at 100–105 °C. The reagent is specific for ketoses and ketose-containing oligosaccharides (sucrose, raffinose), which appear as deep red spots on a white to pink background. After standing few hours pentoses and uronic acids yield strong blue spots. Phosphoric acid can be used instead of trichloroacetic acid[43,46]; in this case, the spots of pentoses and uronic acid appear in a shorter time.

Quantitative Determinations
Quantitative determinations can be carried out either after visualization directly on the paper itself or by elution of the formed dye and photometric determination. Alternatively, the pure separated sugar can be eluted from the paper and quantitative determination carried out using one of the common micromethods.

Direct Determination on the Paper
The direct visual comparison of the spots of the sample with standards of known concentrations is the simplest method. Using only one sample spot only a raw estimation is possible[47,48]. By means of a special method, however, in which two dilutions of the original sample are used together with the original sample and a combination of standard concentrations, visual comparison can achieved with an accuracy of ±10% (the Weisz ring oven technique)[49,50].

The determination of the areas of the spots can be performed either by mechanical planimetry or by weighing. The correlation between the logarithm of the concentrations and the lengths or areas of the spots is linear in most cases[51]. For this purpose the reaction with silver nitrate is the most suitable.

The direct photometry of the spots on the paper can be carried out only with a reflection spectrophotometer. The silver nitrate reaction is also very suitable here due to the great differences between the contrasts of the spots at different concentrations. The relationship between the signal of the photometer and the logarithm of the concentration, however, is not generally linear[36]. This is in contrast to the reaction between aniline–trichloracetic acid where this relation is valid[52]. Greater accuracy is obtained by eluting the dye of the visualization reaction and measuring the concentration of the solution photometrically; the formazane reactions are very suitable in this case.

Determination[36]
Reagent A: 1 M NaOH in methanol
Reagent B: 4% solution of triphenyltetrazolium chloride in methanol. The methanol must be free of acetone, otherwise the color of the background increases.

Just before use, mix reagent A and B in a 1:1 v/v ratio and pour into a flat trough. Draw the dry paper chromatogram, on which the sample and standard sugar solutions have been spotted, through this solution and dry it at room temperature with a fan. Then heat the dry paper in a water saturated atmosphere at 70 °C for 30 minutes; the formation of the formazanes occurs by this procedure. Remove the paper from the hot atmosphere and cut out the spots of formazanes and place in test tubes and extract with methanol–acetic acid 10:1 v/v; after 10 minutes when extraction is complete read the absorbance at 482 nm against a blank. From the calibration curves the concentrations of the separated sugar are evaluated.

Determination of the Sugar in Solution

The paper chromatogram with samples and standards of known concentration are developed and dried; after location of the separated sugar zones by a marker chromatogram they are extracted with water–methanol mixtures. By spraying the paper chromatogram with a solution of boric acid and bromo-cresol purple in methanol the sugars zones can be located without destroying the substances[53].

After extraction, the concentration of the sugars in solution are determined by micromethods as described in Sections 1.1.–1.4. These are either titrimetric procedures [e.g., Cu(II) reaction[54,55], periodic acid oxidation[56], etc.] or photometric procedures (e.g., aniline–phthalic acid[57], 3,5-dinitrosalizylic acid[37] or anthrone–sulfuric acid[26,58]).

2.1.1.2. Paper Chromatography of Sugar Derivatives

2.1.1.2.a. Methyl Glycosides and *O*-Methyl Sugars

For these compounds, which are important in the elucidation of the chemical structures of oligo- and polysaccharides, the following mobile phases have been used for their separation:

1.   *tert.*-Amyl alcohol–1-propanol–water 4:1:1.5 v/v; separation of α- and β-glycosides[59].
2.   1-Butanol–ethanol–water–aqueous ammonium hydroxide (concd) 40:10:49:1 v/v organic phase[60].
3.   Water saturated 1-butanol (development at 37 °C)[61].
4.   Methyl ethyl ketone saturated with 1% ammonium hydroxide[61,62].
5.   2,4,6-Collidine–ethyl acetate–water 2:5:5 v/v; upper phase[63].
6.   1-Butanol–ethanol–water 5:1:4 v/v[64]; upper phase.

The R_G values of *O*-methyl monosaccharides and the R_f values of some 1-*O*-methyl glycosides are listed in Tables 3 and 4, respectively.

For visualization of these compounds some of the reagents for unsubstituted sugars can be used, for example, aniline–phthalic acid[65], *p*-anisidine[66] or *N*-dimethyl-*p*-aminoaniline–HCl–trichloroacetic acid[62].

2.1.1.2.b. Sugar Acetates

The fully acetylated derivatives are weakly water soluble, but they can be separated on hydrophobized chromatographic paper. With the solvent system 1-butanol–pyridine–water 3:1:1 v/v a separation of a mixture of tetra- and pentaacetates of glucose, galactose, xylose and arabinose is possible[67]. This procedure is only of historical interest today, since thin-layer chromatography is a more suitable method for these compounds. For detection, sugar acetates are saponificated with, for example, diethylaminoethanol at 50 °C and visualized with silver nitrate reagent.

2.1.1.3. Sugar Alcohols

The following solvents are used for the development of the paper:

1.   1-Butanol–water saturated[66].
2.   1-Butanol–ethanol–water 4:1.1:1.9 v/v[66] or 4:1:5 v/v[68].
3.   1-Butanol–acetic acid–water 4:1:5 v/v[69].
4.   Ethyl acetate–acetic acid–water 3:1:3 v/v[68].
5.   Benzene–1-butanol–pyridine–water 1:5:3:v/v[66].
6.   1-propanol–ethyl acetate–water 7:1:2 v/v[70].
7.   Nitromethane–acetic acid–ethanol–water saturated with boric acid 8:1:1:1 v/v[71].

Table 3. R_G Values of O-Methyl Sugars

Compound	R_G Values	Compound	R_G Values
Xylose	0.15	**Galactose**	0.07
2-O-Methylxylose	0.38	4-O-Methylgalactose	0.16
2,4-O-Dimethylxylose	0.66	2-O-Methylgalactose	0.23
2,3-O-Dimethylxylose	0.74	2,4-O-Dimethylgalactose	0.41
2,3,4-O-Trimethylxylose	0.94	4,6-O-Dimethylgalactose	0.42
Arabinose	0.12	2,6-O-Dimethylgalactose	0.44
2-O-Methylarabinose	0.38	2,3,4-O-Trimethylgalactose	0.64
2,3-O-Dimethylarabinose	0.64	2,4,6-O-Trimethylgalactose	0.67
2,3,5-O-Trimethylarabinose	0.95	2,3,6-O-Trimethylgalactose	0.71
Ribose	0.21	2,3,4,6-O-Tetramethylgalactose	0.88
Rhamnose	0.30	**Mannose**	0.11
4-O-Methylrhamnose	0.57	4-O-Methylmannose	0.32
3,4-O-Dimethylrhamnose	0.84	2,3-O-Dimethylmannose	0.54
2,3,4-O-Trimethylrhamnose	1.01	4,6-O-Dimethylmannose	0.57
Glucose	0.09	3,4-O-Dimethylmannose	0.58
2-O-Methylglucose	0.22	3,4,6-O-Trimethylmannose	0.79
3-O-Methylglucose	0.26	2,3,6-O-Trimethylmannose	0.81
6-O-Methylglucose	0.27	2,3,4,6-O-Tetramethylmannose	0.96
4,6-O-Dimethylglucose	0.46	**Fructose**	0.12
3,6-O-Dimethylglucose	0.51	3,4-O-Dimethylfructose	0.61
3,4-O-Dimethylglucose	0.52	1,3,4-O-Trimethylfructose	0.83
2,3-O-Dimethylglucose	0.57	3,4,6-O-Trimethylfructose	0.86
2,4,6-O-Trimethylglucose	0.76	1,3,4,6-O-Tetramethylfructose	1.01
2,3,6-O-Trimethylglucose	0.83		
2,3,4-O-Trimethylglucose	0.85		
2,3,4,6-O-Tetramethylglucose	1.00		
2,3,5,6-O-Tetramethylglucose	1.01		

Solvent: 1-Butanol–ethanol–water 5:1:4 v/v; upper phase.
R_G = R_f relating to 2,3,4,6-O-tetramethylglucose (Hirst and Jones).

Table 4. R_f Values of 1-O-Methyl Glycosides

Compound	R_f Values	Compound	R_f Values
Methyl α-D-glucoside	0.58	Methyl β-arabinoside	0.68
Methyl β-D-glucoside	0.63	Methyl α-L-fucoside	0.76
Methyl α-D-galactoside	0.54	Methyl α-L-rhamnoside	0.86
Methyl β-D-galactoside	0.51	Methyl α-D-xyloside	0.75
Methyl α-mannoside	0.68	Methyl β-lactoside	0.33
Methyl α-mannofuranoside	0.74	Methyl β-maltoside	0.37
Methyl α-arabinoside	0.60		

Solvent: tert.-Amyl alcohol–1-Propanol–water 4:1:1.5 v/v (Cifonelli).

Table 5 shows the R_f values of the most common sugar alcohols[72].

Table 5. R_f Values of Sugar Alcohols

Compounds	A	B	C	D	E	G[a]
Ethylene glycol	0.64	-	-	-	0.58	-
Glycerol	0.48	-	-	0.58	0.46	-
Erythrol	0.35	0.31	0.34	0.49	-	-
Adonitol (Ribitol)	0.25	0.28	0.28	0.40	-	2.9
Arabitol	0.22	0.28	-	0.43	-	3.0
Xylitol	0.20	0.24	-	-	-	3.2
Dulcitol (Galactitol)	0.16	0.17	0.18	0.31	0.20	-
Mannitol	0.19	0.20	0.19	0.34	0.22	2.4
Sorbitol	0.18	0.17	0.19	0.31	0.21	1.9
Inositol	0.06	0.07	0.08	0.11	0.07	-

[a]R_f relating to R_f (Glucose) = 1.00.
A =   1-Butanol–acetic acid–water 4:1:5 v/v (Buchanan, Dekker and Long).
B =   Butanol-ethanol–water 4:1:5 v/v (Hackman).
C =   Ethyl acetate–acetic acid–water 3:1:3 v/v (Hackman).
D =   1-Propanol–ethyl acetate, water 7:1:2 v/v (Cerbulis).
E =   Benzene–butanol–pyridine–water 1:5:3:3 v/v (Hough).
G =   Nitromethane–acetic acid–ethanol–water, saturated with boric acid 8:1:1:1 v/v (Robyt).

Visualization

Since sugar alcohols do not contain reactive carbonyl groups they do not react with the common visualization reagents used for mono- and oligosaccharides. Only reagents which attack vicinal hydroxyl groups (e.g., by oxidation) are positive. These are alkaline permanganate, silver nitrate–ammonium hydroxide, lead tetraacetate and sodium periodate.

Boric Acid–Bromocresol Purple[54]

This detection is based on formation of an acid complex between vicinal OH groups and boric acid. This medium-strong acid can be detected by an acid–base indicator.
Reagent: dissolve 40 mg of bromocresol purple in 100mL of ethanol containing 100 mg boric acid and 7.5 mL aqueous 1% sodium tetraborate.
After spraying the developed chromatograms, the sugar alcohol and sugar-containing zones appear as yellow spots on a deep blue background. The color difference fades in a few minutes; the position of the sugar alcohols and other polyhydroxy compounds must be marked with a pencil.

Sodium Periodate–Silver Nitrate[73]

Sugar alcohols (and sugars) are cleaved with periodate and the reaction products containing carbonyl groups are visualized by the alkaline silver nitrate procedure.
Reagent A: 1% solution of sodium periodate in 50% aqueous acetone.
Reagent B: 1 mL saturated silver nitrate is mixed with 100 mL acetone–water 95:5 v/v.
Reagent C: 1% aqueous NaOH.
Reagent D: 5% aqueous sodium thiosulfate solution.

Spray the dry paper first with reagent A and after 2 minutes with reagent B; the nonreducing sugars and sugar alcohols appear as light spots on a brown background; spray with reagent C where the sugar alcohols and nonreducing oligosaccharides turn black as a result of the formation of elementary silver. Finally dip the paper in reagent D for fixation and for removing the excess silver nitrate.

p-Anisidine–Sodium Periodate[74]

This reagent allows the detection of polyols and aldonic acids (and their lactones) in the presence of sugars and uronic acids. The latter compounds react with p-anisidine yielding specific colors; the treatment with periodate hereafter stains the whole paper except for the polyol and aldonic acid containing zones.

Reagent A: 1% solution of p-anisidine in 70% ethanol

Reagent B: 0.1 M sodium periodate: 10 mL are added to 100 mL acetone

Spray the dry paper chromatogram first with reagent A heat 5–10 minutes at 100–105 °C and rinse the sheet with reagent B; the polyols as well as aldonic acids and their lactones appear as white spots on a brown background.

Lead Tetraacetate[69]

This compound cleaves vicinal OH groups with formation of two carbonyl products.

Reagent: 1 g lead tetraacetate is dissolved in 100 mL benzene; the solution is decolored with activated charcoal.

Wet the dry paper chromatogram first with xylene and then spray with the reagent; compounds with vicinal OH groups appear as white spots on a brown background due to hydrolysis of lead tetraacetate to brown lead(IV) oxide.

2.1.1.4. Sugar Acids and Their Lactones and Anhydrides

The following solvents are suitable for this group of compounds:

1.   Pyridine–1-amylalcohol–water 7:6:7 v/v; useful for the separation of galacturonic-, glucuronic- and mannuronic acid[75].

2.   Pyridine–ethyl acetate–acetic acid–water 5:5:1:3 v/v[76].

3.   1-Butanol–water–propionic acid 10:7:5 v/v [77].

4.   1-Butanol–acetic acid–water 3:1:1 v/v[78].

5.   1,1,1-Trimethylethanol (*tert.*-amyl alcohol)–1-propanol–water 4:1:1 v/v[59].

6.   1-Butanol–ethanol–water 4:1:5 v/v[80].

Table 6 lists the R_f values of some common compounds in a few of these solvents. Good separation of uronic acids is obtained with DEAE-cellulose sheets (e.g., Whatman DEAE-81), as illustrated for a mixture of glucuronic acid–galacturonic acid using the solvent system ethyl acetate–acetic acid–water 3:1:1 v/v as the mobile phase. The advantage of the latter system is the complete separation of uronic acids from monosaccharides which have significantly higher R_f values[79].

Visualization

Uronic acids have reducing properties and can be detected and determined by the common sugar reagents (e.g., aniline–phthalic acid, alkaline silver nitrate, etc.). Due to its specificity the naphthoresorcinol–phosphoric acid reagent is recommended[43,46]. The uronic acid containing zones appear as deep blue spots.

Uronic acid lactones and other sugar lactones can be detected by the hydroxamate–Fe(III) reaction[80].

Reagent A: 1 M methanolic hydroxylamine hydrochloride and 1.1 M methanolic KOH are mixed in a 1:1 v/v ratio just before use.

Reagent B: 1.5% solution of $FeCl_3$ in 1% aqueous hydrochloric acid

After development, spray the dry paper chromatogram with reagent A; let it dry at room temperature for 10 minutes and spray with reagent B. The lactones appear as deep purple to violet spots.

Table 6. R_f Values of Sugar Acids and Their Lactones

Compound	A	B	C
D-Galacturonic acid	0.18	-	-
D-Glucuronic acid	0.27	-	-
L-Guluronic acid	0.28	-	-
D-Mannuronic acid	0.35	-	-
2-Ketogluconic acid	0.21	-	-
5-Ketogluconic acid	0.47	-	-
D-Xylono-γ-lactone	-	0.41	-
L-Rhamnono-γ-lactone	-	0.50	0.63
L-Arabono-γ-lactone	-	-	0.59
D-Glucono-γ-lactone	-	0.32	0.56
D-Galactone-γ-lactone	-	0.35	-
D-Mannono-γ-lactone	-	0.25	0.39
D-Glucono-δ-lactone	-	0.22	-

A =   Pyridine–ethyl acetate–acetic acid–water 5:5:1:3 v/v (Fischer and Dörfel).
B =   1-Butanol–ethanol–water 4:1:5 v/v (Abdel-Akher and Smith).
C =   *tert.*-Amyl alcohol–1-propanol–water 4:1:1 v/v (Cifonelli and Smith).

2.1.1.5. Sugar Phosphates

These compounds are of eminent importance for the energy metabolism of biological systems and are present in all living materials.

For their separation by paper chromatography impurities in the paper (mostly cations) must be removed. For this purpose the following procedure is recommended:

Wash the sheets with a solution of 8-hydroxychinoline in a mixture of water and ethanol, followed by treatment with 2 M acetic acid or 1% aqueous oxalic acid and then wash with distilled water[81,82]. Purified paper is also commercially available. The following solvent systems have been approved for the separation of sugar phosphates:

1.   Methanol–formic acid–water 80:15:5 v/v[83].
2.   Methanol–ammonium hydroxide–water 60:10:30 v/v[83].
3.   Ethyl acetate–acetic acid–water 3:3:1 v/v[84].
4.   Methyl cellulose–methyl ethyl ketone–3 M ammonium hydroxide[84].
5.   Ethyl acetate–formamide–pyridine 1:2:1 v/v[84].
6.   *tert.*-Amyl alcohol–water–formic acid (90%) 9:9:3 v/v[82].
7.   Phenol–water 18:7 g/v[81,87].

The R_f values of some sugar phosphates in a few of these solvent systems are listed in Table 7[85].

Table 7. R_f Values of Sugar Phosphates

Compounds	A	B	C	D	E
D-Glucose-1-phosphate	0.27	0.60	0.14	-	0.36
D-Glucose-6-phosphate	0.38	0.48	0.12	0.25	0.29
D-Fructose-6-phosphate	0.34	0.44	0.17	0.28	0.36
D-Fructose-1,6-diphosphate	0.40	0.24	0.08	0.06	0.08
D-Ribose-5-phosphate	-	-	-	0.31	-
D-Ribose-1-phosphate	-	-	0.15	-	0.40
2-Phosphoglyceric acid	0.46	0.18	0.27	-	0.41
3-Phosphoglyceric acid	0.50	0.35	0.23	0.22	0.22
Glyceraldehyde-3-phosphate	-	-	0.07	-	0.19
Glycerol-1-phosphate	-	-	-	-	0.39
Orthophosphate	0.63	0.28	0.33	0.22	0.21
Pyrophosphate	0.46	0.08	-	-	-

A =   Methanol–formic acid–water 80:15:5 v/v (Bandurski).
B =   Methanol–ammonium hydroxide–water 60:10:30 v/v (Bandurski).
C =   Ethyl acetate–acetic acid–water 3:3:1 v/v (Mortimer).
D =   Phenol–water 18:7 v/v (Benson).
E =   Methyl cellosolve–methyl ethyl ketone–3 M ammonium hydroxide 7:2:3 v/v (Mortimer).

A complex mixture of sugar phosphates can be separated advantageously by two dimensional techniques. The following systems have been proposed for this purpose:

1.   1st Direction: ethyl acetate–acetic acid–water 3:3:1 v/v.
        2nd Direction: ethyl acetate–formamide–pyridine 6:4:1 v/v[84].
2.   1st Direction: *tert.*-butanol–formic acid–water 8:3:4 v/v.
        2nd Direction: 1-propanol–25% ammonium hydroxide–water 6:3:1 v/v[86].

With the following system, the formation of sugar phosphates during photosynthesis has been evaluated.

        1st Direction: phenol–water 18:7 w/v.
        2nd Direction: 1-butanol–propionic acid–water 10:5:7 v/v[87].

Visualization

Ammonium Molybdate

The separated sugar phosphates can be visualized with the ammonium molybdate reagent[82]. *Reagent:* 5 mL 60% perchloric acid, 10 mL 1M hydrochloric acid and 25 mL 4% aqueous ammonium molybdate are made up with distilled water to 100 mL.

Spray the dry paper chromatogram and dry it in a warm stream of air; heat the paper for 7 minutes at 85 °C to hydrolyze the phosphate ester groups and afterwards hang the paper in a humid H_2S atmosphere. The sugar phosphates appear as deep blue spots on a slightly bluish background.

Alternatively: irradiate the paper under a UV lamp[83].

Another modification uses an enzymatic cleavage of the phosphate esters by phosphatases according to the following procedure[88]:

Reagent A: 5% solution of $CaCl_2 \cdot 6H_2O$ in 80% ethanol.
Reagent B: solution of alkaline phosphatases in 0.1 M glycine-buffer pH 9.0.
Reagent C: 5% ammonium molybdate in 20% hydrochloric acid.

Reagent D: 50 mg benzidine (or *o*-dianisidine) dissolved in 10 mL acetic acid and diluted with distilled water to 100 mL.

Spray the dry paper chromatogram first with reagent A and evaporate the ethanol. Then spray with reagent B and put the wet paper between two glass plates in a wet atmosphere and incubate for 4 h at 37 °C; the free phosphate ions react with calcium to form insoluble calcium phosphate. Spray with reagent C and after 2 minutes with reagent D. After exposure to NH_3 vapor, the phosphate esters appear as deep blue compounds on a white background.

Ferric Chloride–Sulfosalicylic Acid[89]

Reagent A: 0.1% solution of $FeCl_3 \cdot 6H_2O$ in 80% ethanol.

Reagent B: 1% solution of sulfosalicylic acid in 80% ethanol.

Spray the paper chromatogram first with the reagent A and dry; then spray with reagent B The sugar phosphates appear as white spots on a violet background. The reaction works only at a paper pH of 1.5–2.5 and can be applied only for acid mobile phases.

The sugar components can be visualized according to the reactions on pages 110–112.

Quantitative Determination

The sugar phosphate containing zones are determined by comparison with a reference chromatogram and the untreated zones are cut off; these paper pieces are wet ashed with a mixture of sulfuric acid and perchloric acid and the phosphate content determined by photometry[82,90].

2.1.1.6. Amino Sugars

For chromatographic development only neutral or alkaline solvents are suitable. In acid solvents the amino sugars are unstable.

The following solvents are recommended:

1.   Isoamyl alcohol–pyridine–1% ammonium hydroxide 4:3:10 v/v[91].
2.   Ethyl acetate–pyridine–water–acetic acid 5 :5:3:1 v/v[92].
3.   1-Butanol–pyridine–water 3:2:1.5 v/v[93].

Visualization

The reagent for normal sugars as described on pages 110–112 can be applied (e.g., aniline–phthalic acid, alkaline silver nitrate, etc.) Free amino sugars react like amino acids with the ninhydrin reagent; substituted compounds such as *N*-acetylamino sugars however show a negative reaction.

Specific Reagents for Amino Sugars

Acetylacetone–*p*-dimethylaminobenzaldehyde according to Morgan–Elson (see page 65)[3].

Reagent A: solution of 0.5 mL acetylacetone in 50 mL 1-butanol; just before use add 0.5 mL of a mixture of 5 mL 50% aqueous KOH and 20 mL ethanol to 10 mL of this solution.

Reagent B: 1 g *p*-dimethylaminobenzaldehyde dissolved in 30 mL ethanol and 30 mL concd hydrochloric acid and made up with 1-butanol to 100 mL.

Spray the dry paper chromatogram after development with reagent A and heat for 5 min at 105 °C; spray with reagent B and heat for a further 5 min at 95 °C. The free hexosamines appear as deep cherry-red spots, the acetyl hexosamines as a purple-violet color. The latter compounds give the same color with reagent B alone. Reaction products between sugars and amino acids (glycine, lysine) interfere with the reaction[94].

References

1   Consden, R.; Gordon, A. H.; Martin, A. J. P. *Biochem. J. (London)* **1944**, *38*, 224.
2   Patridge, S. M. *Nature (London)* **1946**, *158*, 270.
3   Patridge, S. M. *Biochem. J.* **1948**, *42*, 238.
4   Jermyn, M. A.; Isherwood, F. A. *Biochem. J.* **1949**, *44*, 402.
5   Linskens, H. F. In *Papierchromatographie in der Botanik*; Linskens, H. F., Ed.; Springer: Berlin, 1959, p 25.
6   Strecker, G.; Lemaire-Poitau, A. *J. Chromatogr.* **1977**, 143, 553.
7   Menzies, I. S. *J. Chromatogr.* **1973**, *81*, 109.
8   Malpress, F. H.; Morrison, A. B. *Nature (London)* **1949**, *164*, 963.
9   Robinson, R.; Morgan, W. Th. *Biochem. J.* **1930**, *24*, 119.
10   Van der Plank, J. E. *Biochem. J.* **1936**, *30*, 457.
11   Somogy, M. *J. Biol. Chem.* **1945**, *160*, 61.
12   Cramer, F. In *Papierchromatographie*; VCH: Weinheim, 1954.
13   Chargaff, E.; Levine, C.; Green, C. *J. Biol. Chem.* **1948**, *175*, 67.
14   De Walley, H. C. *Int. Sugar J.* **1952**, 158.
15   Gordon, H. T.; Thornburg, W.; Werum, L. N. *Anal. Chem.* **1956**, *28*, 849.
16   Evans, E.; Mehl, J. W. *Science* **1951**, *114*, 10.
17   Giovannozzi-Sermani, G. *Nature* **1956**, *177*, 586.
18   Bourne, E. J.; Lees, E. M.; Weigel, H. *J. Chromatogr.* **1963**, *11*, 253.
19   Bailey, R. W.; Pridham, J. B. *Chromatogr. Rev.* **1962**, *4*, 114.
20   Hirst, E. L.; Jones, J. N. K. In *Methods of Plant Analysis*, Vol II; Paech, K.; Tracey, V.N., Eds.; Springer: Berlin, 1955; p 275.
21   Aitken, R. A.; Eddy, B. P.; Ingram, M.; Weurman, *Biochem. J.* **1956**, *64*, 63.
22   Avigad, G. *Biochem. J.* **1959**, *73*, 587.
23   Archibald, A. R.; Manners, D. J. *Biochem. J.* **1959**, *73*, 292.
24   Bayly, R. J.; Bourne, E. J. *Nature (London)* **1953**, *171*, 385.
25   French, D.; Knapp, W. D.; Pazur, H. J. *J. Am. Chem. Soc.* **1950**, *72*, 5150.
26   Dimler, R. J.; Schaefer, W. C.; Wise, C. S.; Rist, C. E. *Anal. Chem.* **1952**, *24*, 1411.
27   Myrbäck, K.; Willstedt, E. *Ark. Kemi* **1953**, *6*, 417.
28   Teichmann, B. *J. Chromatogr.* **1972**, *70*, 99.
29   Feingold, D. S.; Avigad, G.; Hestrin, S. *Biochem. J.* **1956**, *64*, 351.
30   Wallenfels, K.; Bernt, E.; Limberg, G. *Liebigs Ann.* **1953**, *579*, 113; **1953**, *584*, 63.
31   Trevelyan, W. E.; Procter, D. P.; Harrison, J. S. *Nature (London)* **1950**, *166*, 444.
32   Halonen, A. *Paperi ja Puu* **1972**, *3*, 1.
33   Durso, D. F.; Paulson, J. C. *Anal. Chem.* **1958**, *30*, 919.
34   McFarren, E. F.; Brand, K.; Rutkowski, H. R. *Anal. Chem.* **1951**, *23*, 1146.
35   Wallenfels, K.; Bernt, E.; Limberg, G. *Angew. Chem.* **1953**, *65*, 581.
36   Fischer, F. G.; Dörfel, H. *Hoppe-Seyler's Z. Physiol. Chem.* **1954**, *297*, 164.
37   Jeanes, A.; Wise, C. S.; Dimler, R .J. *Anal. Chem.* **1951**, *23*, 415.
38   Weygand, F.; Hofmann, H. *Chem. Ber.* **1950**, *83*, 405.
39   Honda, S.; Matsuda, Y.; Kakeki, K. *J. Chromatogr.* **1979**, *176*, 433.
40   Mowery, D. F. *Anal. Chem.* **1957**, *29*, 1560.
41   Patridge, S. M. *Nature (London)* **1949**, *164*, 443.
42   Robinson, H. M. C.; Rathbun, J. C. *Science* **1958**, *127*, 501.
43   Bryson, J. L.; Mitchell, T. I. *Nature (London)* **1951**, *167*, 864.
44   Horrocks, R. H.; Manning, G. B. *Lancet* **1949**, 1024.
45   Sattler, L.; Zerban, F. W. *Anal. Chem.* **1952**, *24*, 1862.
46   DeWalley, H. C. S.; Albon, N.; Gross, D. *Analyst* **1951**, *76*, 287.
47   Gibbons, G. C.; Boissonnas, R. A. *Helv. Chim. Acta* **1950**, *33*, 1477.
48   Polson, A. *Biochim. Biophys. Acta* **1948**, *2*, 575.
49   Weisz, H. *Mikrochem. Acta (Wien)* **1954**, 7856.
50   Janjic, T.; Celap, M.; Stojkovic, L. *Bull. Soc. Chim. (Belgrad)* **1962**, *27*, 283.
51   Fisher, R. B.; Parsons, D. S.; Morrison, G. A. *Nature (London)* **1948**, *161*, 764.
52   McCready, R. M.; McComb, E. A. *Anal. Chem.* **1954**, *26*, 1645.
53   Gardner, K. J. *Nature (London)* **1955**, *176*, 929.

54   Flood, A. E.; Hirst, E. L.; Jones, J. K. N. *J. Chem. Soc.* **1948**, 1679.
55   Chan, B. K.; Cain, J. C. *J. Chromatogr.* **1965**, *22*, 95.
56   Hirst, E. L.; Jones, J. K. N. *J. Chem. Soc.* **1949**, 1659.
57   Blass, J.; Machebeuf, M.; Nunez, G. *Bull. Soc. Chim. Biol.* **1950**, *32*, 130.
58   Dreywood, R. *Anal. Chem.* **1946**, *18*, 499.
59   Cifonelli, J. A.; Smith, F. *Anal. Chem.* **1954**, *26*, 1132.
60   Hirst, E. L.; Hough, L.; Jones, J. N. K. *J. Chem. Soc.* **1949**, 928.
61   Boggs, L.; Cuendet, L. S.; Ehrenthal, J.; Koch, R.; Smith, E. *Nature* (London) **1950**, *166*, 520.
62   Schaefer, W. C.; von Cleve, J. W. *Anal. Chem.* **1956**, *28*, 1290.
63   Lenz, R. W.; Holmberg, C. V. *Anal. Chem.* **1956**, *28*, 7.
64   Hirst, E. L.; Jones, J. N. K. *Disc. Faraday. Soc.* **1949**, *7*, 268.
65   Bartlett, J. K.; Hough, L.; Jones, J. N. K. *Chem. Ind. (London)* **1951**, 76.
66   Hough, L. *Nature* **1950**, *165*, 400.
67   Micheel, F.; Schneppe, H. *Naturwissenschaften* **1952**, *39*, 380.
68   Hackman, R. H.; Trikojus, V. M. *Biochem. J.* **1952**, *51*, 653.
69   Buchanan, J. G.; Dekker, C. A., Long, A. G. *J. Chem. Soc.* **1950**, 3162.
70   Cerbulis, J. *Anal. Chem.* **1955**, *27*, 1400.
71   Robyt, J. *Carbohydr. Res.* **1975**, *40*, 373.
72   Block, R.; Durrum, E.; Zweig, G. In *A Manual of Paperchromatography and Paper Electrophoresis*; Academic: New York, 1958; p 205.
73   Yamada, T.; Hisamatsu, M.; Taki, M. *J. Chromatogr.* **1975**, *103*, 390.
74   Veiga, L. A.; Chandelier, E. L. *Anal. Biochem.* **1967**, *20*, 419.
75   Masamume, H.; Yoshizawa, Z.; Maki, M.; Tohuku, *J. Expt. Med.* **1951**, *53*, 237.
76   Fischer, F. G.; Dörfel, H. *Hoppe-Seyler's Z. Physiol. Chem.* **1955**, *301*, 224.
77   Macek, K.; Tadra, M. *Chem. Listy* **1952**, *46*, 450.
78   Mapson, L. W.; Isherwood, F. A. *Biochem. J.* **1956**, *64*, 13.
79   Caldes. G.; Prescott, B.; Baker, P. J. *J. Chromatogr.* **1982**, *234*, 264.
80   Abdel-Akher, M.; Smith, F. *J. Am. Chem. Soc.* **1951**, *73*, 5859.
81   Benson, A. A. In *Moderne Methoden der Pflanzenanalyse*, Vol.2; Paech-Tracey, Ed.; Springer: Heidelberg, Berlin, New York, 1955; p 113.
82   Hanes, C. S.; Isherwood, F. A. *Nature (London)* **1949**, *164*, 1107.
83   Bandurski, R. S.; Axelrod, B. *J. Biol. Chem.* **1951**, *193*, 405.
84   Mortimer, R. C. *Can. J. Chem.* **1952**, *30*, 653.
85   Block, R.; Durrum, E.; Zweig, G. In *A Manual of Paperchromatography and Paper electrophoresis*; Academic: New York, 1958; p 199
86   Miethinen, J. K. *Proc. Atom. Conf. Genf* **1958**, *15 P*, 1102.
87   Benson, A. A.; Bassham, I. A.; Calvin, M.; Goodale, T. C.; Haas, V. A.; Stepka, W. *J. Am. Chem. Soc.* **1950**, *72*, 1710.
88   Fletcher, E.; Malpress, F. H. *Nature (London)* **1953**, *171*, 838.
89   Wade, H. E.; Morgan, D. M. *Nature (London)* **1953**, *171*, 529.
90   Berenblum, I.; Chain, E. *Biochem. J.* **1938**, *38*, 295.
91   Raacke-Fels, I. D. *Arch. Biochem. Biophys.* **1953**, *43*, 289.
92   Fischer, F. G.; Nebel, H. G. *Hoppe-Seyler's Z. Physiol Chem.* **1955**, *302*, 10.
93   Payne, W. J.; Kiefer, R. *Arch. Biochem. Biophys.* **1954**, *52*, 1.
94   Immunerz, J.; Vasseur, E. *Nature (London)* **1949**, *165*, 898.

General References

Papierchromatographie in der Botanik; Limskens, H.F., Ed.; Springer: Berlin, 1959.
Hais, M.; Macek, K.; *Handbuch der Papierchromatographie*; VEB Gustav Fischer: Jena, 1958.

2.1.2. Thin-Layer Chromatography

Thin-layer chromatography, described by Ismailov and Schraiber[1], Meinhard and Hall[2], and Kirchner, Miller and Kellner[3,4] was introduced to general analytical practice by Stahl[5-7]. Due to the improved variability of materials, the shorter time for development and the much greater possibilities for visualization of the separated substances this method has replaced paper chromatography in many fields including carbohydrates.

The material of the stationary phase combined in many cases with an inorganic or organic binder is spread on glass plates, on sheets of aluminum or plastic. The spreading of the layers can be carried out by hand with special equipment; for this purpose glass plates with dimensions 200 x 200 x 4 mm are used. Alternatively such plates are commercially available. The procedure of coating such plates depends strongly on the applied material.

Silica Gel

According to the standard procedure, 30 g of the commercial product is slurried with 60–65 mL of water or an aqueous solution of impregnating substance, for example, alkali acetate or alkali phosphate in a mortar with a pestle. The resulting slurry is spread on the glass plates with a special spreading apparatus. The thickness of the layer should vary between 0.25–0.3 mm and for micropreparative scales around 1 mm. Hereafter the plates are predried at room temperature and activated by heating at 110 °C for 30 minutes and then stored in an dessicator over a drying agent. This procedure is valid for adsorption chromatography. Alternatively, the plates are dried completely at room temperature, when the distribution principle should dominate (time: ~12 h) the chromatographic separation. For improvement of the stability of the layers gypsum (silica gel G) or other bonding agents are added.

Kieselguhr (Diatomaceous Earth)

Here the same procedures as for silica gel are valid. This material has a much smaller internal surface than silica gel and the loading capacity is therefore very low.

Cellulose

15 g of commercial material is suspended in 90 mL water and the slurry is homogenized for 30 seconds in a blender. The resulting suspension is spread on the glass plates (thickness: 0.25 mm) which are dried 10 minutes at 100 °C. For microcrystalline cellulose 10 g of the material is homogenized with 45 mL water in a blender, included air removed under vacuum and spread on glass plates with a thickness of 0.5–1 mm. These plates are dried at 80 °C for 30 min.

Development

In contrast to paper chromatography development of thin-layer chromatography plates occurs by the ascending method in appropriate glass chambers. Complete saturation of the atmosphere inside is important to prevent irregularities in the evaporation of the solvent from the layer which cause variations of the R_f values. For this purpose, the walls of the chamber are covered with filter paper, which is soaked with the mobile phase.

2.1.2.1. Separation of Mono- and Oligosaccharides

For the separation of this group of compounds, the following materials are used:

Inorganic materials: silica gel, kieselguhr, aluminum oxide, magnesol.

Organic materials: cellulose, cellulose derivatives, polyamide.

2.1.2.1.a. Separation of Mono- and Oligosaccharides on Non-Impregnated Silica Gel Layers

The separation ability of non-impregnated silica gel is poor and the R_f values are determined primarily by the numbers of -CH(OH) groups; the greater the number, the lower the R_f value.

Table 1.   Separation of Mono- and Oligosaccharides on Non-Impregnated Silica Gel Plates

Compound	A	B	C	D	E	F	G	H
D-Glucose	0.25	0.29	0.48	0.49	0.39	0.53	0.48	0.65
D-Galactose	0.19	0.23	0.38	-	-	0.44	0.40	0.58
D-Mannose	0.32	0.35	0.52	0.55	-	0.56	0.44	0.69
D-Allose	0.24	0.30	0.45	-	0.37	-	-	
D-Altose	0.41	0.43	0.57	-	-	-	-	
D-Talose	0.25	0.35	0.43	-	-	0.56	0.36	
D-Idose	0.43	0.47	0.57	-	0.47	-	-	
D-Gulose	0.30	0.33	0.49	-	-	-	-	
D-Fructose	-	-		0.51	0.38	0.54	-	0.64
L-Sorbose	-	-		-	-	0.61	-	0.69
D(L)-Arabinose	0.31	0.38	0.50	0.57	-	0.61	0.49	0.67
D-Ribose	0.38	0.47	0.54	0.59	-	0.68	0.39	0.72
D-Xylose	0.50	0.51	0.62	0.57	-	0.69	0.59	0.76
D-Lyxose	0.51	0.53	0.62	-	-	-	-	
D-Erythrose	0.63	0.65	0.68	-	0.73	-	-	
D-Threose	0.75	0.71	0.73	-	-			
L-Fucose	-	-	-	-	0.44	0.69	-	0.75
L-Rhamnose	-	-	-	-	0.53	0.79	0.58	0.85
Sucrose	-	-	-	0.25	0.27	0.46	-	0.66
Maltose	-	-	-	0.29	0.18	0.46	-	0.57
Lactose	-	-	-	0.09	-	0.33	-	0.47
Gentiobiose	-	-	-	-	-	0.29	-	-
Cellobiose	-	-	-	0.32	-	0.43	-	0.56
Lactulose	-	-	-	-	-	0.27	-	-
Trehalose	-	-	-	-	-	0.27	-	0.57
Melibiose	-	-	-	0.15	-	0.16	-	0.36
Palatinose	-	-	-	-	-	0.40	-	-
Isomaltose	-	-	-	0.16	-	-	-	-
Raffinose	-	-	-	0.13	0.12	0.14	-	0.42
Melezitose	-	-	-	-	-	-	-	0.56

A =   2-Propanol–ethyl acetate–water 83:11:6 v/v (Papin and Udiman).
B =   1-Butanol–acetone–water 4:5:1 v/v (Papin and Udiman).
C =   Ethanol–2-butanol–water 60:30:10 v/v (Papin and Udiman).
D =   1-Butanol–acetic acid–diethyl ether–water 45:30:15:5 v/v (Hay, Lewis and Smith).
E =   2-Propanol–di-2-propyl ether–65% formic acid 4:4:3 v/v (Micheel and Berendes).
F =   Ethyl acetate–pyridine–water–acetic acid–propionic acid 50:50:10:5:5 v/v (Klaus and Ripphahn).
G =   Ethyl methyl ketone–acetic acid–methanol 60:20:20 v/v (Nemec, Kefurt and Jary).
H =   2-Propanol–acetone–lactic acid 1 M 2:2:1 v/v (Ghebregzabher et al.; Hansen).

Periodic Acid–*p*-Anisidine

The procedure of this reagent, described on page 116 for paper chromatography, is also suitable for silica gel, kieselguhr and cellulose layers.

The sugar alcohol compounds appear as white spots on a brown background. A similar procedure has been described using sodium metaperiodate and benzidine, or *O*-dianisidine or tolidine[100] yielding white spots on a deep blue background. Careful handling is necessary here as benzidine is a carcinogenic substance.

Lead Tetraacetate–Dichlorofluorescein

The lead tetraacetate reacts quickly with vicinal-polyhydroxy compounds at room temperature as described on page 116. By slight heating, the lead tetraacetate converts the 2,7-dichlorofluorescein at the black zones into nonfluorescing compounds. Under UV light only the bands containing polyhydroxy compounds are brightly fluorescent[11,101,102].

Reagent: 5 mL of a saturated solution of lead tetraacetate in pure acetic acid and 5 mL of 2% solution of 2,7-dichlorofluorescein in ethanol is made up with toluene to 100 mL. This reagent is suitable only for inorganic layers (silica gel, alumina, etc.) and is stable for approximately 2–3 hours[11,101,102].

After development, dip the dry plates for 8–10 seconds into the reagent; heat for 3 minutes at 50 °C in a vacuum drying oven or dry with a fan at an elevated temperature; the bands show a bright fluorescence under UV light at 365 nm. The quantitative evaluation of this reaction can be carried out by fluorimetry (excitation: 365 nm, emission: 530 nm).

2.1.2.4. Sugar Acids

This group comprises the oxidation products of sugars such as aldonic acid, alduronic acid, keto-aldonic acid and the related dicarbonic acid (aldaric acid).

2.1.2.4.a. Alduronic Acids

Alduronic acids are the most important groups among these compounds since they are components of many polysaccharides and physiological glycosides. Separation by thin-layer chromatography has been described on silica gel, kieselguhr and on cellulose layers with the following solvent systems:

Support: silica gel, unimpregnated; solvent systems:
1.   Acetone–benzyl alcohol–water–acetic acid 65:22:26:5 v/v[103].
2.   1-Butanol–acetic acid–water 2:1:1[8].

Support: silica gel, impregnated with 0.3 M NaH_2PO_4; solvent systems:
3.   1-Butanol–ethanol–phosphoric acid 1:10:5 v/v[37].
4.   1-Butanol–ethanol–0.1 M hydrochloric acid 1:10:5 v/v[37].

Support: kieselguhr–0.1 M NaH_2PO_4; solvent system:
5.   Acetone–1-butanol–phosphate buffer 40:25:35 v/v[104].

Support: cellulose; with this material the best separations of uronic acids have been obtained; solvent systems:
6.   2-Propanol–pyridine–acetic acid–water 8:8:1:4 v/v[105,106].
7.   Ethyl acetate–formic acid–water 3:1:1 v/v[68].

The visualization of the separated uronic acids can be carried out with sugar reagents since they have the same hemiacetal hydroxyl group at C_1. The best of these is naphthoresorcinol (see pages 135, 136); the uronic acids appear as blue zones on a light background.

The R_f resp. R_G values are listed in Table 19 and they do not differ much from each other. Some of these compounds form lactones with higher R_f values and they can be visualized with the hydroxamate–Fe(III) reaction (see page 117).

Table 19. R_f Values of Alduronic Acid

Compound	A	B	C	D[a]
Glucuronic acid	-	0.37	0.53	0.46
Mannuronic acid	0.36	0.42	0.57	0.56
Galacturonic acid	0.32	0.14	0.42	0.40
Guluronic acid	-	-	0.46	0.53
Glucurono-γ-lactone	0.58	0.51	0.87	-
Mannurono-γ-lactone	0.53	0.57	0.74	1.28

[a] R_G relating to R_f value of glucose = 1.00.
A = 1-Butanol–acetic acid–water 2:1:v/v (silica gel) (Hay, Lewis and Smith).
B = 1-Butanol–ethanol–0.1 M phosphoric acid 1:10:5 v/v (silica gel–0.3 M NaH_2PO_4)
(Ovodov et al.).
C = Acetone–butanol–phosphate buffer 40:25:35 v/v (kieselguhr–0.1 M NaH_2PO_4) (Ernst).
D = 2-Propanol–pyridine–acetic acid–water 8:8:1:4 v/v (cellulose) (Günther and Schweiger).

2.1.2.4.b. Aldonic Acids

These are formed by oxidation of aldoses (e.g., with bromine of iodine). In biological systems they are rather rare such as, for example, gluconic acid in beer, wine and grape juice. This is especially true in the latter two cases when a microbial infection has taken place. Some of these compounds give lactones in acid media which can be separated from the free acids. Thin-layer chromatography has been described using the following conditions.
Support: pure silica gel; solvent systems:
For aldonic acid lactones.
1.   Methyl ethyl ketone–acetic acid–methanol 60:20:20 v/v[12].
2.   1-Butanol–acetic acid–water 2:1:1 v/v[8].
3.   Ethyl acetate–acetone water 40:50:10 v/v[68].
For free acids.
4.   2-Propanol–ethyl acetate–water–concd ammonium hydroxide 35:20:25:25 v/v.
5.   Acetone–ethyl acetate–water–concd ammonium hydroxide 50:20:20:10 v/v.
6.   1-Butanol–acetic acid–water 2:1:1 v/v[8].
Support: silica gel–kieselguhr mixed layers; solvent system:
For aldonic acid lactones.
7.   Benzene–acetic acid–ethanol 2:2:1 v/v[99].
Support: cellulose; solvent systems:
For free acids.
8.   1-Butanol–acetic acid–water 6:1:2:v/v[99].
9.   1-Butanol–formic acid–water 30:5 10 v/v[68].
10.   Ethyl acetate–acetic acid–formic acid–water 18:3:1:4 v/v[53].

In Table 20, the R_f values of some aldonic acids and their lactones are listed for a few solvent systems.

Table 20.   R$_f$ Values of Some Aldonic Acids and Their Lactones

Compound	A	B	C	D	E
D-Erythronic acid	-	-	-	-	0.24
L-Arabonic acid	0.35	-	-	-	0.16
D-Ribonic acid	0.38	-	-	-	-
D-Gluconic acid	0.40	-	-	0.43	-
D-Galactonic acid	0.38	-	-	-	-
D-L-Arabonic acid-γ-lactone	0.64		0.75	-	-
L-Arabonic acid-δ-lactone	0.58	-	-	-	-
D-Ribonic acid-γ-lactone	0.61	0.55	0.73	-	-
D-Xylonic acid-γ-lactone	-	-	0.79	-	-
D-Gluconic acid-γ-lactone	-	0.62	-	-	-
D-Gluconic acid-δ-lactone	0.68	0.59	0.75	-	-
D-Galactonic acid-δ-lactone	0.47	-	-	-	-
D-Galactonic acid-γ-lactone	0.61	-	0.78	-	-
D-Gulonic acid-γ-lactone	0.53	0.35	-	-	-
D-Gulonic acid-δ-lactone	0.43	-	-	-	-
D-Mannonic acid-γ-lactone	-	-	0.62	-	-
L-Rhamnonic acid-γ-lactone	-	-	0.60	-	-

A =   1-Butanol–acetic acid–water 2 :1:1 v/v (silica gel G) (Hay, Lewis and Smith).
B =   Ethyl acetate–acetone–water 4:5:1 v/v (silica gel G) Bancher, Scherz and Kaindl).
C =   Methyl ethyl ketone–acetic acid–methanol 60:20:20 v/v (silica gel G) (Nemec, Kefurt and Jary).
D =   Ethyl acetate–formic acid–water 3:1:1 v/v (cellulose) (Bancher, Scherz and Kaindl).
E =   Ethyl acetate–acetic acid–formic acid–water 18:3:1:4 v/v (cellulose) (Wolfrom, Patin and de Lederkremer).

Visualization

The reagents described for the sugar alcohols can be applied here in the same manner. For aldonic acid lactones the hydroxamate–Fe(III) reaction as described on page 117 is suitable.

2.1.2.5. Sugar Phosphates

On inorganic materials, these very important biologically substances can be separated successfully when the plates have been impregnated before chromatographic development.

Silica Gel

Silica gel H impregnated with 0.5% sodium acetate and 0.5% sodium carboxymethyl cellulose (tylose). 0.5 g sodium acetate and 0.5 g tylose are dissolved in 100 ml distilled water; one part silica gel H is mixed with 2.5 parts of this solution and this is stirred to a homogenous slurry, which is suitable for preparation of the plates These plates are initially maintained for 5–10 minutes at room temperature, then dried at 100–100 °C and stored in an dessicator.

Solvent systems:

1.   Ethyl acetate–acetic acid–water–concd ammonium hydroxide 6:6:2:1 v/v[107].
2.   2-Propanol–concd ammonium hydroxide–water–acetic acid 5:3:1:1 v/v[107].

Visualization

Modified Hanes–Isherwood reagent: Spray the dry plates after development at first with 15% H_2O_2 and heat at 110 °C until excess H_2O_2 has been destroyed; then spray the hot plates with the reagent according to Hanes and Isherwood (see page 118) and irradiate the plates with a UV lamp (Mercury-high presssure lamp) for 20 minutes. The sugar phosphates appear as deep blue spots on a bluish background. The blue color can be intensified by spraying with 2% tin(II) chloride. The bluish background disappears by fumigation with concd ammonium hydroxide[108].

By combination of these two solvent systems, a two-dimensional separation of all the common sugar phosphates is possible.

Table 21. R_f Values of Sugar Phosphates

Compound	A	B	C	D	E	Fª	Gª
3-Phosphoglyceric acid	0.20	0.18	0.68	0.24	0.45	0.36	0.97
2-Phosphoglyceric acid	-	-	0.68	0.24	0.43	0.44	-
Phosphoenolpyruvic acid	0.33	-	0.87	0.23	-	-	-
Glycerol-1-phosphate	0.40	0.21	-	-	-	-	
Glycerol-2-phosphate	0.36	0.26	-	-	-	-	
Glyceraldehyde-3-phosphate	-	-	-	-	0.53	0.45	
Glucose-1-phosphate	0.26	0.19	0.32	0.27	0.70	0.46	1.20
Glucose-6-phosphate	0.26	0.13	0.29	0.17	0.66	0.34	0.56
Fructose-1-phosphate	0.26	0.19	-	-	0.65	0.39	0.59
Fructose-6-phosphate	0.33	0.10	0.41	0.20	0.58	0.37	0.68
Fructose-1,6-diphosphate	0.14	0.06	0.34	0.13	0.29	0.32	0.33
Mannose-1-phosphate	-	-	-	-	-	-	0.90
Mannose-6-phosphate	-	-	-	-	-	-	0.70
Galactose-1-phosphate							0.77

ªR_{Ph} relating to R_f value of inorganic phosphate = 1.00.

A =   Ethyl acetate–acetic acid–water–ammonium hydroxide 6:6:2:1 v/v (silica gel–sodium acetate–tylose) (Bancher and Washüttl).
B =   2-Propanol–concd ammonium hydroxide–water–acetic acid 5:3:1:1 v/v (silica gel–sodium acetate–tylose) (Bancher and Washüttl).
C =   tert.-Amyl alcohol–6% aqueous p-toluene sulfonic acid 6:3 v/v (cellulose) (Waring and Ziporin).
D =   Isobutyric acid–water concd ammonium hydroxide 66:33:1 v/v (cellulose (Waring and Ziporin).
E =   Methanol–water–concd ammonium hydroxide 7:2:1 v/v (cellulose) (Davidson and Drew).
F =   Methanol–acetic acid–water 8:1.5:0.5 v/v (cellulose) (Davidson and Drew).
G =   Ethanol (95%)–0.1 M ammonium tetraborate (pH 9.0) 3:2 v/v (ECTEOLA-cellulose) (Dietrich, Dietrich, and Pontis).

Cellulose

For cellulose layers (thickness: 0.25 mm, activated at 105 °C) the following solvent systems have been recommended.

Solvent systems:
1.   tert.-Amyl alcohol–aqueous 4-toluenesulfonic acid (2 g/30 mL) 60:30 v/v, organic phase[109].
2.   Isobutyric acid–water–concd ammonium hydroxide 66:33:1 v/v[109].
3.   Methanol–acetic acid–water 8:1.5:0.5 v/v[110].
4.   Methanol–concd ammonium hydroxide–water 7:1:2 v/v[110].

The separation of the common sugar phosphates is possible using the two-dimensional method with a combination of either No. 1–2 or 3–4.

ECTEOLA-Cellulose

4 g ECTEOLA-cellulose are slurried in 36 mL 0.004 M aqueous solution of ethylenediamine tetraacetate at pH 7.0 and spread on 20 x 20 cm plates, which are dried at room temperature overnight. Afterwards they are sprayed with 0.1 M ammonium tetraborate pH 9.0 and dried at 50 °C for 30 minutes.

Solvent system:

1.   Ethanol (95%)–0.1 M aqueous ammonium tetraborate pH 9.0 3:2 v/v[111,112].

Visualization

The Hanes–Isherwood reagent (see page 118) is used. In the latter case the modification of Burrows et al.[112] is used.

2.1.2.6. Amino Sugars

These are very important constituents of animal polysaccharides, of glycaminoglycans and glycoproteins. The two most important layers for separation are silica gel and cellulose.

Silica Gel

Support: silica gel activated at 105 °C; solvent systems:

1.   2-Propanol–acetone–1 M lactic acid 4:4:2 v/v[14].
2.   1-Propanol–water 7:1 v/v[113].

Support: silica gel impregnated with 0.2 M NaH_2PO_4, activated at 110–120 °C for 30 minutes; solvent systems:

3.   1-Butanol–acetone–water 4:5:1 v/v[114].
4.   Phenol–water 3:1 v/v[114].

Two-dimensional chromatography using solvent system 3 (1st direction) and solvent system 4 (2nd direction) enables the complete separation of the common N-acetylhexosamines and the corresponding monosaccharides.

Cellulose

On layers of microcrystalline cellulose the separation of the common amino sugars has been achieved with the following combinations.

Solvent systems:

5.   1-Butanol–ethanol–2-propanol–concd ammonium hydroxide–water 2:4:0.5:0.5:1.5 v/v[115].
6.   Ethanol–1-amylalcohol–concd ammonium hydroxide–water 8:2:2:1 v/v[115].
7.   Ethyl acetate–pyridine–acetic acid–water 5:5:1:3 v/v[53].
8.   Ethyl acetate–pyridine–tetrahydrofuran–water 50:22:14:14 v/v[115].

The last solvent system (No. 8) is used for cellulose layers which have been impregnated with borate buffer pH 8 (0.2 M boric acid, 0.05 M NaCl, 0.05 M sodium tetraborate).

Visualization

For this purpose, ninhydrin and the Morgan–Elson reagent (see page 119) or reaction with periodic acid–thiobarbituric acid is used. For the latter, the following procedure has been described[115].

Reagent A: 1.9 g of periodic acid dissolved in 10 mL distilled water; 1 mL of this solution is diluted with acetone to 20 mL.

Reagent B: Sodium arsenite 3.5% in 1 M HCl.

Reagent C: 0.6% Thiobarbituric acid in ethanol.

After development and drying, spray the chromatograms first with reagent A and after 10 minutes with reagent B until the brown iodine color disappears again; after 2 minutes spray with reagent C and heat for 5 minutes at 90 °C. The amino sugars appear with the following colors as listed in Table 22.

Table 22.   Color Reaction of Amino Sugars with Periodic Acid–Thiobarbituric Acid

Compound	Color in Daylight	Color under UV Light
D-Glucosamine	yellow-orange	yellow
D-Galactosamine	light brown	green
N-Acetylglucosamine	red	pink
N-Acetylgalactosamine	red	red

In Table 23, the R_f values of some amino sugars are listed for a few solvent systems.

Table 23. R_f Values of Amino Sugars

Compound	A	B	C	D[a]
2-Amino-2-deoxy-D-glucose (D-Glucosamine)	0.22	0.34	-	1.00
2-Amino-2-deoxy-D-galactose (D-Galactosamine)	0.18	-	-	0.91
3-Amino-3-deoxy-D-mannose	0.24	-	-	-
2-Amino-2-deoxy-D-ribose	0.25	-	-	-
2-Amino-2-deoxy-L-xylose	0.31	-	-	-
2-Amino-2-deoxy-D-lyxose	0.28	-	-	-
N-Acetyl-2-amino-2-deoxy-D-glucose	-	0.79	0.64	1.28
(N-Acetyl-D-glucosamine)	-	-	-	-
N-Acetyl-2-amino-2-deoxy-D-galactose	-	-	0.52	1.24
(N-Acetyl-D-galactosamine)	-	-	-	-

[a] R_G relating to R_f value of D-glucosamine = 1.00.
A =   Pyridine–ethyl acetate–acetic acid–water 5:5:1:3 v/v (cellulose) (Wolfrom, Patin and de Lederkremer).
B =   2-Propanol–acetone–1 M lactic acid 4:4:2 v/v (silica gel) (Hansen).
C =   1-Butanol–acetone–water 4:5 :1 v/v (silica gel) (Hotta and Kurokawa).
D =   1-Butanol–ethanol–2-propanol–concd ammonium hydroxide 2:4:0.5:1.5 v/v (cellulose) (Günther and Schweiger).

For the amino sugars, derivatization before separation by thin-layer chromatography has been described. The modification occurs either at the semiacetal hydroxyl at C-1 of the sugar or at the free amino group.

Reaction with 7-Amino-4-methylcoumarin

For the first example, a fluorescence labeling technique was developed for labeling *N*-acetylglucosamine and its oligomers using 7-amino-4-methylcoumarin (AMC) and the reductive amination with sodium cyanoborohydride according to the following scheme[116].

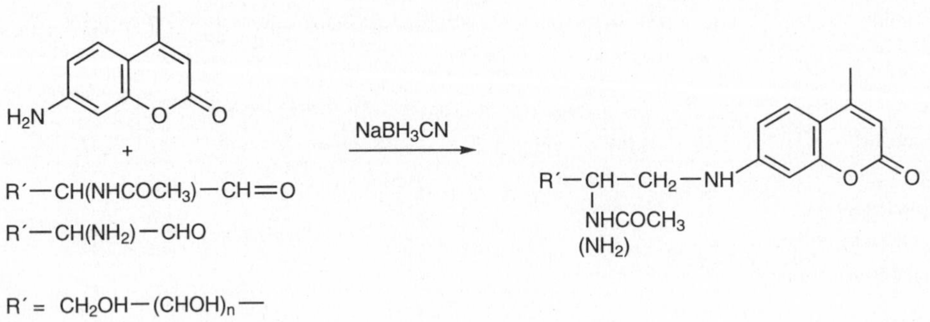

Reaction between 7-Amino-4-methylcoumarin and Amino Sugars

Microscale Procedure

In a screw-cap conical tube dissolve 1 nmol of the sugar in 50 µl methanol containing 3 nmol AMC; add 2 µL of acetic acid and heat the mixture for 2 hours at 40 °C; then add 50 µL of 0.1 mM solution of NaBH$_3$CN and keep overnight at 40 °C; then evaporate the mixture to dryness under nitrogen, dissolve the residue in 1 mL of water and clean the solution by passage through a small column containing mix bed ion exchanger (Biorad AG 501); elute at first with water and then with ammonium hydroxide. Evaporate the latter eluate to dryness and separate the labeled compound by thin-layer chromatography using the following systems.

Support: silica gel; solvent system:

1.   Chloroform–methanol–water 10:8:2 v/v.

Support: octadecylsilane bonded silica gel; solvent system:

2.   Acetonitrile–water 95:5 v/v.

The separated substances show intense fluorescence at excitations of 360–400 nm with emission wavelengths of 430–445 nm.

Reaction with 1-Fluoro-2,4-dinitrobenzene

This reagent reacts with the free amino groups of the amino sugars yielding the corresponding 2,4-dinitro derivatives but the conversion is not quantitative since formation of byproducts occurs[117]. Better results are obtained with the corresponding amino deoxyalditols, formed by the reduction of the amino sugars with sodium tetrahydroborate. The separation of the isolated DNP-amino sugars and DNP-aminodeoxyalditol can be carried out under the following conditions.

Support: polyamide; solvent systems:

1.   Formic acid–water 1:4–1:49 v/v.

2.   Chlorobenzene–acetic acid containing 2.9 mM EDTA 5:1 v/v.

3.   Chlorobenzene–acetic acid containing 2.9 mM–1-pentanol–5% (w/v) benzeneboronic acid in methanol 50:3:8:4 v/v.

Polyamide appears to be a more inert support than silica gel. The latter material is known to catalyze decomposition of the compounds on its surface especially when it is dry and activated.

2.1.2.7. Sugar Osazones

These compounds are formed by reaction of reducing sugars with phenylhydrazine (see page 41) in weakly acidic media. Since they are scarcely soluble in water, reducing sugars can be isolated as their osazone derivatives from aqueous solutions and characterized without interference from other water soluble compounds.

These sugar osazones can be separated by thin-layer chromatography according to the following procedures.

Support: silica gel. On pure silica gel good separations of the sugar osazones has been achieved with the following solvent combinations. The osazones are dissolved in ethanol for spotting on the plates.
Solvent systems:
1. Benzene–acetone 20:25 v/v[68].
2. 1-Butanol–hexane 40:40 v/v[68].
3. Chloroform–dimethylformamide 70:10 v/v[68].

The separation occurs according to the number of C-atoms and only to a small extent according to the stereochemical configuration.

Support: kieselguhr, impregnated with 0.05 M sodium tetraborate; solvent system:
4. Chloroform–dioxane–tetrahydrofuran–0.1 M sodium tetraborate 40:20:20:1.5 v/v[118].

Table 24. R_f Values of Sugar Osazones

Compound	A	B	C	D	E	F
Glyoxal-bis-phenylhydrazone	0.77	1.00	1.00	-	-	-
Methylglyoxal-bis-phenylhydrazone	0.85	1.00	1.00	-	-	-
Glycerinaldehyde-osazone	0.65	1.00	0.82	-	-	-
Erythrose-osazone	0.45	0.75	0.40	-	-	-
Xylose-osazone	0.21	0.46	0.15	0.72	-	-
Arabinose-osazone	0.27	0.52	0.17	0.91	0.65	-
Glucose-osazone	0.09	0.26	0.07	0.39	0.22	-
Galactose-osazone	0.08	0.22	0.05	0.52	0.29	-
Sorbose-osazone	0.08	0.19	0.05	0.21	-	-
Rhamnose-osazone	-	-	-	-	0.78	-
Lactose-osazone	-	-	-	0.02	-	0.50
Maltose-osazone	-	-	-	0.12	-	0.57

A = Benzene–acetone 20:25 v/v (silica gel) (Bancher, Scherz and Kaindl).
B = 1-Butanol–hexane 40:40 v/v (silica gel) (Bancher, Scherz and Kaindl).
C = Chloroform–dimethylformamide 70:10 v/v (silica gel) (Bancher, Scherz and Kaindl).
D = Chloroform–dioxane–tetrahydrofuran–0.1 M sodium tetraborate 40:20:20:1.5 v/v (kieselguhr–0.05 M sodium tetraborate) (Rink and Herrmann).
E = Chloroform–acetone–ethanol (95%) 50:50:30 v/v (Silene EF) (Per Tore).
F = Chloroform–acetone–ethanol–water 50:50:30:5 v/v (Silene EF) (Per Tore).

Support: silene EF (calcium silicate); solvent systems:

5.　　Chloroform–acetone–ethanol 50:50:30 v/v (for osazones of pentoses)[119].

6.　　Chloroform–acetone–ethanol–water 50:50:30:5 v/v (for osazones of hexoses and disaccharides)[119].

Support: Polyamide; solvent systems:

7.　　Benzene–dimethylformamide 97:3 v/v (for osazones of monosaccharides)[120].

8.　　Pyridine–water 15:85 v/v (for osazones of oligosaccharides)[120].

Visualization

The deep yellow osazones are visible on the plates after separation but the colors are unstable and disappear after a short time. Stable colors are obtained by spraying the dry plates after separation with a solution of diazotized sulfanilic acid in 1 M Na_2CO_3. The yellow spots of the osazones turn a deep brown/brown-red color[120].

References

1　　Ismailow, N. A.; Schraiber, M. S. *Farmatsya* **1938**, *3*, 1.

2　　Meinhard, J. E.; Hall, N. F. *Anal. Chem.* **1949**, *21*, 185.

3　　Kirchner, J. G.; Miller, J. M.; Keller, G. J. *Anal. Chem.* **1951**, *23*, 420.

4　　Kirchner, J. G.; Miller, J. M. J. *Anal. Chem.* **1954**, *26*, 2002.

5　　Stahl, E. *Pharmazie* **1956**, *11*, 633.

6　　Stahl, E. *Chem.-Ztg* **1958**, *82*, 323.

7　　Stahl, E. *Parfüm u. Kosmetik* **1958**, *39*, 564.

8　　Hay, G. W.; Lewis, B. A.; Smith, F. J. *Chromatogr.* **1963**, *11*, 479.

9　　Micheel, F.; Berendes, O. *Mikrochim. Acta* **1963**, 519.

10　　Becker, M.; Shefner, A. M. *Nature* **1964**, *202*, 803.

11　　Klaus, R.; Ripphahn, J. J. *Chromatogr.* **1982**, *244*, 99.

12　　Nemec, J.; Kefurt, K.; Jary, J. J. *Chromatogr.* **1967**, *26*, 116.

13　　Ghebregzabher, M.; Rufini, S.; Monaldi, B.; Lato, L. J. *Chromatogr.* **1976**, *127*, 133.

14　　Hansen, S. A. J. *Chromatogr.* **1975**, *105*, 388.

15　　Papin, I. P.; Udiman, M. J. *Chromatogr.* **1979**, *170*, 490.

16　　Conain, C.; Gallo, I.; Capitano, M. A. *Biochim. Biol. Sper.* **1965**, *4*, 217.

17　　Bancher, E.; Scherz, H.; Kaindl, K. *Mikrochim. Acta* **1964**, 652.

18　　Pataki, G. In *Dünnschichtchromatographie in der Aminosäuren und Peptidchemie*; Walter de Gruyter: Berlin, 1966; p 23.

19　　Bancher, E.; Scherz, H.; Prey, V. *Mikrochim. Acta* **1963**, 712.

20　　Scherz, H.; Rücker, W.; Bancher, E. *Mikrochim. Acta* **1965**, 876.

21　　Scherz, H. unpublished.

22　　Siegenthaler, U.; Ritter, W. *Mittl. Geb. Lebensm. Hyg.* **1977**, *68*, 448

23　　Gauch, R.; Leuenberger, U.; Baumgartner, E. J. *Chromatogr.* **1979**, *174*, 195

24　　Bailey, D. J. *Chromatogr.* **1977**, *130*, 431

25　　Ghebregzabher, M.; Rufini, S.; Sapia, G. M.; Lato, M. J. *Chromatogr.* **1979**, *180*, 1.

26　　Nurok, D.; Zlatkis, A. J. *Chromatogr.* **1977**, *142*, 449.

27　　Würsch, P.; Roulet, P. J. *Chromatogr.* **1982**, *244*, 177.

28　　Covacevich, M. T.; Richards, G. N. J. *Chromatogr.* **1976**, *129*, 420.

29　　Schäffler, K. J.; Morel du Boil, P. G. J. *Chromatogr.* **1972**, *72*, 212.

30　　Schur, F.; Pfenninger, H.; Narciss, L. *Schweiz. Brauerei Rdsch.* **1973**, *84*, 93.

31　　Pastuska, G. *Fresenius' Z. Anal. Chem.* **1961**, *179*, 427.

32　　Prey, V.; Berbalk, H.; Kausz, M. *Mikrochim Acta* **1961**, 968.

33　　Jacin, M.; Mishkin, A. R. J. *Chromatogr.* **1965**, *18*, 170.

34   Lato, M.; Brunelli, B.; Ciuffini, C.; Mezzetti, T. *J. Chromatogr.* **1968**, *39*, 26.
35   Martinez, J.; Olano, A. *Chromatographia* **1981**, *14*, 621.
36   de Zeeuw, R. A.; Dull, G. G. *J. Chromatogr.* **1975**, *110*, 279.
37   Ovodov, Y. S.; Evtushenko, E. V.; Vaskovsky, V. E.; Ovodova, R. G.; Soloveva, T. F. *J. Chromatogr.* **1967**, *26*, 111.
38   Hansen, S. A. *J. Chromatogr.* **1975**, *107*, 224.
39   Ragazzi, E.; Veronese, G. *Farmaco (Pavia) Ed. Prat.* **1963**, *18*, 152.
40   Kremer, B. *J. Chromatogr.* **1978**, *166*, 335.
41   Lato, M.; Brunelli, B.; Ciuffini, G.; Mezzetti, T. *J. Chromatogr.* **1969**, *39*, 407.
42   Weidemann, G.; Fischer, W. *Hoppe-Seyler's Z. Physiol. Chem.* **1964**, *336*, 189.
43   Pifferi, P. G. *Anal. Chem.* **1965**, *37*, 925.
44   Adachi, S. *J. Chromatogr.* **1965**, *17*, 295.
45   Doner, L. W.; Biller, L. M. *J. Chromatogr.* **1984**, *287*, 391.
46   Weill, C.; Hanke, P. *Anal. Chem.* **1962**, *34*, 1736.
47   Shannon, J. C.; Creech, R. G. *J. Chromatogr.* **1969**, *44*, 307.
48   Collins, F. W.; Chandorkar, K. R. *J. Chromatogr.* **1971**, *56*, 163.
49   Stahl, E.; Kaltenbach, U. *J. Chromatogr.* **1961**, *5*, 351.
50   Prey, V.; Scherz, H.; Bancher, E. *Mikrochim. Acta* **1963**, 567.
51   Waldi, D. *J. Chromatogr.* **1965**, *18*, 417.
52   Baltes, W.; Liesk, J.; Domesle, A. *Chem. Mikrobiol. Technol. Lebensm.* **1973**, 2, 92.
53   Wolfrom, M. L.; Patin, D. L.; de Lederkremer, R. *J. Chromatogr.* **1965**, *17*, 488.
54   Spitschan, R. *J. Chromatogr.* **1971**, *61*, 169.
55   Vomhof, D. W.; Tucker, T. C. *J. Chromatogr.* **1965**, *17*, 300.
56   Hoton-Dorge, M. *J. Chromatogr.* **1976**, *116*, 417.
57   Schweiger, A. *J. Chromatogr.* **1962**, *9*, 374.
58   Raadsveld, C. W.; Klomp, H. *J. Chromatogr.* **1971**, *57*, 99.
59   Métraux, J. P. *J. Chromatogr.* **1982**, *237*, 525.
60   Horikoshi, T.; Koga, T.; Hamada, S, *J. Chromatogr.* **1987**, *416*, 353.
61   Briggs, J.; Chambers, J.; Finch, P.; Slaiding, I. N.; Weigel, H. *Carbohydr. Res.* **1980**, *78*, 365.
62   Per Tore, J. *J. Chromatogr.* **1963**, *12*, 413.
63   Marais, J. *J. Chromatogr.* **1967**, *27*, 321.
64   Affonso, A. *J. Chromatogr.* **1967**, *27*, 324.
65   Stehlik, G. In *Chromatographie, Elektrophorese Symp. Academic Europ.*; Bruxelles, 1969; pp 524–528.
66   Buchan, J. L.; Savage, R. L. *Analyst* **1952**, *77*, 401.
67   Kocourek, J.; Ticha, H.; Kostir, J. *J. Chromatogr.* **1966**, *24*, 117.
68   Bancher, E.; Scherz, H.; Kaindl, K. *Mikrochim. Acta* **1964**, 1044.
69   McNally, S.; Overend, W. G. *J. Chromatogr.* **1966**, *21*, 160.
70   Scherz, H. *Z. Lebensm. Unters. Forsch.* **1985**, *181*, 40.
71   Bial, R. *Dtsch. Med. Wochenschr.* **1903**, *29*, 253, 477.
72   Dische, Z. *Mikrochim. Acta* **1937**, 2, 13.
73   Bailey, R. W.; Bourne, E. J. *J. Chromatogr.* **1960**, *4*, 206.
74   Walkey, J. W.; Tillman, J. *J. Chromatogr.* **1977**, *132*, 172.
75   Vomhof, D.; Truitt, J.; Tucker, T. *J. Chromatogr.* **1966**, *21*, 335.
76   Raadsveld, C. W.; Klomp, H. *J. Chromatogr.* **1971**, *57*, 103.
77   Jones, J. K.; Pridham, J. B. *Nature* **1953**, *172*, 161.
78   Hsu, F.; Nurok, D.; Zlatkis, A. *J. Chromatogr.* **1978**, *158*, 411.
79   Morcol, T.; Velander, W. H. *Anal. Biochem.* **1991**, *195*, 153.
80   Weisz, H. *Mikrochim. Acta* **1954**, 785.
81   Scherz, H. *Mikrochim. Acta* **1967**, 490.
82   Scherz, H. Stehlik, G.; Bancher, E.; Kaindl, K. *Chromatogr. Rev.* **1968**, *10*, 1.
83   Wing, E.; BeMiller, J. N. In *Methods in Carbohydrate Chemistry*, Vol VI; Wolfrom, M., Ed.; Academic: New York, London, 1972; p 42ff.
84   Robards, K.; Whitelaw, M. *J. Chromatogr.* **1986**, *373*, 81.
85   Churms, S. *J. Chromatogr.* **1990**, *500*, 555.
86   Avigad, G. *J. Chromatogr.* **1977**, *139*, 343.
87   Jen-Kun Lin; Shan-Shou Wu *Anal. Chem.* **1987**, *59*, 1320.

88   Rosenfelder, G.; Mörgelin, M.; Jui-Yoa Chang; Schönenberger, C.A.; Braun, D.G., Towbin, H. *Anal. Biochem.* **1985**, *147*, 156
89   Wei Tong Wang; Le Donne, N.C.; Ackermann, B.; Sweeley, C.C. *Anal. Biochem.* **1984**, *141*, 366.
90   Prey, V.; Berbalk, H. Kausz, M. *Mikrochim. Acta* **1962**, 449.
91   Sinner, M. *J. Chromatogr.* **1976**, *121*, 122.
92   Gee, M. *Anal. Chem.* **1963**, *35*, 350.
93   Wing, R. E.; Collins, C. L.; BeMiller, J.N. *J. Chromatogr.* **1968**, *32*, 303.
94   Tate, M. E.; Bishop, C.T. *Can. J. Chem.* **1962**, *40*, 1043.
95   Deferrari, J. O.; de Lederkremer, R.M.; Matsuhiro, B.; Sproviero, J. *J. Chromatogr.* **1962**, *9*, 283.
96   Abdel-Akher, M. Smith, F. *J. Am. Chem. Soc.* **1951**, *73*, 5859.
97   Kremer, B. P. *J. Chromatogr.* **1975**, *110*, 171.
98   Israeli, Y.; Morel, J. P.; Morel-Desrosiers, N. *Carbohyd. Res.* **1994**, *263*, 25.
99   Haldorsen, K. M. *J. Chromatogr.* **1977**, *134*, 467.
100   Lajunen, K.; Purokoski, S.; Pitkänen, E. *J. Chromatogr.* **1980**, *187*, 455.
101   Tanner, H. Z. *Obst–Weinbau* **1967**, *103*, 610.
102   Gübitz, G.; Frei, R.W.; Bethke, H. *J. Chromatogr.* **1976**, *117*, 337.
103   Fey, R.; Mack, J. *Lebensmittelchemie u. gerichtl. Chemie* **1981**, *35*, 50.
104   Ernst, W. *Anal. Chim. Acta* **1968**, *40*, 161.
105   Friese, P. *Fresenius' Z. Anal. Chem.* **1980**, *301*, 389.
106   Günther, H.; Schweiger, A. *J. Chromatogr.* **1968**, *34*, 498.
107   Bancher, E.; Washüttl, J. *Microchim. Acta* **1967**, 223.
108   Washüttl, J.; Bancher, E. *Microchim. Acta* **1967**, 395.
109   Waring, P.; Ziporin, Z. *J. Chromatogr.* **1964**, *15*, 168.
110   Davidson, I. W. F.; Drew, W.G. *J. Chromatogr.* **1966**, *21*, 319.
111   Dietrich, C. P.; Dietrich, S. M. C.; Pontis, H. G. *J. Chromatogr.* **1964**, *15*, 277.
112   Burrows, S., Grylls, F. S.; Harrison, J.S. *Nature* **1952**, *170*, 800.
113   Gal, A. *Anal. Biochem.* **1968**, *24*, 452.
114   Hotta, K.; Kurokawa, M. *Anal. Biochem.* **1968**, *26*, 472.
115   Günther, H.; Schweiger, A. *J. Chromatogr.* **1965**, *17*, 602.
116   Prakash, C.; Vijay, I.K. *Anal. Biochem.* **1983**, *128*, 41
117   Talieri, M. J.; Kilic, N.; Thompson, J. S. *J. Chromatogr.* **1981**, *206*, 353.
118   Rink, M.; Herrmann, S. *J. Chromatogr.* **1963**, *12*, 415.
119   Per Tore, J. *Anal. Biochem.* **1964**, *7*, 123.
120   Haas, H. J.; Seeliger, A. *J. Chromatogr.* **1964**, *13*, 573

General References

CRC Handbook of Chromatography; Eds. G. Zweig, J. Sherma., Vol I. Carbohydrates; Ed. S. Churms; CRC: Boca Raton Florida, 1982.
Dünnschichtchromatographie, Ein Laboratoriumsbuch; Ed. E. Stahl, 2nd ed; Springer: Berlin, Heidelberg, New York, 1967.

2.1.3. Column (High-Pressure) Liquid Chromatography

Column chromatographic separations of carbohydrates can be grouped either on the basis of the mechanism of separation, (e.g., distribution chromatography, adsorption chromatography, ion exchange chromatography, gel exclusion chromatography), the type of the separation [e.g., low-pressure liquid chromatography (LPLC), high-pressure liquid chromatography (HPLC), gel permeation chromatography (GPC)], or according to the carbohydrates to be separated (e.g., monosaccharides, oligosaccharides, polysaccharides or complex mixtures of carbohydrate derivatives).
In the following section the category dealing with the type of separation is discussed in detail.

2.1.3.1. Low-Pressure Liquid Chromatography of Carbohydrates
Khym and Zill[1] demonstrated the first anion exchange separation of sugars as borate complexes[2] on Dowex-1, a strong base anion exchange resin with a particle size of 200–400 mesh (16/mesh = diameter in mm), by low-pressure liquid chromatography[3–6].
The separation of carbohydrates on deactivated coal was published in 1950[7,8] using water or water/alcohol mixtures as eluant[9].
The separation of carbohydrates on cellulose powder based on the distribution chromatography separation mechanisms has also been described[10].

2.1.3.2. High-Pressure Liquid Chromatography (HPLC) of Carbohydrates
Although the advent of low-pressure liquid chromatography was a decisive step forward in carbohydrate analysis, this has been superseded by high-pressure liquid chromatography, one of the most widely used separation techniques in this field. This has led to the development of stationary phases with different matrices, porosity and surface modifications[11,12].

2.1.3.2.1. Separation of Carbohydrates Using Unmodified Silica Stationary Phases
The separation of carbohydrates on unmodified silica gel is almost impossible due to the interactions between the free silanol groups of the silica gel and the hydroxyl groups of the carbohydrates[13,14]. Some papers have demonstrated separations of underivatized carbohydrates on pure silica gel such as LiChrosorb Si60 using a mixture of formate, methanol and water as well as acetonitrile and water (9:1, v/v) as the mobile phase[15–18]. For this reason, unmodified silica gels are used mainly in adsorption chromatography for the precolumn separation of carbohydrates derivatized with agents such as p-nitrobenzoate. This has been shown to offer the additional advantage of enhanced sensitivity[19–22].

2.1.3.2.2. Separation of Carbohydrates Using Chemically Modified Silica Stationary Phases

Amine-Bonded Stationary Phases
Several papers have dealt with the separation of carbohydrates on chemically modified silica gels, in particular amine-bonded phases[23–29]. There are several explanations for the mechanisms involved in the separation of carbohydrates on these phases. Firstly, the separation is accomplished by the distribution of the carbohydrates between the stationary phase and the mobile phase, the latter consisting almost exclusively of acetonitrile and water. The composition of the eluant has a significant effect on the retention behavior[30] although the relatively high content (> 70%) of acetonitrile may cause problems with regards to solubility. A second disputed mechanism proposes that retention is caused by the formation of hydrogen bonds between the hydroxyl groups of the carbohydrates and the amino groups on the stationary phase and that the mechanism of separation is based on adsorption as

well as on distribution and surface tension[31,32]. Figure 1 shows a separation of carbohydrates on an amino stationary phase.

A disadvantage of these stationary phases is the limited column life due to hydrolysis of the alkaline matrix and the formation of Schiff bases between the amino group and the sugar molecules or other carboxyl groups in the system[33,34]. By means of in situ preparation of amine-bonded stationary phases a higher stability may be obtained by the permanent regeneration of the particle surface[35]. Furthermore, different amines such as piperazine and tetraethylenepentamine (TEPA) have been compared[36,37]. It is also possible to separate more carbohydrates on an in situ modified phase, because a mobile phase with a higher water content may be used, improving the solubility of weakly soluble carbohydrates. Although, theoretically, in situ modifications are easy to carry out, it has proven time consuming to obtain a stable baseline under experimental conditions and hence reproducibility is rather poor.

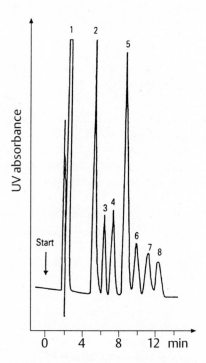

Figure 1.   Separation of Monosaccharides on a Chemically Bonded Amino Stationary Phase
Column: silica gel aminopropyl stationary phase, 125 x 4.6 mm i.d., 5 µm (Macherey, Nagel & Co, Düren, Germany); *mobile phase*: acetonitrile–water 75:25 v/v; *flow rate*: 2mL/min; *detection*: UV, 188 nm (Binder[26]; with permission)
1: solvent, **2**: rhamnose, **3**: xylose, **4**: arabinose, **5**: fructose, **6**: mannose, **7**: glucose, **8**: galactose

Diol- and Polyol-Cyano-Bonded Stationary Phases

As in the case of amine-bonded stationary phases a mixture of acetonitrile and water (85:15, v/v) is used as the mobile phase with this type of stationary phase. This again causes problems with solubility. However, in comparison to amine-bonded phases stability is increased because Schiff bases cannot be formed. If it is desirable to separate anomers, triethylamine has to be added[38,39].

Silica gel modified with cyano-bonded groups has also been used successfully[40–42].

2.1.3.2.3. Reversed-Phase Stationary Phases

As mentioned before, unmodified silica is only of limited use in the analysis of carbohydrates because of its polarity. However, alkylated or phenyl-bonded silica gels may be applied successfully to the chromatographic separation of less polar sugars. Consequently, octadecyl silica allows the analysis of gluco-oligomers with water as the mobile phase[43]. In this case, the mechanism of separation is the differences in molecular weight. Reversed-phase LC offers several advantages for the analysis of carbohydrates, the abolition of toxic solvents as well as the easy coupling with structure defining analytical methods such as mass spectrometry. Column temperature is another important factor with regard to separation efficiency. Although resolution of oligosaccharides is better at lower temperatures (5 °C), a considerable increase in analysis time has to be accepted. Moreover, the length of the alkyl chain bonded to the silica particles as well as pore size have a significant effect on separation efficiency[44–47].

2.1.3.2.4. Separation of Carbohydrates Using Pressure Stable Cation Exchangers

A. Silica-Based Cation Exchange Stationary Phases

Separations using cation exchangers based on silica gel are not described in the literature as often as separations on polymer-based cation exchangers. Generally, separations on silica phases resemble

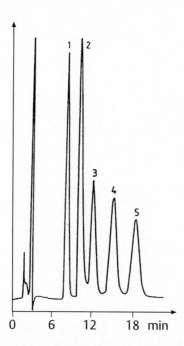

Figure 2.   Separation of Mono- and Oligocarbohydrates on 5 μm RSiL-Cat in the Ca^{2+} Form
Mobile phase: water–acetonitrile 75:25 v/v with 0.075% triethylamine (Verzele, Simoens and Van Damme[56]; with permission)
1: glucose, **2**: fructose, **3**: maltose, **4**: lactose, **5**: maltotriose[56]

those on polymer-based phases, but at room temperature, faster analyses are possible with silica phases. The chromatographic distributions show polysaccharides eluting before monosaccharides. The efficiency is improved considerably by eluting with acetonitrile–water mixtures. The relative elution times with water are caused by a size exclusion effect or by mixed chromatographic effects. With increasing acetonitrile content anomeric separation becomes more apparent[48-55].

Figure 2 shows a typical chromatogram for a cation exchanger based on silica gel with an acetonitrile–water mixture as the mobile phase. Although the structurally similar pairs lactose–maltose and glucose–fructose are well separated, the separation of fructose–sucrose is poor. Addition of methanol to the acetonitrile–water mobile phase reduces the retention time of most carbohydrates (Figure 3).

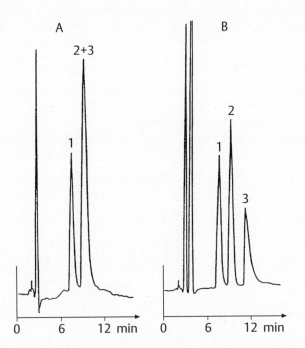

Figure 3.   Separation of Monosaccharides on 5 µm RSiL-Cat in the Ca²⁺ Form
Column: 250 x 4.6 mm; *flow rate*: 1 mL/min; *mobile phase*: A: acetonitrile–water 75:25 with 0.075% TEA and B: acetonitrile–water–methanol 68:22:10 with 0.075% triethylamine
1: glucose, **2**: sucrose with fructose in A
2: sucrose, **3**: fructose in B. Note the specific retardation of fructose in B triethylamine (Verzele, Simoens and Van Damme[56]; with permission)

As a result of high pressure resistance higher flow rates can be used on silica-based rather than polymer-based cation exchangers, which sometimes results in a faster separation. This can, however, depend on the cross-linking of the polymer. In spite of this fact there is a tendency to use higher cross-linked cation exchange resins. One important problem with silica-based cation exchangers is the dissolution of the matrix in alkaline medium.

On the other hand, silica gel itself has cation exchange properties due to the ionization of silanol groups at pH >3. For example in situ Cu^{2+}-coated silica gel leads to a system which allows the separation of carbohydrates (Figure 4)[57]

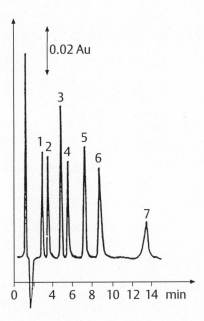

Figure 4.   HPLC Chromatogram of Carbohydrates Using a Cu(II) Modified Silica Gel Stationary Phase *Column*: 200 x 4.8 mm i.d., packing Cu(II) modified silica gel 5 μm; *mobile phase*: acetonitrile–water 72:25 containing ammonia 1.5 M and Cu(II) 0.02×10^{-3} M; *flow rate*: 2 mL/min; *temperature*: 20 °C; *detection*: UV, 254 nm (Leonard, Guyon and Fabiani[57]; with permission)
1: rhamnose, **2**: xylose. **3**: fructose, **4**: glucose, **5**: saccharose, **6**: maltose, **7**: lactose

A comparison of strongly acidic cation exchangers with sulfonic acid bound either to an organic polymer matrix, e.g. polystyrene–divinylbenzene, or to a silica matrix shows that better results can be achieved using a polymer matrix material[58–60]. Figure 5 shows a schematic overview of the retention times of sugars using different HPLC stationary phases.

B. Polymer-Based Cation Exchange Stationary Phases

Cation exchange resins based on polystyrene–divinylbenzene matrices are excellent for the separation of carbohydrates and are among the stationary phases most commonly applied. Other polymers such as polyacrylates are much less convenient because of the lack of an affinity between the π-electron systems of the components on the polystyrene matrix.

Polymeric cation exchanger resins, therefore, are usually based on finely sized spherical beads of styrene–divinylbenzene copolymers with a mean diameter of 8–12 μm to which sulfonate groups have been attached[61].

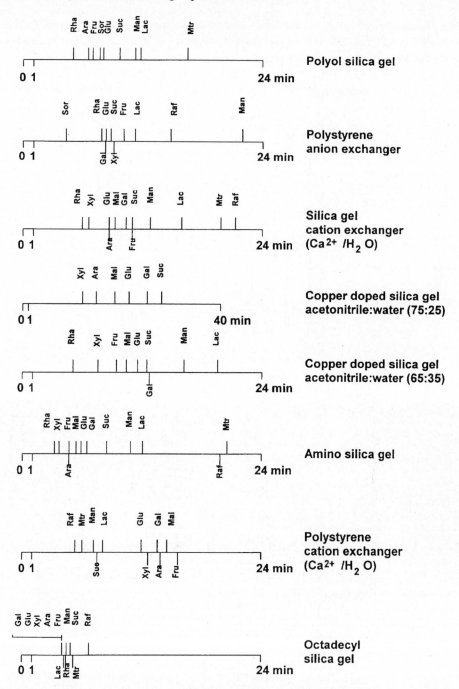

Figure 5.   Schematic Representation of Sugar Retention Times
Glu: glucose, **Fru**: fructose, **Suc**: sucrose, **Man**: mannose, **Lac**: lactose, etc. (Verzele, Simoens and Van Damme[56]; with permission)

PS–DVB particles are normally produced by emulsion or suspension polymerization giving a wide size distribution. To obtain particles with a narrow size distribution a sieving process such as sedimentation or centrifugation has to be carried out. A yield of about 30% is normal.

By using a swelling step, as described by Ugelstad[62], during the synthesis it is possible to produce monosized particles. In this process an emulsion polymerization leads to a particle diameter of about 1 μm. In a second step water insoluble molecules with a low molecular weight (e.g., 1-chlorodecane) are transferred into the particles by diffusion. In the third step monomers such as styrene and/or divinylbenzene are mixed with the swelled particles into which they diffuse. It is possible to select the porosity of the particles and Figure 6 shows the differences between porous and nonporous polystyrene–divinylbenzene particles.

In the case of producing nonporous particles, which could be useful for polysaccharide analysis, only monomers are used in the growing process.

Porous beads are produced in an inert dilutent such as toluene which is incorporated into the matrix of the growing particles. In the last polymerization step, when the temperature is increased, the inert dilutent comes out of the particles, different pore sizes can be obtained (Figure 7). By varying the inert dilutent, the proportion between inert dilutent and monomer, and the temperature of the polymerization step, the pore size and the specific surface area can be adjusted. These matrices are suitable for HPLC and solid-phase extraction procedures [63–65].

The PS–DVB particles however, differ in their degree of cross-linking (Table 1). While lower cross-linked resins are used to optimize the separation of oligosaccharides, higher cross-linked resins are mainly applied to high resolution separations of mono- and disaccharides (Figure 8). Depending on the cross-linking such particles are more pressure and pH stable than silica-based materials.

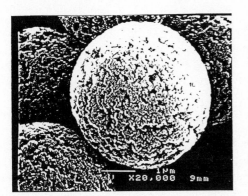 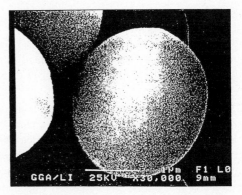

Figure 6.   Porous and Nonporous Polystyrene–Divinylbenzene (PS–DVB) Copolymer Matrices and Their Characteristic Surface Consistences: Porous PS–DVB: 5 μm, 350 m^2/g, Nonporous PS–DVB: 2.5 μm, 4 m^2/g

Table 1.   Types of Sulfonated Styrene–Divinylbenzene Copolymer Resins

Counter Ion	Ca^{2+}	Ca^{2+}	Pb^{2+}	Ag$^+$	Ag$^+$	H$^+$	H$^+$
Cross-linkage (%)	8	4	8	4	6	8	2
Particle size (μm)	9	25	9	25	11	9	5–17
pH range	5–9	5–9	5–9	6–8	6–8	1–3	1–3

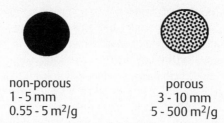

non-porous
1 - 5 mm
0.55 - 5 m²/g

porous
3 - 10 mm
5 - 500 m²/g

Figure 7.   Size and Specific Surface Area of Nonporous and Porous PS–DVB Beads

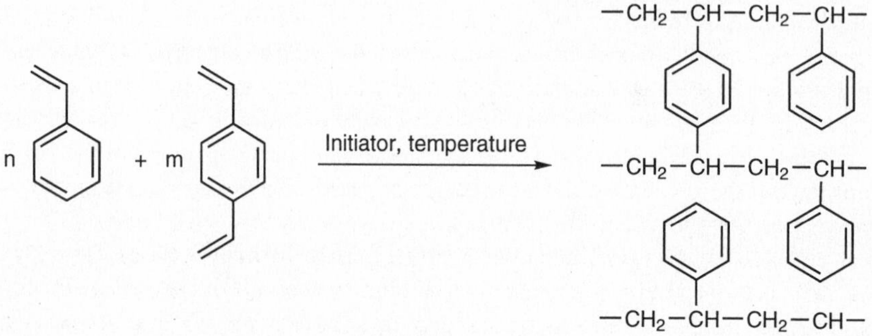

Figure 8.   Structure of Polystyrene–Divinylbenzene Resin

Uncharged carbohydrates, present at low pH are adsorbed by the polymer resin, whereas a charged form which is present at high pH (pH > 11) remains unretained.

Therefore ionic groups such as $-SO_3^-$, modify the hydrophobic resin. Furthermore, a loading of sulfonated ion exchangers with inorganic counter ions affects the retention of nonionized carbohydrates dramatically.

Polymeric cation exchangers are therefore usually based on finely sized spherical beads of styrene–divinylbenzene copolymers with a mean diameter of 8–12 μm to which sulfonate groups have been attached to produce a strong acid cation exchange resin. The particles, however, differ in their degree of cross-linking. While lower cross-linked resins are used to optimize the separation of oligosaccharides, higher cross-linked resins are mainly applied to high resolution separations of mono- and disaccharides. Furthermore, selectivity depends strongly on the counter ion, typically a proton or a metal, which has been bound to the cation exchange resin. Stationary phases with a cross-linking degree of 8% divinylbenzene are fairly pressure stable and, hence, may be used under high-performance liquid chromatographic conditions. If the degree of cross-linking is less than 4% pressure stability may not suffice at higher flow rates. While dilute mineral acids such as sulfuric acid are used in the case of a protonated cation exchange resin[66], deionized water is the mobile phase of choice with metal loaded cation exchange resins[67,68]. Organic modifiers such as acetonitrile may be added to the eluant up to a concentration of 30% without causing any damage to the polystyrene-based support matrix. Even additives such as triethylamine which raises the pH and increases the degree of mutarotation of the reducing carbohydrates may be added[69].

Coupling of Different Loaded Polymer-Based Cation Exchangers

Ion-moderated high-performance liquid chromatography, which depends on the different counter ion loading has been employed to separate monosaccharides, aldehydes, ketones, alcohols (e.g., H^+, Pb^{2+} counter ion) and oligosaccharides (e.g., Ca^{2+}, Ag^+ counter ion loading).

Stationary phases prepared from different ion exchange resins allow good separations of oligosaccharides, whereas monosaccharides and sugar degradation products are poorly resolved. On the other hand, some of the stationary phases show a good ability to separate monosaccharides and further degradation products, but fail to separate oligomeric sugars.

By combining appropriate ion-exchange resins with a single mobile phase, their separation specificities can be united to provide a system that permits the above substance groups to be separated in one analytical run. To attain this objective, it is possible to make a series connection of two columns with an initial separation step on a primary column and further separation of the eluted fractions on the second column using the same mobile phases. In this arrangement the lengths of the columns and the stationary phases can be varied to achieve an acceptable elution order[70].

Depending on the degree of cross-linking and the type of counter ion cation exchange resins differ in their potential field of application. The protonated form with a cross-linking degree of 8% is highly suited to the separation of alcohols, ketones, aldehydes, sugar acids and carbohydrates[66], while the same resin loaded with Ca^{2+} or Pb^{2+} has been applied successfully to the separation of mono- and oligosaccharides. Ag^+ is the preferred counter ion for the analysis of oligosaccharides[70]. Hence, depending on the counter ion, some cation exchange resins may be more suited to the separation of oligosaccharides rather than monosaccharides and sugar degradation products, whereas others such as the calcium form are more appropriate for the analysis of monosaccharides and various sugar degradation products such as aldehydes and ketones. However, by means of coupling various differently loaded cation exchange resin columns it becomes possible to analyze in a single run both oligo- as well as monosaccharides and their degradation products[70]. Figure 9, for instance, shows the separation of various saccharides and degradation products by means of a calcium-loaded sulfonated poly(styrene–divinylbenzene) cation exchange resin column (A: cross-linkage 7.5%, 8μm, 300 x 7.8 mm I.G., Spherogel Carbohydrate N, Beckman) coupled to a silver-loaded cation exchange resin column (B: Aminex cross-linkage 4%, 24 μm, HPX 42A, Bio-Rad, Richmond, USA) at a column temperature of 95 °C with deionized water as mobile phase. The flow rate was 1.2 mL/min, and detection of the eluted analytes was accomplished by means of a refractive index detector.

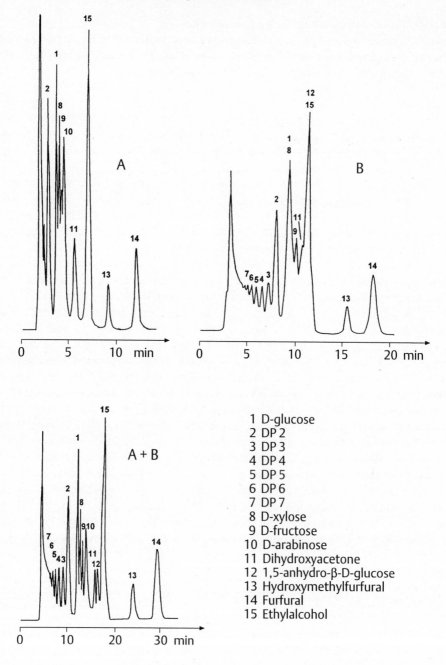

Figure 9.   Separation of Mono- and Oligosaccharides Using Coupled Cation Exchangers (Bonn[70]; with permision)

C. Separation Mechanisms

Due to the importance of cation exchange resins in the analysis of carbohydrates, it is important to discuss the mechanisms of separation including the so-called ion moderated partition chromatography mechanism[71–73].

Ion Exchange

Ion exchange chromatography makes use of the different ionic interactions of the sample molecules with the charged functional groups of the supporting matrix. In principle, the exchange of an ionic species (E^+) of opposite charge to the fixed ionic charge group (R^-) on the copolymer (P) through another counter ion (A^+) is based on a stoichiometric reaction:

$$P–R^-E^+ + A^+ \longrightarrow P–R^-A^+ + E^+$$

This equilibrium can be used to separate cations on a cation exchange resin by elution chromatography. Provided that the aqueous eluant contains the electrolyte E^+Y^- and the stationary phase attains its $P–R^-E^+$ form, the two injected sample ions A^+ and B^+ will compete with the eluant cation E^+ for the cation exchange sites R^- on the copolymer P. The separation of A^+ and B^+ is accomplished through their different ion exchange equilibria, which determine the retention times.

Ion Exclusion

The ion exclusion phenomenon can be explained as the repulsion of ions. As long as no other interactions with the stationary phase occur, ionic substances are separated according to their degree of dissociation. Organic acids, for instance, are differently dissociated depending on their pK_a values as well as the pH of the eluant. Based on the Donnan equilibrium only neutral and undissociated organic acid molecules may penetrate into the matrix, whereas completely dissociated molecules are repulsed. Therefore, a weak acid is retained longer than a strong acid. In principle, retention time should be a function of the pK_a value of the sugar acid investigated. Tanaka et al.[74] were able to find such a correlation for inorganic and low molecular weight aliphatic acids. In the case of longer alkyl chains, however, additional effects such as adsorption and size exclusion determine retention.

Size Exclusion

The size exclusion mechanism is based on the physical exclusion of molecules too large to deeply penetrate the pore structure of the resin. This property depends on the degree of cross-linking of the polystyrene–divinylbenzene matrix[75,76]. The extent of sieving is determined both by the Stokes radius of the solute as well as by the pore size or pore diameter of the beads. The latter is a function of cross-linkage and swelling of the support matrix. The actual separation occurs by repeated diffusion of the sample molecules into and out of the pores of the beads. Sample molecules with sizes greater than the pore diameter of the support matrix cannot enter the pores, and are excluded and rapidly eluted from the column in the void volume. Molecules with sizes smaller than the pore diameter enter the pores and differentially elute in volumes greater than the void volume.

Ligand Exchange and Counter Ions

The selectivity of the cation exchange resin is determined to a great extent by the type and the load of the counter ion. Resins for the analysis of carbohydrates are usually loaded with calcium, silver or lead. Since a change of the counter ion usually causes a tremendous change in volume, prepacked columns cannot be loaded with a different counter ion in situ. For the same reason, the counter ion should not be changed in order to guarantee an optimum of separation efficiency.

Proton-loaded resins are usually run with dilute sulfuric acid, which guarantees permanent regeneration. An intensive study of the chromatographic behavior of 63 substances (acids, aldehydes, ketones, alcohols and carbohydrates) on a sulfonated 8% cross-linked polystyrene–divinylbenzene cation exchanger with proton loading (HPX87H, BioRAD) is described by Pecina et al.[66]. Figure 10 shows the high separation efficiency of a sulfonated 8% cross-linked PS–DVB cation exchanger.

The increase in retention times with increasing chain length observed with linear acids, aldehydes and alcohols is caused mainly by a reversed-phase mechanism. Other influences determining the elution order are the positions of the functional groups and branching points in the chain: compounds containing a terminal functional group are more retarded than secondary or tertiary isomers; branched chain compounds have shorter retention times than the corresponding straight chain isomers.

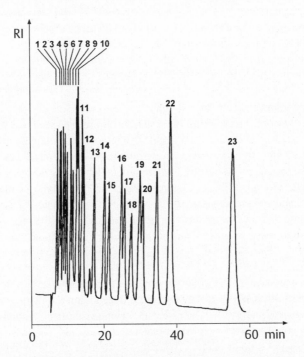

Figure 10. Optimized Separation of a Mixture of Alcohols, Aldehydes, Ketones, Acids and Carbohydrates
Column: HPX87H (300 x 7.8 mm i.d.); *column temperature*: 70 °C; *mobile phase*: 0.005 M sulfuric acid; *flow rate*: 0.7 mL/min; *refractive index detection*: [17]
1: cellobiose, **2**: 2-oxoglutaric acid, **3**: D-galacturonic acid, **4**: D-glucose, **5**: D-galactose, **6**: dulcitol, **7**: glyceraldehyde, **8**: glycolic acid, **9**: dihydroxyacetone, **10**: formic acid, **11**: acetic acid, **12**: levulinic acid, **13**: acetaldehyde, **14**: acetone, **15**: propanal, **16**: *tert.*-butanol, **17**: 1-propanol, **18**: butanal, **19**: 2-butanol, **20**: isobutanol, **21**: 1-butanol, **22**: furfural, **23**: 5-methylfurfural (Pecina, Bonn, Burtscher and Bobleter[66]; with permission)

This effect on elution order is demonstrated by the isomeric butanols and propanols: *tert.*-butanol < isobutanol < 2-butanol < 1-butanol; 2-propanol < 1-propanol. For the linear butanediol isomers the following elution order is observed: 1,2- < 1,3- < 1,4-butanediol.

The propanediol isomers, however, have nearly identical retention times in the temperature range investigated. The relationship between the position of a functional group and the retention time was also found with carbonyl compounds: aldehydes are more retarded than the corresponding ketones. Thus, acetone is eluted earlier than propanal, and methyl ketone earlier than butanal.

The influence of an increasing number of identical functional groups in a given compound on the elution order is exemplified by alcohols derived from 1-butanol: *meso*-erythritol < 1,2,4-butanetriol < butanediol isomers < 1-butanol. The type of functional group also determines the capacity factors of compounds with identical chain length. The elution order RCO_2H < RCHO/RCOR' < ROH is clearly demonstrated by the series propanoic acid < acetone < propanal < propanol isomers and acetic acid < acetaldehyde < ethanol. These sequences remain unchanged, when a carboxylic, carbonyl or hydroxy group is introduced into a molecule already containing a certain functionality. Thus, when acetic acid is regarded as the basic unit, the following elution order applies: oxalic < glyoxylic–glycolic < acetic acid.

By introducing additional functional groups (e.g., OH) into a molecule already containing one other (e.g., CO_2H), analogous effects are observed: glyceric acid is eluted earlier than lactic acid, which in turn has a shorter retention time than propanoic acid. The same applies to carbohydrates: hexoses and pentoses have shorter retention times than the corresponding deoxysugars, which have one hydroxy group less.

Elution of the separated compounds normally occurs in order of decreasing size if there are no additional interactions between the solute and the stationary phase. In this case a linear correlation between elution volume and the logarithm of molecular weight will be obtained.

With regard to sugar oligomers, several studies have shown linear calibration curves for metal-loaded cation exchange resins[77–79]. In the case of calcium- and silver-loaded resins especially size exclusion is the dominating mechanism of separation of sugar oligomers. Separation efficiency is influenced both by the counter ion as well as by cross-linking[80,81]. It has been shown that silver-loaded cation exchange resins result in the separation of a greater number of oligosaccharides. The greater retention and the enhanced resolution are brought about by the formation of strong silver monodentate complexes with oligosaccharides, which are much stronger than the bidentate complexes formed between calcium and oligosaccharides[82]. Furthermore, by varying the amount of silver placed on the resin, a level of 70 to 80% was found to yield optimum resolution. Beyond that level, a marked reduction in resolution was observed. The role of the metal counter ion in size exclusion chromatography on styrene–divinylbenzene copolymers is its impact on the swelling behavior of the resin. Furthermore, ligand exchange seems to play an additional role.

In the separation of oligosaccharides stronger complexes are formed with silver as the counter ion than, for example, Ca^{2+}. In consequence a greater number of oligosaccharides up to a dp (degree of polymerization) of 15 can be separated[81]. The amount of silver loading on the sulfonated PS–DVB resin shows an influence on the oligosaccharide separation Figure 11. The optimum silver loading has been found at approx. 70%.

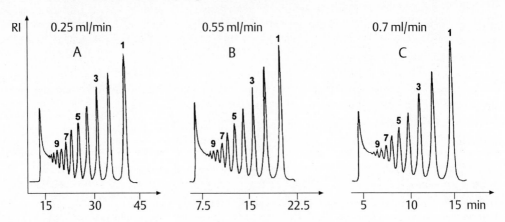

Figure 11.   Chromatograms of an Acid-Hydrolyzed Corn Syrup at Various Flow Rates
Column: (300 mm x 7.8 mm i.d.) Hamilton HC-40 5% cross-linked resin (10–15 µm) 71% silver
form; *flow rate*: A: 0.25 mL/min, B: 0.55 mL/min and C: 0.70 mL/min; numbers over peaks indicate
degree of polymerization (DP); H indicates higher excluded oligosaccharides (Scobell and Brobst[81];
with permission).

Ion Moderated Partition Chromatography

The metal counter ions bound to the cation exchange resin are able to bind sample molecules through
coordination complexes, which results in their separation as a result of differences in the strength of
their coordination complexes and their degree of solubility in the mobile phase[83]. In the analysis of
oligo- and monosaccharides especially on cation exchange resins using aqueous eluants it has been
shown that loading resins with a metal counter ion results in improved selectivity compared to a resin
in its protonated form. In contrast to oligosaccharides that are mainly retained due to size exclusion,
ligand exchange is the dominating mechanism of separation in the case of monosaccharides. The
stability of the coordination complexes depends both on the type of counter ion as well as on the
configuration of the hydroxyl groups of the polyols investigated[82]. Goulding has investigated several
counter ions, including the monovalent cations Li^+, Na^+, Rb^+ and Ag^+, as well as the divalent cations
Mg^{2+}, Sr^{2+}, Ba^{2+}, Cd^{2+} and the trivalent La^{3+}. Capacity factors k', are given in Table 2.

On the basis of NMR spectroscopic and electrophoretic investigations, it can be concluded that the
configuration of the hydroxyl group at C-3 of the pyranose ring can form relatively strong chelate
complexes when they are in the position ax–eq–ax (I: stability constants in the range of 1–5 mol^{-1}). On
the other hand, relatively unstable complexes are formed, when the hydroxyl groups have assumed the
position eq–ax–eq (Figure 12).

In II the complex can only be formed from a pair of ax–eq (cis) hydroxyl groups which means it is
weaker with a stability constant of approx. 0.1 mol^{-1}, the sequence eq–ax–eq (III) of the three hydroxyl
groups on the other hand can only form weaker bidentate chelates.

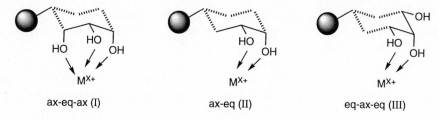

ax-eq-ax (I)   ax-eq (II)   eq-ax-eq (III)

Figure 12. Chelates Formed by Pyranose Rings and Metal Cations

Table 2. Capacity Factors (k') of Carbohydrates on Aminex A-5 Eluting with Water[82]

Counter Ion	Sucrose	Glucose	Galactose	Mannose	Talose	Fructose
$Tris^+$ 0.10	0.35	0.35	0.35	d	–	0.35
Li^+	0.20	0.35	0.50	0.45	–	0.45
Na^+	0.20	0.40	0.55	0.70	0.50	0.65
K^+	0.40	0.70	0.80^a 1.10^a	–	–	1.05
Rb^+	0.40	0.75^b	0.75^b 1.00^a	0.75^b 1.25^a	0.95^b	0.90
Ag^+	0.45	0.75	0.85	0.95^b	1.7	1.15
Ti^+	0.45	0.70	0.85	–	–	1.10
Mg^{2+}	0.15	0.30	0.40	0.40	–	0.35
Ca^{2+} 0.15	0.30^a	0.45^a 0.45^a 0.30^a	0.45^a 0.65^a 0.40^a	3.8 0.65^a –	1.15^c –	1.10^c
Sr^{2+}		0.20	0.45^b	0.55^b		
Ba^{2+} 0.25	0.40^b	0.45^b	0.70^b	–	1.00^c	
Cd^{2+} 0.15	0.25	0.35	–	–	0.50	
La^{3+} 0.10	0.15	0.25	0.30	4.0	0.35	

Additional results

Glucose: k' = 1.25 (Rb^+), 1.15 (Ag^+), 1.3 (Ca^{2+}), 1.0 (La^{3+}).

Galacitol (sorbitol): k' = 0.60 (Rb^+), 2.1 (Ca^{2+}).

Xylose: k' 0.45^a, 0.60^a (Ca^{2+}).

α-D-Galactosyl-1,2-glycerol: k' = 0.50 (Li^+), 0.30 (Na^+), 0.40 (Rb^+).

^{a}α,β-Forms. ^{b}Asymmetric peak, partial separation. ^{c}Skew peak. ^{d}Not measured.

Interactions with the Sulfonate Groups

The sulfonate groups attached to the styrene–divinylbenzene copolymers cause retention of the sample molecules in a twofold way. Firstly, analytes may be adsorbed by dipole–dipole interactions or hydrogen bonds. Secondly, the intraparticle bound water may assume a more polar structure than outside the matrix due to the impact of the sulfonate groups. Under such conditions, sample compounds may be separated according to their varying degrees of preference for the more intraparticle phase over the less polar mobile phase. Both the support matrix and the sulfonate groups can be considered inert carriers of a polar fluid film. However, a clear distinction between the two mechanisms is not possible. Nevertheless, both adsorption and partitioning are the main forces in the separation of carbohydrates on sulfonated styrene–divinylbenzene copolymers when water or dilute sulfuric acid are employed as eluants.

Interactions with the Support Matrix
The support matrix of a sulfonated poly(styrene–divinylbenzene) resin is a nonpolar aromatic mesh that permits the reversed phase partitioning of sample molecules between the polar mobile phase and the nonpolar resin backbone. This mechanism predominates in separations involving molecules such as aliphatic acids and alcohols as well as phenols[72]. Retention is greatest in the case of aqueous eluants. Elution, however, can be accelerated by the addition of organic solvents. A linear relation is usually observed between the retention times and the logarithm of the content of organic modifier in the eluant. An increase in column temperature also decreases retention times. A linear correlation between log k and 1/t can be found. The slope of the straight line in the Van't–Hoff diagram corresponds to the standard enthalpy for the transition of the compound from the mobile to the stationary phase[84–86]. Within an homologous series, however, the capacity factors increase exponentially with increasing carbon chain length.

Variables Affecting the Liquid Chromatographic Separation of Carbohydrates on Cation Exchangers
The aforementioned separation mechanisms involved in ion moderated partition chromatography on sulfonated only (styrene–divinylbenzene) copolymers are additionally influenced by a great variety of variables such as particle size, cross-linkage, type of counter ion, column temperature, composition of the eluant and flow rate.

Particle Size
While the number of theoretical plates of the column increases with decreasing particle size, the pressure built up by the stationary phase and the eluant flow increases. Due to the elastic structure of the resins, however, a certain pressure may not be exceeded without the risk of generating a void volume in the column head through compression of the column packing material. At a given particle size, column pressure is related directly to the flow rate of the eluant. Therefore, an upper pressure limit is given in the case of prepacked columns.

Cross-Linkage
The cross-linkage of a resin is one of the most important parameters, because separation can only take place, when the sample molecules penetrate at least partially into the matrix. For this reason, cation exchange resins used for the analysis of oligosaccharides are usually less cross-linked than those designed for the separation of monomers.

The use of silver as the counter ion leads to a particularly high resolution of gluco-oligomers, with an optimum at 70% loading and 4% divinylbenzene cross-linking. In addition to size exclusion, ligand exchange and matrix adsorption contribute to this separation.

The possibility of obtaining analogous separations with identical resolutions with other materials has been investigated by the use of a weakly cross-linked support and substitution of Ag^+ by H^+ in order to eliminate ligand exchange in favor of the size-exclusion mechanism. The separation capacity, however, was maintained. An additional comparison was drawn with regard to the influence of silver loading on a 2% cross-linked material[87,88].

The low degree of cross-linking (2%) of sulfonated poly(styrene–divinylbenzene) copolymer (hydrogen form) causes the exchange material to be less stable towards high pressure. A major advantage of hydrogen-loaded ion exchanger lies in the continuous regeneration, in contrast to the 70% silver-loaded columns, which can be operated only with deionized water as the mobile phase. Through the use of sulfuric acid and water, respectively, as the mobile phase and refractive index detection, the sensitivity in the determination of mono-, di- and oligosaccharides reaches the nanogram range (Figure 13).

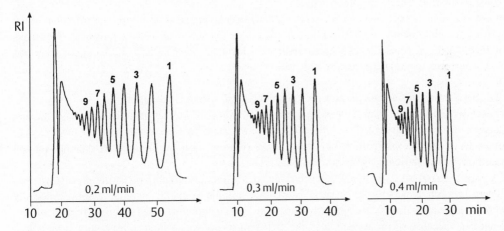

Figure 13. Separation of Oligosaccharides Derived from an Enzymatic Hydrolysate of Starch
Column: 2% cross-linked hydrogen-loaded sulfonated poly(styrene–divinylbenzene) ion exchanger; *mobile phase*: 0.005 M sulfuric acid; *flow rate*: see chromatograms; *temperature*: 76 °C; the numbers on the peaks indicate DP values (Derler, Hörmeyer and Bonn[87]; with permission)

Column Temperature

Column temperature is an important factor in optimizing carbohydrate separations. On the one side, the rate of diffusion of the sample solute into the particle is a temperature-dependent process. At elevated temperatures the distribution equilibrium between the mobile phase and stationary phase is reached more rapidly, This results in a minimum of band broadening as well as a maximum number of theoretical plates. Column temperatures effect mainly reversed phase partitioning and size exclusion.

Eluant

To separate carbohydrates on sulfonated cation exchange resins loaded either with H^+ or various metals often only water or diluted sulfuric acid are used as eluants. However, the speed of elution may be controlled by the addition of water soluble organic modifiers. However, the amount of organic modifier that can be added to the mobile phase is limited, because it causes a change in the swelling of the support matrix. The preferred modifier for sulfonated poly(styrene–divinylbenzene) resins is acetonitrile. In the case of a sulfonated poly(styrene–divinylbenzene) cation exchanger in H^+-form (HPX-87H), it may be added up to a relative concentration of 40% (v/v). In contrast, even a few per cent of methanol leads to a rapid decrease in separation efficiency.

The effect exerted by the organic modifier on retention times depends on the dominating mechanism affecting separation. Compounds, which are separated mainly on the basis of reversed phase partitioning, are eluted more rapidly as expected upon addition of acetonitrile to the mobile phase. But hydrophilic samples such as carbohydrates, which are normally separated by normal phase partitioning, may be retained even slightly longer by the addition of a nonpolar modifier because they are attracted more towards the comparatively polar matrix.

In the case of protonated cation exchange resins, the pH may be varied in addition. The pH value of the eluant effects predominantly the ion exchange and the ion exclusion of weak electrolytes. However, when 0.005 M sulfuric acid is used as eluant, organic acids are present in their undissociated form and hence no effect on retention times can be expected.

2.1.3.2.5. Separation of Carbohydrates on Pressure Stable Anion Exchangers

Anion exchange stationary phases are available both on silica as well as on polymers[81–92]. The first applications of anion exchange materials to carbohydrate analysis were performed on low-pressure systems and used complex formation. Whereas the use of borate allowed the separation of neutral and acidic carbohydrates, the analysis times were longer than 1 hour[93].

Since pressure stable silica gels are hydrolyzable under alkaline conditions, porous poly(styrene–divinylbenzene) copolymers are the material of choice for anion exchange chromatography, especially in combination with pulsed amperometric detection. Even nonporous poly(styrene–divinylbenzene) resins are used because of the reduction in the diffusion of the solute molecules It should be noted, however, that carbohydrates may be degraded under alkaline conditions and higher temperatures. This is in particular true for the Lobry de Bruyn van Ekenstein reaction[94].

A. Silica-Based Stationary Phases

For the separation of carbohydrates and their derivatives ion-exchange chromatography with silica-based anion exchangers has been used. There are two basic types of anion exchangers: those comprising strong base anion exchange groups such as quaternary ammonium and those bearing weak base anion exchange groups such as primary amines. In most cases both functional groups are bonded to 5 or 10 µm spherical silica. As mentioned previously, aminopropyl silica packings, used in normal phase partition chromatography, may also act as weak base anion exchangers under certain conditions.

Examples of silica-based strongly basic anionic exchangers are Nucleosil 10 SB (Macherey & Nagel), Zorbax SAX (DuPont) or Partisil-10 SAX (Whatman). Several applications have been carried out on these materials. Voragen et al.[95] separated uronic acids and oligogalacturonic acids on Zorbax SAX and Nucleosil 10 SB. The uronic acids were eluted with 0.7 M acetic acid. The separation of unsaturated oligogalacturonic acids was performed with 0.3 M sodium acetate buffer, pH 5.4, up to hexamers and 0.4 M sodium acetate buffer, pH 5.4, up to octamers (Figure 14).

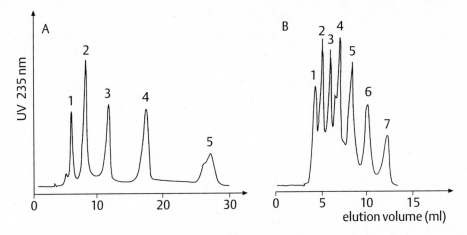

Figure 14.   Separation of Unsaturated Oligogalacturonic Acid on a Nucleosil 10 SB or Zorbax SAX Column
Flow rate: 1 mL/min; *temperature*: 40 °C; *detection*: UV, 235 nm; A: separation of oligomeres up to hexamers with 0.3 M sodium acetate buffer (pH 5.4) as eluant and B: separation of oligomers up to octamers with 0.4 M sodium acetate buffer (pH 5.4) as eluant; U-di = unsaturated digalacturonic acid; U-tri = unsaturated trigalacturonic acid, etc. (Voragen, Schols, Devries and Pilnik[95]; with permission)
1: U-di, **2**: U-tri, **3**: U-Tetra, **4**: U-penta, **5**: U-hexa, **6**: U-hepta, **7**: U-octa

The most important disadvantage of silica-based strongly basic anionic exchangers is their rapid deterioration, even when guard columns are used. Thus, their lifetime is comparatively short.

Examples of silica-based weakly basic anionic exchangers are Ultrasil-NH_2 (Beckman) or Lichrosorb 10 NH_2 (Merck). Voragen et al.[95] used Lichrosorb 10 NH_2 for the separation of unsaturated oligogalacturonic acids. The unsaturated oligogalacturonic acids were separated with 0.11 M sodium acetate buffer, pH 7.5, as eluant (Figure 15).

Compared with the results obtained with the strongly basic anionic exchanger the results were inferior. The same unsaturated oligogalacturonic acids could only be separated up to the pentamers. Another application, performed by Nebinger et al.[96], is the analysis of even- and odd-numbered hyaluronate oligosaccharides. The simultaneous separation of even- and odd-numbered hyaluronate degradation products, containing 2–8 sugar residues, was carried out on a Ultrasil-NH_2 column using 0.1 M KH_2PO_4, pH 4.75, as eluant (Figure 16). Compared with silica-based strongly basic anionic exchangers, the weakly basic species are more stable[97,98].

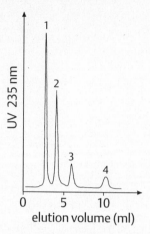

Figure 15.   Separation of Unsaturated Oligogalacturonic Acids on a LiChrosorb 10 NH₂ Column
Eluant: 0.11 M sodium acetate buffer (pH 7.5); *flow rate*: 1.5 mL/min; *temperature*: 40 °C; *detection*: UV, 235 nm; separation of enzyme (PAL) digest of pectate (Voragen, Schols, Devries and Pilnik[95]; with permission)
1: U-di, **2**: U-tri, **3**: U-Tetra, **4**:U-penta

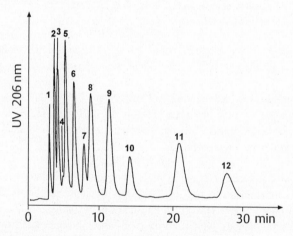

Figure 16.   Simultaneous Identification of Even- and Odd-Numbered Hyaluronate Degradation Products
Column: Ultrasil-NH₂, 250 x 4.6 mm; *eluant*: 0.1 M KH₂PO₄, pH 4.75; *flow rate*: 1 mL/min; *detection*: UV, 206 nm; GlcUA = D-glucuronic acid; GlcNA = N-Acetyl-D-Glucosamine; a mixture of 1.25 nm of each sugar was applied to the column and eluted
1: GlcNAc, **2**: βGlcNAc-1-4βGlcUA1-3GlcNAc, **3**: βGlcUA1-3GlcNAc, **4**: GlcUA, **5**: βGlcNAcl-4(βGlcUA1-3βGlcNAcl-4)₂, **6**: (βGlcUA1-3βGlcNAcl-4)₂ or (βGlcNAcl-4βGlcUA1-3)₂, **7**: βGlcUA1-3βGlcNAcl-46lcUA, **8**: βGlcNAcl-4(βGlcUA1-3βGlcNAcl-4)₃, **9**: (βGlcUA1-3βGlcNAcl-4)₃ or (βGlcNAcl-4βGlcUA1-3)₃, **10**: βGlcUA1-3(βGlcNAcl-4βGlcUA1-3)₂, **11**: βGlcUA1-3(βGlcNAcl-4)₄ or (βGlcNAcl-4βGlcUA1-3)₄, **12**: βGlcUA1-3(βGlcNAcl-4βGlcUA1-3)₃ (Nebinger, Koel, Franz and Werries[96]; with permission)

B. Polymer-Based Anion Exchange Stationary Phases

High-performance anion exchange chromatography (HPAEC) using polymer-based stationary phases and high pH eluants in combination with pulsed amperometric detection (PAD) is a powerful tool for carbohydrate separation[99–101]. Elution at high pH allows separation of carbohydrates as their oxyanions, and PAD is characterized by high sensitivity and relative specificity for compounds with hydroxyl groups. Pellicular supports containing latex particles with a superficial thin layer of strong anion exchanger are currently employed for HPAEC-PAD of mono-, oligo- and polysaccharides. The use of a macroporous poly(N,N,N-trimethylammoniummethylstyrene–divinylbenzene) strong anion exchanger has also been reported[90].

Dionex have introduced a range of polymeric nonporous Micro Bead TM pellicular resins[102,103].

Figure 17 shows a pellicular anion exchange resin. These beads are characterized by rapid mass transport, fast diffusion, high pH stability and good mechanical stability (>4000 psi). The resin composition consists either of a vinylbenzyl chloride/divinylbenzene macroporous substrate fully functionalized with an alkyl quaternary ammonium group (Carbo Pac MA1) or a polystyrene–divinylbenzene substrate agglomerated with 350 nm Micro Bead quaternary amine functionalized latex (Carbo Pac PA1) or an ethylvinylbenzene–divinylbenzene substrate agglomerated with 350 nm Micro Bead quaternary amine functionalized latex (Carbo Pac PA-100).

Neutral monosaccharides act as weak acids as seen by their pK_a values. At high pH, they are partially ionized allowing anion exchange chromatography to be performed.

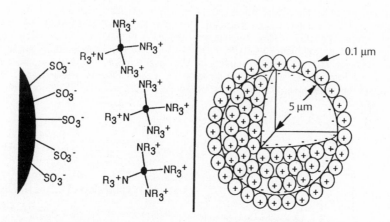

Figure 17.   Pellicular Anion Exchange Resin Bead (Dionex[102]; with permission)

In order to optimize baseline stability and detector sensitivity in many carbohydrate analyses, it is necessary to add a strong base to the eluant stream post-column[104]. This is required in two main instances: (a) when using eluant with concentration below 75 mM NaOH 300–500 mM sodium hydroxide is generally required to optimize PAD (pulsed amperometric detection) sensitivity and minimize baseline drift and (b) during gradient analyses of carbohydrates involving a change in pH (Figure 18).

Pulsed amperometry is sensitive to changes in eluant pH. To minimize baseline shifts the post-column addition of NaOH is recommended. Figure 19 shows a separation of monosaccharides on a Dionex CarboPac PA1.

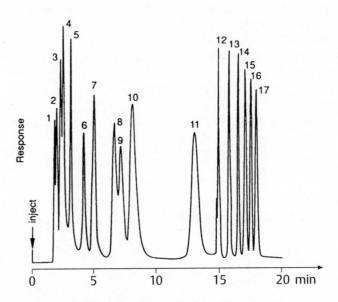

Figure 18.   Separation of Carbohydrates on an Anion Exchanger Stationary Phase
Electrode: Au; *column*: Dionex CarboPac PA1; *eluants*: (I) 100 mM NaOH, (II) 50 mM NaOH/500 mM NaOAc, (III) H$_2$O; *elution*: isocratic with 50% I/50% III (0–6 min), linear gradient to 50% I/50% II (6–15 min); isocratic with 50% I/50% II (15–21 min); *post-column reagent*: 0.40 M NaOH
1: inositol, **2**: xylitol, **3**: sorbitol, **4**: mannitol, **5**: fucose, **6**: rhamnose, **7**: arabinose, **8**: glucose, **9**: xylose, **10**: fructose, **11**: sucrose, **12**: maltose, **13**: maltotriaose, **14**: maltotetraose, **15**: maltopentaose, **16**: maltohexaose, **17**: maltoheptaose (Johnson and La Course[104]; with permission)

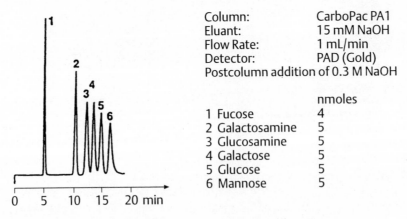

Column:	CarboPac PA1
Eluant:	15 mM NaOH
Flow Rate:	1 mL/min
Detector:	PAD (Gold)
Postcolumn addition of 0.3 M NaOH	

	nmoles
1 Fucose	4
2 Galactosamine	5
3 Glucosamine	5
4 Galactose	5
5 Glucose	5
6 Mannose	5

Figure 19.   Separation of Neutral and Basic Monosaccharides (Dionex[102]; with permission)

A new anion exchange stationary phase based on a nonporous highly cross-linked polystyrene–divinylbenzene matrix (2.21 µm, 7m^2/g) derivatized by direct nitration of the spherical particles, followed by reduction and quaternization with iodomethane has been described[105,106].

Separation of oligosaccharides by anion exchange chromatography is strongly affected by oligosaccharide acidity and by the accessibility of oxyanions to functional groups attached to the stationary phase. The disaccharides trehalose, isomaltose, gentiobiose, nigerose, and maltose are similar in structure (they are all composed of two D-glucosyl residues), and differ only in the configuration of their glucosidic bonds. These disaccharides follow the elution order shown in HPAEC using a currently available strong anion exchange resin. The differences in the retention times can be related either to the different acidity of the substituted hydroxyl groups or to the different configuration of the glucosidic bond, which alters their orientation when adsorbed by the stationary phase. The greater retention time of the trisaccharide maltotriose compared to the trisaccharide *iso*-maltotriose is another example of the selectivity of the stationary phase for isometric forms.

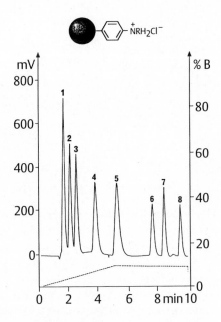

Figure 20. Separation of Mono- and Oligosaccharides
Stationary phase: quart. amino derivatized PS–DVB (2.1 µm); *mobile phase*: gradient elution 50 mM NaOH (A), 1 M NaOAc (B); *detection*: PAD, E_1 = 100 mV (t_1 = 300 ms), E_2 = 600 mV (t_2 = 120 ms), E_3 = –6mV (t_3 = 300 ms) (Corradini, Corradini, Huber and Bonn[106]; with permission). Using the same analytical system, the separation of a sample of starch hydrolysate up to DP 21 could be achieved within 30 minutes
1: glucose (3.45 ppm), **2**: turanose (4.01 ppm), **3**: maltose (4.07 ppm), **4**: panose (5.24 ppm), **5**: maltotriose (8.33 ppm), **6**: laminaritriose (6.72 ppm), **7**: maltotetraose (6.60 ppm), **8**: maltopentaose (6.32 ppm)

Figure 20 shows the separation of mono- and oligosaccharides, whereas Figure 21 reports the separation of a sample of starch hydrolysate, marketed by Fluka as Dextrin 20, containing over 20 linear homologous malto-oligosaccharides. The number of each peak in Figure 21 corresponds to the number of glucose residues in the linear malto-oligosaccharide.

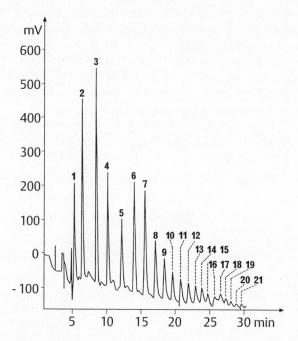

Figure 21.   Separation of Maltodextrines
          Separation conditions: see Figure 20 (Corradini, Corradini, Huber and Bonn[105]; with permission)

Anion Exchange Chromatography of Carbohydrates as Borate Complexes

In their fundamental work on the separations of sugars by ion exchange, Khym and Zill[1] showed, for instance, the resolution of fructose from galactose and glucose or the separation of arabinose from xylose and glucose. They used a Dowex-1 resin (40–80 μm strong base anion exchanger) for the separation and an orcinol (3,5-dihydroxytoluene)- as well as anthrone (9,10-dihydro-9-oxoanthracene) tests for visualization of eluted carbohydrates. Although Khym and Zill were not able to analyze all commonly occurring hexoses and pentoses in one run, improvements in the method and type of anion exchange resins allowed faster and more selective separations. Syamanda et al.[107] reduced analysis time by application of a positive chloride ion concentration gradient for faster elution of carbohydrates. Kesler[3] demonstrated the separation of 15 carbohydrates and 3 aldehydes within 7 hours using a purified and sized Dowex-1-X8 (5–40 μm) anion exchange resin.

With the advent of high-performance liquid chromatography especially high resolution separation became possible[108,109]. The separation of a 16-component standard mixture of carbohydrates under gradient conditions in less than 4 hours is shown in Figure 22.

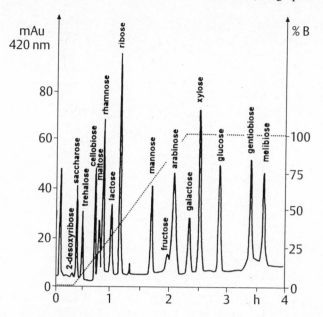

Figure 22. Separation of Carbohydrates as Borate Complexes by Anion Exchange Chromatography
Column: CA-X4 resin, 190 x 6 mm i.d.; *gradient*: 0.07–0.6 M borate, pH 8.0–10.5; *flow rate*:
1mL/min; *pressure*: 4.5 MPa: *temperature*: 60 °C; *detection*: orcinol–sulfuric acid, photometric
detection at 420 nm; *sample size*: each peak represents 16 nmol monosaccharide and 8 nmol
disaccharide (Bauer and Voelter[108]; with permission)

Elution was accomplished by a combination of linear gradients ranging from 0.07 M borate buffer (pH
8.0) to 0.5 M borate buffer (pH 10.5) on a DA-X4F anion exchange resin, which consists of 4% cross-
linked polystyrene particles of particle size 11±1 μm. For on-line detection, carbohydrates were
degraded with concentrated sulfuric acid to furfurals and hydroxymethylfurfurals and reacted with
orcinol to yield a dye, whose absorbance was measured photometrically at 420 nm. The kinetics of dye
formation can be deduced from Figure 23. Different residence times in the reaction bath were achieved
through different coil length. Reaction times between 3 and 4 minutes were found to be sufficient for
stable and reproducible quantitation.

Spectrophotometric post-column labeling has played an important role in monitoring carbohydrates in
borate complex anion exchange chromatography because carbohydrates have no characteristic
absorption in the UV/visible range. Measurement of the refractive index is insensitive due to
temperature instability or the composition of the mobile phase. Labeling methods with phenolic
compounds such as orcinol are not practical because they require the use of corrosive acids. Sinner and
Puls[110] described the detection of reducing carbohydrates by reduction of added copper(II) to copper(I)
which can be visualized as the 2,2'-bicinchoninate complex. New advances for borate complex anion
exchange chromatography of carbohydrates came from the group of Honda[111] in the early 1980s. He
introduced 2-cyanoacetamide as a new sensitive fluorogenic reagent for reducing carbohydrates. Under
stepwise elution with 0.2 M (pH 7.2) and 0.5 M (pH 9.6) borate buffers Honda et al. described the
automated analysis of mono- and disaccharides using the Hitachi No. 2633 resin of pellicular,
quaternary ammonium type having an average diameter of 11 μm (Figure 24).

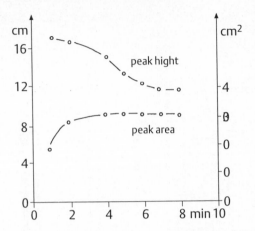

Figure 23. Time Dependency of Color Development of Sucrose with Orcinol–Sulfuric acid
Flow rates: 1 mL/min borate buffer and 1.1 mL/min sulfuric acid (95–97%) containing 1 g/L orcinol; *temperature*: 100 °C; *inner diameter of coil*: 0.7 mm (Bauer and Voelter[108]; with permission)

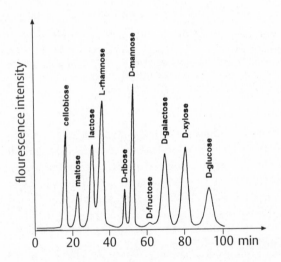

Figure 24. Separation of 10 Mono- and Disaccharides with Fluorimetric Detection of 2-Cyanoacetamide Derivatives
Column: Hitachi No. 2633 resin (11µm, quaternary ammonium); *eluants*: 0.2 M borate (pH 7.2) for 22 min, then stepwise change to 0.5 M borate (pH 9.6); *reagent solution*: 2-cyanoacetamide in 0.3 M borate buffer (pH 7.5); *post-column reaction temperature*: 100 °C; *excitation maximum*: 331 nm; *emission maximum*: 383 nm; *sample size*: 25 nmol of each carbohydrate (Honda, Matsuda, Takahashi and Kakehi[111]; with permission)

Using a purified sample of 2-cyanoacetamide, the mass detection limit for most aldoses was found at the 0.1 to 0.01 nmol level[112]. Post-column derivatization with 2-cyanoacetamide derivatives was also shown to be useful for UV detection at 180 nm [113]. The lower detection limit was 1 nmol for all aldoses, which is one or two orders of magnitude higher than that of fluorimetric monitoring. Table 3 shows a comparison of relative molar absorbance with relative fluorescence intensities. It is noteworthy that for different carbohydrates the range of absorbance is narrower (55–230) than that of fluorescence (5–335).

Table 3. Sensitivities of Aldoses Relative to Glucose

| | Relative Molar Response | |
	UV absorbance	Fluorescence
Glucose	100	100
Rhamnose	230	57
Lyxose	209	92
Ribose	55	5
Mannose	118	17
Fucose	165	21
Arabinose	192	63
Galactose	223	120
Xylose	189	33

In a further study, Honda et al.[114] also demonstrated the possibility of electrochemical detection of 2-cyanoacetamide derivatives of carbohydrates. Studies of the mechanism of the reaction between 2-cyanoacetamide and reducing carbohydrates suggests the intermediate formation of conjugated diene compounds, which are the compounds considered to be electrochemically oxidized. The detection limit for glucose is about 20 pmol and the linearity of the detector signal under optimized conditions (flow rate, borate concentration, reaction temperature, pH, reagent concentration, reaction time) is observed between 50 and 2000 pmol.

C. Separation Mechanisms

Anion Exchange
Since carbohydrates are weak acids with pK_a values ranging from 12 to 14 (e.g., glucose 12.35, xylose 12.29), they can only dissociate partially or completely under strong alkaline conditions[115]. On an anion exchange material, polyols are eluted with aqueous sodium hydroxide in decreasing order of alcohols, mono-, di- and oligosaccharides. Analysis is not disturbed by co-occuring nonionic and basic molecules, which are eluted in the void volume. However, detection of carbohydrates under such conditions poses a problem. In the meantime it has been possible to replace the derivatization procedure with pulsed amperometric detection[116]. Since carbohydrates are present as oxy-anions under strongly alkaline conditions, complexation with borate is not necessary. Actually the differential dissociation of the hydroxyl groups suffices to separate carbohydrates on a strong anion exchange resin. Anomer hydroxyl groups, for instance, dissociate more easily than others. The use of pulsed amperometric detection has solved the problems associated with conventional amperometric detection, namely the irreversible coating of electrodes with oxidation products. Today therefore gold electrodes are mainly in use for pulsed amperometric detection and in most instances the pH value of a 0.1 M sodium hydroxide solution is enough.

Use of Borate Ions in Carbohydrate Chromatography
The potential of borate esters of polyhydroxy compounds such as aldoses, ketoses and polyalcohols to

allow their effective separation has long been known in the literature. Coleman and Miller[117] were the first to observe the electrophoretic migration of a D-glucose-borate complex. Neutral carbohydrates can readily be separated as negatively charged borate complexes on strongly basic anion exchangers. Khym and Zill demonstrated the first anion exchange separation of sugars on Dowex-1, a strong-base anion exchange resin. In aqueous solution, an equilibrium between the borate ion and a diol function can be postulated resulting in the formation of 1:1 or 1:2 complexes.

Migration and retention behavior and consequently separation selectivity are governed by the stability of the complexes between the polyhydroxy compound and the borate ion. Böseken and coworkers studied the stability of esters of boric acid and borate using conductometry and polarimetry[118]. Van Duin et al.[119] and Bell et al.[120] employed ^{11}B NMR spectroscopy to estimate the relative stability of borate esters. Van Duin investigated the influence of the vicinity (α,β- or α,γ-), the configuration (threo/erythro and syn/anti), and the number of hydroxyl groups (2–5) as well as the presence of substituents on complex stability. Recently, the migration behavior of borate complexes of polyol derivatives obtained by reductive animation of sugars and uronic acids has also been studied by capillary zone electrophoresis[121–123].

2.1.3.3. Gel Permeation Chromatography (GPC) of Carbohydrates

Gel permeation chromatography takes place according to molecular size and is therefore important for the separation and purification of carbohydrates in the higher molecular range[124]. Different gels, e.g. agarose, dextrane or polyacrylamide, are used as stationary phases[125–130]. Oligo-carbohydrates, especially gluco and malto oligomers with a degree of polymerization up to 20, can be separated and isolated[131]. Separations are time consuming because of the soft gel material. The introduction of higher cross-linked polymers (TSK-Gels, Shodex Ottpak, Shodex Ionpak) which perform with the same separation mechanism, allows shorter analysis times using water and organic solvents as the mobile phase[132–136].

2.1.3.4. Methods for the Detection of Carbohydrates by HPLC

Several different detection systems are available for the liquid chromatographic analysis of carbohydrates. They can be divided into two groups: bulk property and solute property detectors. The former determines the change in the overall physical properties of the mobile phase. Three main types belong to this group: refractive index, conductivity and dielectric constant monitors. The second group consists of ultraviolet (UV) adsorption, radioactive, polarographic, fluorescence, mass, and mass spectrometric detectors.

Refractive Index Detection

The differential refractometer is one of the most important detection systems for carbohydrate analysis. The refractometer measures the difference between the refractive index of the reference solution and the mobile phase. The detection limit of mono- and oligosaccharides is dependent on the stationary phase and lies in the range of 3–5 µg[137–144].

UV Absorption Detection

The UV absorption detector is sensitive for solutes which have a reasonable absorption band in the UV region. The adsorption maximum for carbohydrates is in the range 188–192 nm and the detection limit lies in the range of 1–10 µg[145,146].

Fluorescence Detection System

For sensitive detection of carbohydrates simple HPLC methods with fluorescence detection has been described for the analysis of neutral and amino sugar residues and similarly for sialic acids in glycoproteins. Monosaccharides can be labeled by reductive amination, for example, with anthranilic acid in the presence of sodium cyanoborohydride, and the derivatives separated in the femtomol range[147,148].

Radiation Detection System

Radiation detectors are of special interest, especially in biology or pharmacy when quantitative analysis of degradation or transformation products of certain compounds is necessary in a very low detection range. When radioactively labeled lignocellulosic biomass, e.g. wood cultivated in a $^{14}CO_2$-gassed plant growth chamber, or commercially available U-^{14}C-labeled carbohydrates are subjected to a chemical degradation the resulting labeled compounds can be isolated by HPLC (Figure 25)[149].

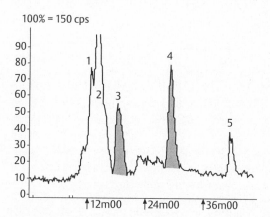

Figure 25. Semi-Preparative Separation of ^{14}C-Labeled Sugar Degradation Products in Poplar Wood Hydrolysate
Column: H-loaded ion exchanger; *mobile phase*: water; *flow rate*: 1.0 mL/min; *column temperature*: 85 °C; *detection*: solid scintillation; *injection volume*: 1 mL
1: cellobiose, **2**: glucose, **3**: methyl glyoxal, **4**: HMF, **5**: furfural (Bonn[149]; with permission)

The most important method of detection for the radioactive labeled compounds is β-scintillation counting off- and on-line which uses low-energy beta emitters, such as C-14 or H-3. Efficiencies of 55% for C-14 and 2% for H-3 have been described.

Mass Detection

By using mass detectors both reducing and nonreducing carbohydrates as well as oligosaccharides can be detected. The limit of detection of this method is comparable to that of refractive index. Low volatile mobile phase systems, e.g. aqueous ionic solutions, reduce the applicability of this method. The principle of mass detection is based on the spraying of the eluant and subsequent transportation through a tube by a gas stream. As a consequence, the mobile phase is evaporated and the sample particles pass through a light beam which is scattered and then detected[150–152].

Electrochemical Detection

Electrochemical detection in flowing systems is usually achieved by controlling the potential of the working electrode at a fixed potential value, corresponding to the limiting current plateau region of the compounds of interest. The current is measured as a function of time as the compounds undergo an oxidation or reduction. This technique is called flow amperometry. Amperometric detection for liquid chromatography results in sharp current peaks reflecting the concentration profiles of the eluted compounds as they pass through the detector. Accordingly, the magnitude of the peak current serves as a measure of concentration. Controlled-potential detectors are characterized by a remarkable sensitivity (down to the pg level), high selectivity (towards electroactive species), wide linear range, low dead volumes, fast response, and relatively low cost [153]. The choice of electrode material is more critical in LCEC than in the usual electroanalytical experiment, primarily due to the mechanical ruggedness and long-term stability required. Several experimental parameters such as composition of the electrolyte solution, chemical and physical properties of the electroactive species of interest, and the potential range investigated, have to be taken into account. The most commonly used working electrodes are glassy carbon, carbon paste, mercury, and platinum electrodes, however, gold, copper, nickel and more recently modified electrodes have been employed as well in LCEC.

Aldehyde and terminal alcohol moieties in carbohydrates can be oxidized at gold electrodes in alkaline media and at platinum electrodes under all pH conditions. The anodic response of these compounds is electrocatalytic with direct participation of the electrode surface within the oxidation mechanism. However, due to irreversible adsorption of surface-active species, including intermediate oxidation products as well as solution impurities, the electrode becomes fouled and the anodic current decays to zero. Thus carbohydrates usually cannot be detected by direct constant-potential amperometry at conventional noble metal electrodes. Various other metal electrodes, such as nickel[154-157], copper[158], nickel–copper, and nickel–chromium–iron alloys[159] have been investigated for the constant potential amperometric detection of carbohydrates following chromatographic separations. Chemically modified electrodes, such as cobalt phthalocyanine containing carbon paste[160], cupric salts on glassy carbon (GC) substrates[161,162], copper microparticles in Nafion coatings[163], or nickel salts on GC substrates[164], have also been reported as electrochemical detectors. The redox mechanism of carbohydrates at metallic electrodes, such as copper and nickel, is as complicated as the oxidation at noble metal electrodes. The oxidation of carbohydrates in strongly alkaline solutions is supposed to be electrocatalytic, based on the reaction of electrochemically generated M(III)–oxyhydroxide complexes with the carbohydrates leading to oxidation products and reformation of M(II) species. Though many electrode materials have been investigated, the main problems encountered during amperometric measurements of carbohydrates at metal electrodes at constant potential could not be solved. Loss of sensitivity and reproducibility, either due to electrode fouling or surface corrosion, has to be faced during long term use of these electrodes.

More recently, Johnson and co-workers[165,166] have introduced pulsed electrochemical detection (PED) at gold or platinum electrodes, a technique which has become very popular not only for the LC–EC detection of carbohydrates[167-169], but also for the determination of other aliphatic compounds such as alcohols, amines and sulfur compounds. This detection mode overcomes the problem of lost activity of noble metal electrodes (electrode fouling) associated with the fixed-potential detection of carbohydrates by the use of a triple-pulse waveform (Figure 26), combining anodic and cathodic polarizations. Figure 26 shows the potential–time (E-t) waveforms most commonly applied in pulsed electrochemical detection (PED). For the liquid chromatographic detection of carbohydrates a pulsed waveform of the type illustrated in Figure 26A and 26B is most frequently employed.

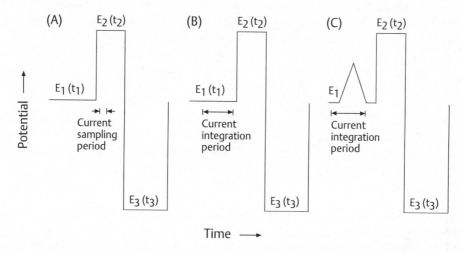

Figure 26. Potential–Time (E-t) Waveforms
Processes: E_1: anodic detection, E_2: oxidative cleaning, E_3: cathodic reactivation; *waveforms*: A: pulsed amperometric detection (PAD) with a short current sampling period (e. g., 16.7 ms), B: PAD with a long current integrated sampling period and C: integrated PAD with a long current integration period during a fast triangular staircase scan (Johnson and La Course[165]; with permission)

In Figure 26A, for a detection time period (t_1) ($t_1 = t_{del} + t_{int}$) a constant potential E_1 is applied, where oxidation of the compounds takes place. After a delay (t_{del}) the current is sampled for a short time (t_{int}) (e.g., 16.7 ms) at the end of the pulse, where the charging current has decayed to a minimum. The potential is then stepped to E_2, where the electrode undergoes oxidative cleaning during a time period of t_2, followed by reductive regeneration of the electrode surface at potential E_3 (t_3).

The fundamental method of current measurement in PED instrumentation is by electronic integration of electrode current during the period t_{int}. When the output voltage signal is directly proportional to the average current (i. e., $i_{avg} = q/t_{int}$ Cs^{-1}) this style of signal presentation has resulted in the common name 'pulsed amperometric detection' (PAD), whereas the names 'pulsed coulometric detection' (PCD) and 'integrated pulsed amperometric detection' (IPCD) have been suggested for signal presentations using the current integral (i. e., $q = \int i\, dt$ C). Figure 26B shows the method known as 'pulsed coulometric detection' (PCD). In this case, the current is integrated over a longer period and the time period chosen is an integral number (m) of cycles of the line voltage (i. e., $t_{int} = m$ 16.7 ms in U.S.A. and $t_{int} = m$ 20.0 ms in Europe) with typical total integration times of greater than 200 ms. The use of this type of waveform eliminates the most common electrical interference [60 Hz (U.S.A.) and 50 Hz (Europe)] encountered in pulsed electrochemical detection. Thus, lower detection limits for most compounds can be reached.

Optimization of all PED potential and time parameters is a complex task, which can be achieved either by the 'trial and error' method based on multiple injections in LC–PED with manual adjustments to each waveform parameter between injections, or according to a recently published automated procedure, based on pulsed voltammetry at hydrodynamic electrodes such as the rotated disk electrode[170,171]. The parameters are optimized with the intent of maximizing the signal-to-noise ratio (S/N) whilst maintaining a reasonably high value of waveform frequency $[f = 1/(t_1 + t_2 + t_3)]$. Typically the waveforms are executed at a frequency of 1–2 Hz. In Figure 27 the effect of different integration time periods (t_{int}) on the LC–PED response of some carbohydrates is demonstrated.

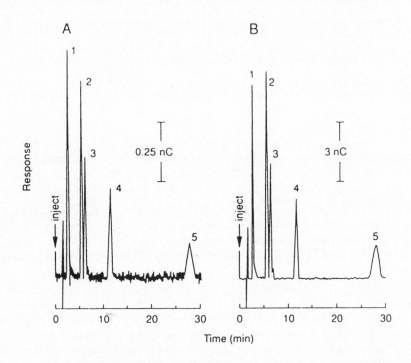

Figure 27   Comparison of LC–PED Results for Two Values of Integration Time Period (t_{int}) for the PED Waveform in Figure 26A and 26B.
Electrode: Au; *column*: Dionex CarboPac PA1; *eluant*: isocratic with 0.10 M NaOH; *sample volume*: 25 µL; *waveform*: E_1 = +0.15 V vs. Ag/AgCl (t_{del} = 520 ms, t_{int} = variable), E_2 = +0.70 V (t_2 = 120 ms), E_3 = -0.30 V (t_3 = 360 ms); *values of t_{int}*: A: 17 ms and B: 200 ms; *peaks* (0.1 nmol ea.):
1: D-glucitol, **2**: D-glucose, **3**: D-fructose, **4**: sucrose, **5**: maltose (reproduced with permission from ref. 171)

Further representative LC–PED results for carbohydrates and other compounds such as alcohols and sulfur compounds are summarized in ref. 172.

Pre-Column Derivatization

With this method carbohydrates are derivatized before the chromatographic separation[173]. In addition to perbenzoates[174,175] and dansylhydrazones[176], 2-aminopyridine and derivatives[177,178] are used for the derivatization of mono- and oligosaccharides. These methods have proved to be especially useful for samples containing proteins and fats which may interfere with the carbohydrates.

Post-Column Derivatization
This method of detection is used when the concentration of the respective carbohydrates is too low for the previously described detection methods. The carbohydrates are first separated and then derivatized to yield easily detectable products[179–181].

On-Line Coupling of HPLC with Mass Spectrometry
HPLC–MS Interfaces
An ideal detector system for HPLC should combine high detection sensitivity with maximum identification capability[182]. Of all detector systems used with HPLC separations, the mass spectrometer most probably comes closest to being the ideal detector, because it is sensitive at the attomole range and provides information concerning molecular mass, atomic composition, and molecular structure.

Since the mass spectrometer has to be operated under high vacuum conditions, the on-line coupling of HPLC with mass spectrometry (MS) requires the use of an interface to handle the high gas loads that are produced by vaporization of the HPLC effluent. The earliest attempts to couple liquid chromatography with mass spectrometry (HPLC–MS) involved direct liquid introduction[183] as well as moving belt transfer interfaces[184] and could not achieve widespread acceptance due to problems with interfacing and memory effects. The thermospray interface (TSP) was the first commercially successful HPLC–MS interface[185]. In the TSP process, the column effluent flows through an electrically heated vaporizer and forms a supersonic jet of fine droplets. Sample ions are either preformed in the aqueous buffer or generated by gas phase reactions with a plasma that is produced with a discharge electrode. After evaporation of the solvent, sample ions are electrically deflected for mass analysis into the sampling cone of the mass spectrometer situated perpendicular to the vapor flow.
The first useful desorptive ionization process for the analysis of large, polar, and thermally labile molecules was fast atom bombardment (FAB) (Barber et al.[186]) and Caprioli et al.[187,188] developed the continuous flow FAB probe (CF–FAB). This is based on a ca. 1–20 µL/min flow of a mixture of the sample together with a liquid matrix, typically glycerol, into the source of a mass spectrometer with subsequent atom bombardment of the liquid as it flows over the surface of a target. Positive and negative analyte ions are sputtered from the matrix and are mass analyzed[189].

The development of the electrospray ionization (ESI) technique[190] and the closely related atmospheric pressure chemical ionization technique (APCI)[191] has had a major impact on the practical applicability of HPLC–MS. In ESI, the mobile phase passes through a metal capillary that is maintained at a potential between 2 and 5 kV. A charged aerosol is formed at the tip of the capillary and, as solvent is removed by the vacuum system, the charge density on the surface increases to such an extent that charge repulsion exceeds the surface tension and smaller droplets are formed. At a certain point the dimensions of the droplets become so small that ions are directly ejected into the gas phase and can be mass analyzed[192]. The APCI interface comprises a heated nebulizer similar to the thermospray interface. Sample ions are generated through chemical gas phase reactions with positive or negative reactant ions that are formed in the APCI source from a corona discharge[193].

Applications of HPLC–MS in the Analysis of Carbohydrates
Various HPLC techniques have been developed to facilitate carbohydrate analyses in an effort to improve our understanding of their role in biochemistry. Reversed-phase[194,195] normal-phase[196,197] and cation exchange[198,199] stationary phases have been applied successfully to the separation of mono- and oligosaccharides.

More recently, high-performance anion exchange chromatography (HPAEC) at high pH with pellicular strong-base anion exchangers has enabled fast and high-resolution separations of carbohydrates[200,201]. A major problem in interfacing HPAEC with mass spectrometry is the presence of nonvolatile buffer components such as sodium hydroxide or sodium acetate which have to be removed before entering the mass spectrometer. Simpson and Fenselau[202] used a micromembrane suppressor to remove alkaline salts from the mobile phase after the chromatographic separation.

Figure 28 depicts the schematic diagram of the integrated HPAEC–TSP–MS system. A gradient of aqueous sodium hydroxide was formed by a binary gradient system and separation of carbohydrates was accomplished on a 250 × 4.6 mm I.D. column packed with a 10 µm polymeric anion exchanger. After passing the UV detector, which monitored the acetylated carbohydrates at 220 nm, the column effluent was neutralized in the micromembrane suppressor which was continuously regenerated with 35 mM sulfuric acid at a flow rate of 10 mL/min. Due to a maximum operating pressure of the suppressor of 100 psi and a back pressure of the TSP interface between 600 and 700 psi, a booster pump had to be inserted between the membrane suppressor and the TSP interface.

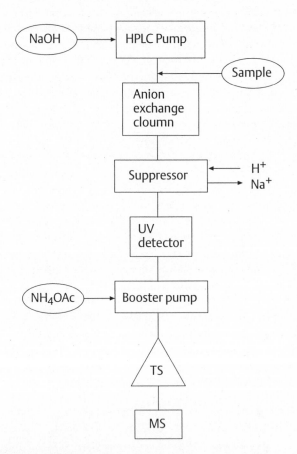

Figure 28. Integrated HPAEC–TSP–MS System
TS: thermospray interface, MS: mass spectrometer (Simpson, Fenselau, Hardy, Townsend, Lee and Cotter[202]; with permission)

The chromatogram of a gradient separation of carbohydrate standards shows that the two epimeric monosaccharides N-acetylglucosamine and N-acetylmannosamine can easily be separated in less than 15 minutes (Figure 29). A good correlation between the 220 nm UV trace and the total ion chromatogram can be seen. Gradient elution results in improved peak shape and lower detection limits of carbohydrates in comparison to isocratic separations.

Some of the primary features of the N-acetyllactosamine mass spectrum obtained by gradient HPAEC–MS analysis (Figure 30) are $[MH–18]^+$ at m/z 366, $[MH–3x18]^+$ at m/z 330. $[MH–4x18]^+$ at m/z 312, a protonated N-acetylglucosamine fragment $[GlcNAcH]^+$ at m/z 222, $[GlcNAcH–18]^+$ at m/z 204, $[GlcNAcH–2x18]^+$ at m/z 186, $[GlcNAcH–3x18]^+$ at m/z 168, an ammonium adduct of galactose $[GalNH_4]^+$ at m/z 198, and $[GalNH_4–18]^+$ at m/z 180. Each loss of 18 probably corresponds to loss of water.

Boulenguer et al.[203] separated permethylated oligosaccharides in a fused silica capillary column (250 mm × 0.32 mm i.d.) packed with a 5 μm octadecylsilica stationary phase with a gradient of 0–17% methanol in water/10% thioglycerol at a flow rate of 5 μL/min. The column was connected directly to the capillary of a CF–FAB probe. Separation efficiency of the capillary column was impaired only slightly by band broadening in the CF–FAB probe as illustrated by the minor peak deformation of eluted oligosaccharides from hen ovomucoid.

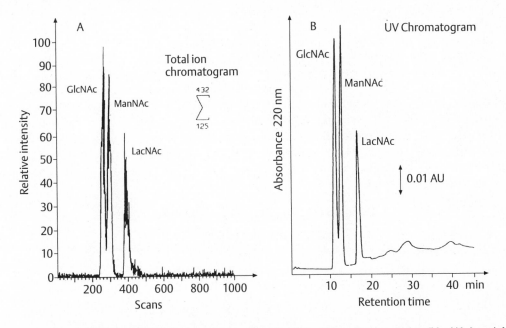

Figure 29.   Gradient Separation of N-Acetylglucosamine (GlcNAc), N-Acetylmannosamine (ManNAc) and N-Acetyllactosamine (LacNAc)
Column: Dionex AS-6 analytical anion exchange column (10 μm, 250 × 4.6 mm i.d.); *eluant*: A: 10 mM NaOH and B: 100 mM NaOH; *gradient*: 8 min at 0% B, 0–20% B in 4 min, 20–100% B in 11 min; *flow rate*: 1 mL/min; *sample size*: 100 μg each; *vaporizer temperature*: 190 °C; *detection*: A: TSP–MS, 125–432 amu and B: UV at 220 nm (Simpson, Fenselau, Hardy, Townsend, Lee and Cotter[202]; with permission)

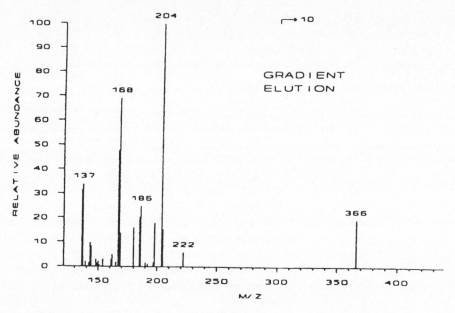

Figure 30. Thermospray Mass Spectrum of N-acetyllactosamine (100 µg) (Simpson, Fenselau, Hardy, Townsend, Lee and Cotter[202]; with permission)

The chloride attachment mechanism was utilized for coupling HPLC with MS through the APCI interface[204]. Chromatography was carried out on a 250×4 mm i.d. column packed with 7 µm NH_2–silica. The mobile phase consisted of a mixture of methanol or acetonitrile with a low amount of chloroform. A strong base is formed in the APCI source from the polar solvent molecules by corona discharge (Scheme I), followed by ion molecule reactions. The strong base reacts with chloroform and generates chloride ions (Scheme II) which can associate with neutral molecules containing polar functionalities such as hydroxyl or carboxyl groups. Under atmospheric conditions these associates become stabilized by removal of excess internal energy by collisions, resulting in stable ions with hydrogen bonding (Scheme III).

Scheme I, Strong Base Generation

$$ROH + e^- \xrightarrow{\text{discharge}} RO^- + H$$

Scheme II, Chloride Generation

$$HCCl_3 + RO^- \longrightarrow CCl_3^- + ROH$$
$$CCl_3^- \longrightarrow CCl_2 + Cl^-$$
$$CCl_2 + H_2O \longrightarrow CO + 2\,HCl$$

Scheme III, Association

$$R'OH + Cl^- \longrightarrow (R'OH\cdots Cl)^{-*} \xrightarrow{\text{collisional stabilization}} (R'OH\cdots Cl)^-$$

The negative ion mass spectra of underivatized monosaccharides and some oligosaccharides show dominant quasi-molecular ions [M+Cl]⁻, few fragment ions attributed to sugar ring fragmentation products, and few cluster ions [M+Cl+nH₂O]⁻ (n = 1 or 2), [M+Cl+mCH₃OH]⁻ (n = 1 or 2), [M+Cl+nH₂O+mCH₃OH]⁻ (n, m = 1, 2 or 2, 1].

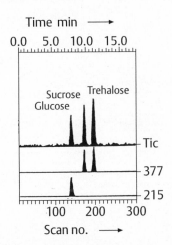

Figure 31. Separation of Carbohydrates by HPLC–APCI–MS
Column: Lichrosorb NH₂ column (7 µm, 250 × 4 mm i.d.); *eluant*: acetonitrile–water–chloroform 60:37:3; *flow rate*: 1 mL/min; *vaporizer temperature*: 170 °C; *corona discharge current*: 5 µA; *detection*: negative ion APCI–MS, 1–800 amu in 4 s (Kato and Numajiri[204]; with permission)

Figure 31 illustrates a chromatogram of the separation of glucose, sucrose, and trehalose using an eluant of acetonitrile–water–chloroform (60:37:3) at a flow rate of 1 mL/min. Microgram quantities of carbohydrates yielded satisfactory molecular ion intensities. The chromatogram shows that glucose eluted first, followed by sucrose and trehalose. The [Glucose+Cl]⁻ adduct ion was detected at m/z 215 whereas sucrose and trehalose provided isobaric chloride adduct molecular ions at m/z 377, but they could easily be determined on the basis of their retention times.

Mass spectrometric ionization techniques as described above generate mainly molecular or quasi-molecular ions with only little fragmentation. One of several strategies to obtain structural information from fragment ions is collisionally induced dissociation (CID) combined with tandem mass spectrometry (MS/MS). A typical fragmentation may be represented as shown in Scheme IV, where M_{PC} is the mass of a precursor ion, M_{PD} that of a product ion, and M_N that of an ejected neutral particle. Thus, MS–MS experiments essentially consist of measuring M_{PC} and then finding which ions (M_{PD}) are obtained from its fragmentation, or measuring the product ion (M_{PD}) and finding which ion (M_{PC}) is its precursor. The interpretation of MS–MS spectra affords structural information about a given analyte and is based on correlation of the collisionally induced dissociation process with structural elements of the molecule.

Scheme IV

$$M_{PC} \xrightarrow{\text{CID}} M_{PD} + M_N$$

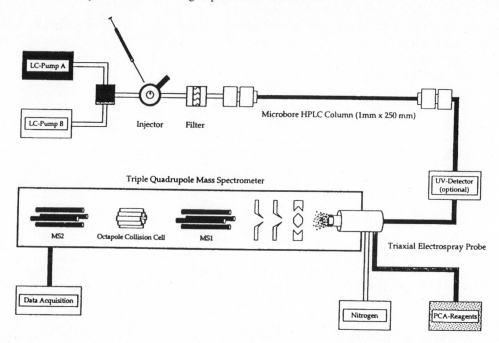

Figure 32. Setup for HPLC–MS–MS Including Post-Column Addition of Metal Chlorides for In-Source Formation of Carbohydrate-Metal Ion Complexes (Kohler and Leary[205]; with permission)

Kohler and Leary described a technique for on-line post-column addition (PCA) of metal chlorides using a triaxial electrospray probe in order to form carbohydrate-metal ion complexes that could be analyzed by positive ion ESI–MS or ESI–MS–MS[205]. The metal chloride solution (LiCl, NaCl, KCl, RbCl, CsCl, $CoCl_2$) was added to a carbohydrate sample directly within the ion source of the mass spectrometer. The multiple step process of chromatographic separation, post-column complexation and tandem mass spectrometry was implemented by an HPLC–MS–MS system using a triaxial electrospray probe (Figure 32).

The PCA of metal chlorides entailed considerable enhancements in detection sensitivity of carbohydrates and assisted, in the case of cobalt chloride, in structural analysis of oligosaccharide by MS–MS. The most significant increase in intensity was observed for the lithium and sodium complexes of cellobiose. The signal of the $[M+Li]^+$ ion was found to be 70 times more abundant than the $[M+H]^+$ signal and detection limits of 10 pmol were realized in HPLC–MS runs. The applicability of this technique for structure elucidation was demonstrated by microbore HPLC, in-source PCA and ESI–MS–MS of oligosaccharide test mixtures. Figure 33 shows the reconstructed ion chromatograms of cellobiose and di-*N*-acetylchitobiose. The MS–MS spectrum corresponding to cellobiose after CID is depicted in Figure 34. CID of the $[M+Co–H]^+$ complexes yields cross ring fragmentation which allows determination of specific linkages. For cellobiose, losses of $C_2H_4O_2$ (m/z 340) and $C_4H_8O_4$ (m/z 280) are observed which are characteristic of a 1,4-linkage. A glycosidic cleavage ion is observed at m/z 220.

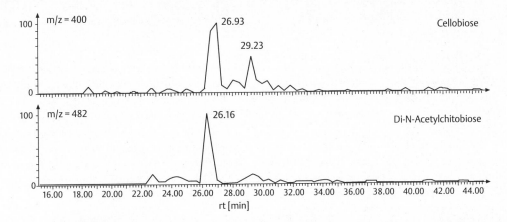

Figure 33. Reconstructed Ion Chromatograms of 1 nmol Cellobiose and 1 nmol Di-*N*-acetylchitobiose with Post-column Addition of CoCl₂

Column: Spherisorb s5NH₂ (5 µm, 250 × 1 mm i.d.); *eluant*: A: acetonitrile and B: water; *gradient*: 0–50% B in 10 min, 50–88% B in 10 min, 100% B for 30 min; *flow rate*: 20 µL/min; *post-column addition of CoCl₂ (c = 1nmol/µm)*: flow rate 1.67 µL/min; *detection*: positive ion ESI–MS, 380–940 amu in 2.5 s (Kohler and Leary[205]; with permission)

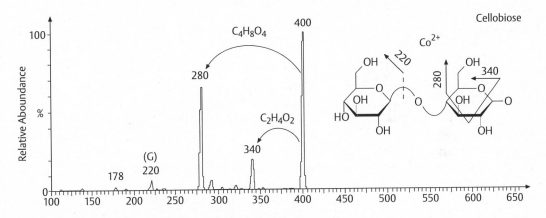

Figure 34. On-Line HPLC–MS–MS Spectrum of Cellobiose

Scan range: 100–400 amu in 5 s; *collision gas*: 10^{-3} mbar argon; *collision energy*: 20 eV (Kohler and Leary[205]; with permission)

2.1.3.5. Preparative Isolation of Carbohydrates

The preparative isolation of carbohydrates is of particular interest in the fields of biotechnology and gene technology for the purpose of structural elucidation of important polysaccharides. In most cases analytical techniques form the basis of the preparative isolation of carbohydrates from various matrices. Thus, well characterized stationary phases are used. Pure water is the preferred mobile phase because direct isolation is part of the chromatographic separation. If necessary, highly volatile mobile phase additives should be used to enable a fast and easy removal. One critical question to be considered is the amount of carbohydrate to be isolated. In biochemical applications micro- and milligram amounts are sufficient to enable a structural elucidation of the respective carbohydrates, e.g.

nuclear magnetic resonance spectroscopy. Thus, analytical scale instruments are suitable for micropreparative purposes.

Figure 35 depicts the isolation of gluco-oligosaccharides obtained from a biomass hydrolysate. These compounds may serve as standard substances as they can only be obtained with difficulty by synthesis. In addition, radioactively marked derivatives can be isolated semi-preparatively by this method[166]. If larger amounts have to be prepared, as in the gram or kilogram range, the simple application of analytical equipment is not suitable. This is necessary in the pharmaceutical and food industry where flow rates in the range of liters per minutes have to ensured. Furthermore, stationary phases are applied in dimensions of 2 x 30cm and above with resin material also used for analytical scale applications. For preparative purposes column dimensions can reach 3 x 230cm with 40 kg of stationary phase and a flow rate of several liters per minute[206–210].

Likewise, displacement chromatographic methods using analytical instrumentation also enable the isolation of specific components in the gram range as an alternative to elution chromatography[211–213].

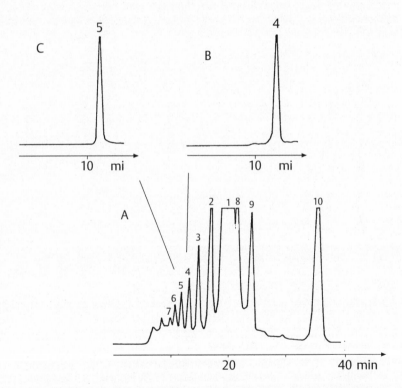

Figure 35. A: Semi-Preparative HPLC Separation of Gluco-Oligomers, 1,6-Anhydro-ß-D-glucose and Hydroxymethylfufural from Poplar Wood Hydrolysates. B: Isolated Cellotetraose Fraction. C: Isolated Cellopentaose Fraction
Column: 70% Ag-loaded ion exchanger; *mobile phase*: water; *flow rate*: 1.0 mL/min; *column temperature*: 85 °C; *detection*: refractive index
1: glucose, 2: cellobiose, 3: cellotriose, 4–7: gluco-oligomers, numbers indicate d.p., 8: fructose, 9: 1,6-anhydro-β-D-glucose, 10: hydroxymethylfurfural (Bonn[149]; with permission)

References

1   Khym, J. X.; Zill, L. P. *J. Am. Chem. Soc.* **1950**, *74*, 2090.
2   Isbell, H. S.; Brewster, J. F.; Holz, N. B.; Frish, H. L. *J. Res. Natl Bur. Stand.* **1948**, *40*, 129.
3   Kesler, R. B. *Anal.Chem.* **1967**, *39*, 1416.
4   Walborg, E. F.; Lantz, R. S. *Anal.Biochem.* **1986**, *22*, 123.
5   Havlicek, J.; Samuelson, O. *Anal. Chem.* **1968**, *47*, 123.
6   Jandera, P.; Churacek, J. *J. Chromatogr.* **1974**, *98*, 79.
7   Baker, S. A.; Bourne, J. J.; Theander, O. *J. Chem. Soc.* **1955**, 4276.
8   Whistler, R. C.; Durso, D. F. *J. Am. Chem. Soc.* **1950**, *72*, 677.
9   Whistler, R. C.; BeMiller, J. N. In *Methods in Carbohydrate Chemistry*, Vol. 1; Whistler, R. L.; Wofram, M. L., Eds.; Academic: New York, London, 1962; p 42 ff.
10   Thoma, J. A.; Wright, H. B.; French, D. *Arch. Biochem. Biophys* **1950**, *85*, 152.
11   Scherz, H.; Bonn, G. K. *Encyclopedia of Analytical Science*, Vol. 1; Townshend, A., Ed.; Academic: New York, 1995.
12   *Carbohydrate Analysis, J. Chromatogr.*; Ziaol El Rassi, Ed.; **1995**.
13   Ennor, K. S.; Honeyman, J.; Shaw, C. J.; Stening, T. C. *J. Chem. Soc.* **1968**, 2921.
14   Bonmahraz, M.; Davydor, V.; Kiselev, A. *Chromatographia* **1982**, *15*, 751.
15   Rocca, J. L.; Rouchouse, A. *J. Chromatogr.* **1976**, *117*, 216.
16   McGinnis, G. D.; Fang, P. *J. Chromatogr.* **1978**, *153*, 107.
17   Olst, H. v.; Joosten, G. E. H. *J. Liq. Chromatogr.* **1979**, *2*, 111.
18   Aitzetmüller, K. *Chromatographia* **1980**, *13*, 432.
19   Nachtmann, F.; Budna, K. W. *J. Chromatogr.* **1977**, *136*, 279.
20   Verhaar, L. A. T.; Kuster, B. F. M. *J. Chromatogr.* **1981**, *220*, 313.
21   Wheals, B. B.; White, P. C. *J. Chromatogr.* **1980**, *189*, 414.
22   Robards, K.; Whitelaw, M. *J. Chromatogr.* **1986**, *373*, 81.
23   Schwarzenbach, R. *J. Chromatogr.* **1975**, *117*, 206.
24   Jones, A. D.; Burns, J. W.; Sellings, S. E.; Cos, J. A. *J. Chromatogr.* **1977**, *144*, 169.
25   Linden, J. C.; Lawhead, C. L. *J. Chromatogr.* **1975**, *117*, 211.
26   Binder, H. *J. Chromatogr.* **1980**, *189*, 414.
27   Orth, P.; Engelhardt, H. *Chromatographia* **1982**, *15*, 91.
28   Wheals, B. B.; White, P. C. *J. Chromatogr.* **1979**, 176.
29   Porsch, B. *J. Chromatogr.* **1982** *253* 49.
30   Verhaar, L. A. T.; Kuster, B. F. M. *J. Chromatogr.* **1982**, *234*, 57.
31   Novolov, Z. L.; Meagher, M. M.; Reilly, P. I. *J. Chromatogr.* **1985**, *319*, 51.
32   Yost, R. W.; Ettre, L. S.; Coulon, R. D. *Practical Liquid Chromatography An Introduction*; Perkin-Elmer: Norwalk, C.T., 1980; p 79.
33   Verhaar, L. A. T.; Kuster, B. F. M. *J. Chromatogr.* **1981**, *220*, 313.
34   Brons, E.; Olieman, J. *J. Chromatogr.* **1983**, *159*, 79.
35   Hendrix, D. L.; Lee, R. E.; Baust, J.; James, H. *J. Chromatogr.* **1981**, *210*, 45.
36   Aitzetmüller, K. *J. Chromatogr.* **1978**, *156*, 354.
37   Baust, J. G.; Lee, R. E.; Rojas, R. R.; Hendrix, D. L.; Friday, D.; James, J. *J. Chromatogr.* **1983**, *261*, 65.
38   Verzele, M.; Simoens, G.; van Damme, F. *Chromatographia* **1978**, *23*, 292.
39   Verzele, M.; van Damme, F. *J. Chromatogr.* **1986**, *362*, 23.
40   Pirisino, J. F. In *Food Constituents and Food Residues–Their Chromatographic Determination*; Lawrence, J. F., Ed; Dekker: New York, 1984; p 159ff.
41   Rabel, F. M.; Caputo, A. G.; Butts, E. T. *J. Chromatogr.* **1976**, *126*, 731.
42   Porsch, P. *J. Chromatogr.* **1982**, *253*, 49.
43   Rajakylä, E. *J. Chromatogr.* **1986**, *353*, 1.
44   Beynon, P. J.; Collins, P. M.; Gardiner, D.; Overend, W. E. *Carbohydr. Res.* **1968**, *6*, 431.
45   Heyrand, A.; Rinando, M. *J. Liq. Chromatogr.* **1980**, *3*, 721.
46   Cheetham, W. H.; Sirimanne, P.; Day, P. R. *J. Chromatogr.* **1981**, *207*, 439.
47   Verhaar, L. A. T.; Kuster, B. F. M.; Claessens, H. A. *J. Chromatogr.* **1984**, *284*, 1.
48   Wheaton, R. M.; Baumann, W. C. *Ann. N. Y. Acad. Sci.* **1953**, *57*, 159–176.
49   Huber, C.; Bonn, G. K. In *Carbohydrate Analysis, J. Chromatogr.*, Library Vol. 58; Ziad El. Rassi, Ed.; 1995; p 147ff.

50   Jandera, P.; Churacek, J. *J. Chromatogr.* **1973**, *86*, 351.
51   Jandera, P.; Churacek, J. *J. Chromatogr.* **1974**, *98*, 55.
52   Churms, S. *J. Chromatogr.* **1992**, *51B*, 230.
53   Hicks, K. B. *Adv. Carbohydr. Chem. Biochem.* **1988**, *46*, 17.
54   Lee, Y. C. *Anal. Biochem.* **1990**, *189*, 151.
55   Johnson, D. C.; Dobberpuhl, D.; Roberts, R.; Vandeberg, P. *J. Chromatogr.* **1993**, *640*, 79.
56   Verzele, M.; Simoens, G.; van Damme, F. *Chromatographia* **1987**, *23*, 292.
57   Leonhard, J. L.; Guyon, F.; Fabiani, P. *Chromatographia* **1984**, *18*, 600.
58   Vratny, P.; Frei, R.; Brinkmann, U. A.; Nielsen, N. W. *J. Chromatogr.* **1984**, 355.
59   Domard, A.; Rinaudo, M. *Polym. Commun.* **1984**, *25*, 55.
60   Robards, K.; Whitelaw, M. *J. Chromatgr.* **1986**, *373*, 81.
61   Dorfner, K. In *Ion Exchangers*; Dorfner, K., Ed.; Walter de Gruyter: Berlin, New York, 1991; p 189.
62   Ugelstad, J.; Mork, D. C.; Kaggerud, K. H.; Ellingsen, T.; Berge, A. *Adv. Colloid.Inter. Sci* **1980**, *13*, 101.
63   Huber, C.; Oefner, P.; Bonn, G. *J. Chromatogr.* **1992**, *599*, 113.
64   Huber, C.; Oefner, P.; Bonn, G. *Nucl. Acids Res.* **1993**, *21*, 106.
65   Huber, C.; Oefner, P.; Bonn, G. *Anal. Biochem.* **1993**, *212*, 351.
66   Pecina, R.; Bonn, G.; Burtscher, E.; Bobleter, O. *J. Chromatogr.* **1984**, *287*, 245.
67   Bonn, G.; Pecina, R.; Burtscher, E.; Bobleter, O. *J. Chromatogr.* **1984**, *287*, 215.
68   Bonn, G.; Bobleter, O. *J. Chromatographia* **1984**, *8*, 445.
69   Verhaar, L. A. T.; Kuster, B. F. M. *J. Chromatogr.* **1981**, *210*, 279.
70   Bonn, G. *J. Chromatogr.* **1985**, *322*, 411.
71   Wood, R.; Cumming, L.; Jupille, T. *J. Chromatogr. Sci.* **1980**, *18*, 551.
72   Jupille, T.; Gray, M.; Black, B.; Gould, M. *Am. Lab.* **1981**.
73   Küper, B. *Chem. Rundschau* **1982**, *37*, 6.
74   Tanaka, K.; Ishizuka, T.; Sunahara, T. *J. Chromatgr.* **1979**, *174*, 153.
75   Deuel, H.; Solm, J.; Angyas-Weisz, L. *Helv. Chim. Acta* **1950**, *33*, 2171.
76   Deuel, H.; Solm, J.; Angyas-Weisz, L. *Helv. Chim. Acta* **1951**, *34*, 1849.
77   Fitt, L.; Hassler, W.; Just, D. *J. Chromatogr.* **1980**, *187*, 381.
78   Brunt, K. *J. Chromatgr.* **1982**, *246*, 145.
79   Schmidt, J.; John, M.; Wandrey, C. *J. Chromatgr.* **1982**, *213*, 151.
80   Scobell, H. H; Brobst, K.; Steele, E. *Cereal. Chem.* **1977**, *54*, 905.
81   Scobell, H. H; Brobst, K. *J. Chromatogr.* **1981**, *212*, 51.
82   Goulding R. W. *J. Chromatogr.* **1975**, *103*, 229.
83   Davankov, V.; Semechkin, A. V. *J. Chromatogr.* **1977**, *141*, 313.
84   Colin, H.; Diez-Masa, J. C.; Guiochon, G.; Czajkowsk, T.; Miedziak, I. *J. Chromatogr.* **1978**, *167*, 41.
85   Vigh, G.; Varga-Puchony, Z. *J. Chromatogr.* **1980**, *196*, 1.
86   Vrátny, P.; Coupek, J.; Vozka, S.; Hostomska, Z. *J. Chromatogr.* **1983**, *254*, 143.
87   Derler, H.; Hörmeyer, H. F.; Bonn, G. *J. Chromatogr.* **1988**, *440*, 281.
88   Hicks, K. B.; Hotchkiss, A. T., Jr *J. Chromatogr.* **1988**, *441*, 382.
89   Olechno, J.; Carter, S.; Edwards, W. T.; Gitte, D. G.; Townsend, R. R.; Lee, Y. C.; Hardy, M. R. *Techniques in Protein Chemistry*; Academic: San Diego, 1989; p 364.
90   Lee, D. P.; Bunker, M. T. *J. Chromatogr.Sci.* **1989**, *27*, 496.
91   Lee, Y. C. *Anal. Biochem.* **1990**, *189*, 151.
92   Lacourse, W. R. *Analusis* **1993**, 21, 181.
93   Honda, S.; Takahashi, M.; Kakehi, K.; Ganno, S. *Anal. Biochem.***1981**, *113*, 130.
94   Bonn, G.; Binder, H.; Leonhard, H.; Bobleter, O. *Monatsh. Chem.* **1985**, *116*, 961.
95   Voragen, A. G. J.; Schols, H. A.; Devries, J. A.; Pilnik, W. *J. Chromatogr.* **1982**, 244, 327.
96   Nebinger, P.; Koel, M.; Franz, A.; Werries, E. *J. Chromatogr.* **1983**, *265*, 19.
97   Churms, S. *J. Chromatogr.* **1992**, *51B*, 246–248.
98   Churms, S. C. *J. Chromatogr.* **1990**, *500*, 555.
99   Hardy, M. R.; Townsend, R. R. *Proc. Natl. Acad. Sci* USA **1988**, *85*, 3289.
100   Lee, Y. L. *Anal. Biochem.* **1990**, *189*, 151.
101   Johnson, D.C.; Dobberpuhl, D.; Roberts, R.; Vandeberg, P. *J. Chromatogr.* **1993**, *640*, 79.
102   Dionex Sunnyvale, CA, USA; Technical Note 1993 LPN 0342101 10 M 493.
103   Weiß, J. *Ionenchromatographie*; VCH: Weinheim, New York, 1991.
104   Johnson, D.C.; La Course, W. R. *Anal. Chem.* **1990**, *62*, 589 A.
105   Corradini, C.; Corradini, D.; Huber, C.; Bonn, G. *J. Chromatogr.* **1994**, *685*, 213.

106   Corradini, C.; Corradini, D.; Huber, C. G.; Bonn, G. K. *Chromatographia* **1995**, *41*, 511.

107   Syamanda, R.; Staples, R. C.; Bolck, R. J. *Contrib. Boyce Tompson Inst.* **1962**, *21*, 363.

108   Bauer, H.; Voelter, W. *Chromatographia* **1976**, *9*, 433.

109   Voelter, W.; Bauer, H. *J. Chromatogr.* **1976**, *126*, 693.

110   Sinner, M.; Puls, J. *J. Chromatogr.* **1978**, *156*, 197.

111   Honda, S.; Matsuda, Y.; Takahashi, M.; Kakehi, K. *Anal. Chem.* **1980**, *52*, 1079.

112   Honda, S.; Takahashi, M.; Kakehi, K.; Ganno, S. *Anal. Biochem.* **1981**, *113*, 130.

113   Honda, S.; Takahashi, M.; Nishimura, Y.; Kakehi, K.; Ganno, S. *Anal. Biochem.* **1981**, *118*, 162.

114   Honda, S.; Konishi, T.; Suzuki, S. *J. Chromatogr.* **1984**, *299*, 245.

115   Rendleman, J. A. *Advances in Chemistry*, Vol. 117 1973; American Chemical Society: Washington DC; p 51ff.

116   Rocklin, R. D.; Pohl, C. A. *J. Liq. Chromatogr.* **1983**, *6*, 1577.

117   Coleman, G. H.; Miller, A. *Proc. Iowa Acad. Sci.* **1942**, *49*, 257.

118   Böseken, J. *Adv. Carbohydr. Chem.* **1949**, *4*, 189.

119   Van Duin, M.; Peters, J. A.; Kieboom, A. P. G.; Van Bekkum, H. *Tetrahedron* **1985**, *41*, 3411.

120   Bell, C. F.; Beauchamp, R. D.; Sort, E. L. *J. Chromatogr.* **1989**, *185*, 39.

121   Vorndran, A. E.; Grill, E.; Huber, C. G.; Oefner, P. J.; Bonn, G. K. *Chromatographia* **1992**, *34*, 109.

122   Oefner, P. J.; Vorndran, A. E.; Grill, E.; Huber, C. G.; Bonn, G. K. *Chromatographia* **1992**, *34*, 308.

123   Grill, E.; Huber, C. G.; Oefner, P. J.; Vorndran, A. E.; Bonn, G. K. *Electrophoresis* **1993**, *14*, 1004.

124   Churms, S. C. *Adv. Carbohydr. Chem. Biochem.* **1970**, *25*, 13.

125   Dellweg, H.; Johan, M.; Trenel, G. *J. Chromatogr.* **1971**, *57*, 89.

126   Brown, W.; Andersson, Ö. *J. Chromatogr.* **1971**, *57*, 255.

127   Brown, W. *J. Chromatogr.* **1970**, *52*, 273.

128   John, M.; Trenel, G.; Dellweg, H. *J. Chromatogr.* **1969**, *42*, 197.

129   Schmidt, F.; Enevoldsen, B. S. *Carbohydr. Res.* **1978**, *61*, 197.

130   Sabbagh, N.; Fagerson, I. S. *J. Chromatogr.* **1973**, 86, 184.

131   Heyrand, A.; Rinaudo, M. *J. Chromatogr.* **1978**, *166*, 149.

132   Apffel, J. A. *J. Chromatogr.* **1981**, *206*, 43.

133   Noel, D.; Hanai T.; D' Amboise, M. *J. Liq.Chromatogr.* **1979**, *2*, 1325.

134   Fukano, K.; Komija, K.; Sasaki, H.; Hashimoto, T. *J. Chromatogr.* **1978**, *166*, 47.

135   Kato, Y.; Sasaki, H.; Aiura, M.; Hashimoto, T. *J. Chromatogr.* **1978**, *153*, 546.

136   Schwald, W.; Concin, R.; Bonn, G.; Bobleter, O. *Chromatographia* **1985**, *20*, 35.

137   Rabel, F. M.; Caputo, A. G.; Butts, E. T. *J. Chromatogr.* **1976**, *126*, 731.

138   Linden, J. C.; Lawhead, C. L. *J. Chromatogr.* **1975**, *105*, 125.

139   McGinnis, G. D.; Fang, P. *J. Chromatogr.* **1978**, *153*, 107.

140   Rocca, J. L.; Rouchouse, A. *J. Chromatogr.* **1976**, *117*, 216.

141   Baust, J. G.; Lee, R. E.; Rojas, R. R.; Hendrix, D. L.; Friday, D.; James, H. *J. Chromatogr.* **1983**, *261*, 65.

142   Slavin, J. L.; Marlett, J. A. *J. Agric. Food. Chem.* **1983**, *31*, 467.

143   Palmer, J. K.; Brandes, W. B. *J. Agric Food Chem.* **1974**, *22*, 709.

144   Ferreira, M. R.; Leitao, R.; Ferreira, M. A. *Rev. Port. Farm.* **1995**, *45*, 75.

145   Hettinger, J.; Majors, R. E. *Varion Instruments Applications* **1976**, *10*, 6.

146   Hughes, D. E. *J. Chromatogr.* **1985**, *331*, 183.

147   Anumula, R. K. *Methods Protein Struct. Anal.*; Plenum: New York, 1994; pp 195–206.

148   Rooyakkers van Eijk, D. R.; Deutz, N. E. *J. Chromatogr.* **1996**, *730*, 99.

149   Bonn, G. *J. Chromatogr.* **1987**, *387*, 393.

150   Macrae, R.; Dick, J. *J. Chromatogr.* **1981**, *210*, 138.

151   Macrae, R.; Tugo, L. C.; Dick, J.; *Chromatographia* **1982**, *15*, 476.

152   Bonn, G.; Bobleter, O. *Chromatographia* **1984**, *18*, 445.

153   Kissinger, P. T.; Heinemann, W. R. *Laboratory Techniques in Electroanalytical Chemistry*, 2nd Ed.; Marcel Dekker: New York, 1996.

154   Buchberger, W.; Winsauer, K.; Breitwieser, Ch. *Fresenius' Z. Anal. Chem.* **1983**, *315*, 315.

155   Buchberger, W. *Chromatographia* **1990**, *30*, 57.

156   Stitz, A.; Buchberger, W. *Fresenius' Z. Anal. Chem.* **1991**, *339*, 55.

157   Stitz, A.; Buchberger, W. *Electroanalysis* **1994**, *6*, 251.

158   Luo, Z. M.; Baldwin, R. P. *J. Electroanal. Chem.* **1995**, *387*, 87.

159   Marioli, J. M.; Kuwana, T. *Electroanalysis* **1993**, *5* 11.

160   Santos, L. M.; Baldwin, R. P. *Anal. Chem.* **1987**, *59*, 1766.

161   Prabhu, S. V.; Baldwin, R. P. *Anal. Chem.* **1989**, *61*, 852.

162   Prabhu, S. V., Baldwin, R. P. *Anal. Chem.* **1989**, *61*, 2258.
163   Zadeii, J. M.; Marioli, J. M.; Kuwana, T. *Anal. Chem.* **1991**, *63*, 649.
164   Casella, I. G.; Desimoni, E.; Salvi, A. M. *Anal. Chim. Acta* **1991**, *243*, 61.
165   Johnson, D. C.; LaCourse, W. R. *Anal. Chem.* **1990**, *62*, 589A.
166   LaCourse, W. R.; Johnson, D. C. *Electroanalysis* **1992**, *4*, 367.
167   Roberts, R. E.; Johnson, D. C. *Electroanalysis* **1994**, *6*, 193.
168   Roberts, R. E.; Johnson, D. C. *Electroanalysis* **1994**, *6*, 269.
169   Elliot, P. T.; Olsen, S. A.; Tallman, D. E. *Electroanalysis* **1996**, *8*, 443.
170   LaCourse, W. R.; Johnson, D. C. *Anal. Chem.* **1993**, *65*, 50.
171   LaCourse, W. R.; Johnson, D. C. *Carbohydr. Res.* **1991**, *215*, 159.
172   Johnson, D. C., LaCourse, W. R. In *Carbohydrate Analysis, J. Chromatogr.*, Library Vol 58; Ziad el Rassi, Ed.; Elsevier: Amsterdam, 1995.
173   Majors, R. E.; Barth, H.G.; Lochmüller, C. H. *Anal.Chem.* **1984**, *56*, 329.
174   Galensa, R. Z. *Lebensm. Unters. Forsch.* **1984**, *178*, 199.
175   Lehrfeld, J. *J. Chromatogr.* **1976**, *120*, 141.
176   Hull, S. R.; Turco, S. J. *Anal. Biochem.* **1985**, *146*, 143.
177   Takemoto, H.; Hase, S.; Ikenaka, T. *Anal. Biochem.* **1985**, *145*, 245.
178   Kwon Hyokjoon; Kim; Joon *J. Liq. Chromatogr.* **1995**, *18*, 1437.
179   Frei, R. W.; Jansen, H.; Brinkman, U. A. T. *Anal. Chem.* **1985**, *57*, 1529.
180   Hicks, K. B. *Adv. Carbohydr. Chem. Biochem.* **1988**, *4*, 67.
181   Vratny, P.; Brinkman, U. A. T.; Frei, R. W. *Anal. Chem.* **1985**, *57*, 224.
182   Tomer, K. B. In *HPLC Detection; Newer Methods*; Patonay, G., Ed.; VCH: New York, 1992; pp 163–195.
183   Games, D. F. In *Soft Ionization Biological Mass Spectrometry*; Morris, H. R., Ed.; Heyden: London, 1981; p 54.
184   Santikarn, S.; Ritter, G.-R.; Reinhold, V. N. *J. Charbohydr. Chem.* **1987**, *6*, 141.
185   Blakeley, C. R.; Vestal, M. L. *Anal. Chem.* **1985**, *55*, 750.
186   Barber, M.; Bordoli, R. S.; Sedwick, R. D.; Tyler, A. N. *J. Chem. Soc., Chem. Commun.* **1981**, 325.
187   Caprioli, R. M.; Fan, T.; Cottrell, J. S. *Anal. Chem.* **1986**, *58*, 2949.
188   Caprioli, R. M. *Anal. Chem.* **1990**, *62*, 477a.
189   Caprioli, R. M.; DaGue, B. B.; Wilson, K. *J. Chromatogr. Sci.* **1988**, *26*, 640.
190   Whitehouse, C.; Dreyer, R. N.; Yamashita, M.; Fenn, J. B. *Anal. Chem.* **1985**, *57*, 675.
191   French, J. B.; Thomson, B. A.; Davidson, W. R.; Reid, N. M.; Buckley, J. A. *In Mass Spectrometry in the Environmental Sciences*; Safe, S.; Karasek, F. W.; Hutzinger, O., Eds.; Pergamon: 1982.
192   Fenn, J. B. *J. Am. Soc. Mass Spectrom.* **1994**, *4*, 524.
193   Kambara, H.; Kanomata, I. *Mass Spectroscopy* **1976**, *24*, 229.
194   Rajakyla, E. *J. Chromatogr.* **1986**, *353*, 1.
195   Tomiya, N.; Kurono, M.; Ishihara, H.; Tejima, S.; Endo, S.; Arata, Y.; Takahashi, N. *Anal. Biochem.* **1987**, *163*, 489.
196   Wight, A. W.; van Niekerk, P. J. *Food Chem.* **1983**, *10*, 211.
197   Koizumi, K.; Utamura, T.; Kubota, Y.; Hizukiri, S. *J. Chromatogr.* **1987**, *409*, 396.
198   Verzele, M.; Simoens, G.; van Damme, F. *Chromatographia* **1987**, *23*, 292.
199   Derler, H.; Hörmeyer, H. F.; Bonn, G. *J. Chromatogr.* **1988**, *440*, 281.
200   Technical Note TN 20; Dionex; LPN 03421-01 489.
201   Hardy, M. R.; Townsend, R. R. *Proc. Natl. Acad. Sci. U.S.A.* **1988**, *85*, 3289.
202   Simpson, R. C.; Fenselau, C. C.; Hardy, M.; Townsend, R.; Lee, Y. C.; Cotter, R. *Anal. Chem.* **1990**, *62*, 248.
203   Boulenguer, P.; Leroy, Y.; Alonso, J. M.; Montreuil, J.; Ricart, G.; Colbert, C.; Duquet, D.; Dewaele, C.; Fournet, B. *Anal. Biochem.* **1988**, *168*, 164.
204   Kato, Y.; Numajiri, Y. *J. Chromatogr.* **1991**, *562*, 81.
205   Kohler, M.; Leary, J. A. *Anal. Chem.* **1995**, *67*, 3501.
206   Lamblin, E.; Klein, A.; Boersma, A.; Din, N.; Roussel, P. *Carbohydr. Res.* **1983**, *118*, 1.
207   Verzele, M.; Dewaele, D.; Mag, L. C.; **1985**, *3*, 22.
208   Clark, P. I.; Narasimhan, S.; Williams, J. M.; Clamp, J. R. *Carbohydr. Res.* **1983**, *118*, 147.
209   Hsieh, P.; Rosner, M. R.; Robbins, P. W. *J. Biol. Chem.* **1983**, *258*, 2548.
210   Thiem, J.; Sievers, A. *Fresenius' Z. Anal. Chem.* **1980**, *304*, 369.
211   Lamblin, G.; Boersma, A.; Klein, A.; Roussel, P.; Halbeek, H.; Vliegenhart, J. F. G. *J. Biol. Chem.* **1984**, *259*, 9051.
212   Frenz, J.; Horvath, Cs.; AIChE, J.; **1985**, *31*, 400.
213   Hostettmann, K.; Hostettmann, M.; Marston, A. *Preparative Chromatography Techniques; Application in Natural Product Isolation*; Springer: Berlin, 1986.

2.1.4. Gas Chromatography

Whilst paper and thin-layer chromatography are the methods of choice for the qualitative and semi-quantitative analysis of sugars, gas chromatography (GC) is preferable for the quantitative determination of sugars, especially in such cases where a multicomponent mixture is present (e.g., hydrolysis products of polysaccharides or glycoproteins).

Mono- and oligosaccharides are nonvolatile and must, therefore, be converted into volatile derivatives prior to analysis by gas chromatography. The following derivatives are suitable for this purpose:

1.   *O*-Methyl ethers.
2.   *O*-Trimethylsilyl ethers.
3.   *O*-Acetyl esters.
4.   *O*-Trifluoroacetyl esters.
5.   *O*-Butane borates.

2.1.4.1. Separation of Mono- and Oligosaccharides as Their per-*O*-Methylated Derivatives

The separation of a series of sugars as their per-*O*-methylated derivatives (D-xylose, L-arabinose, D-mannose, D-glucose, and D-galactose)[1] by gas chromatography was carried out for the first time on a 1.6 m Celite column coated with 25% apiezone.

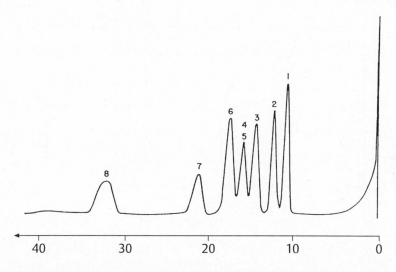

Figure 1.   Gas Chromatogram of the Separation of the Totally Methylated Glucose and Fructose Isomers
Column: L: 1.52 m, i.d.: 2 mm; *stationary phase*: 5% diethylene glycol succinate; *temperature*: 136 °C
1: methyl 2,3,4,6-tetra-*O*-methyl-β-D-glucopyranoside, **2**: methyl 1,3,4,6-tetra-*O*-methyl-α-D-fructopyranoside, **3**: methyl 1,3,4,6-tetra-*O*-methyl-β-D-fructopyranoside, **4**: methyl 2,3,4,6-tetra-*O*-methyl-α-D-glucopyranoside, **5**: methyl 1,3,4,6-tetra-*O*-methyl-α-D-fructopyranoside, **6**: methyl 2,3,5,6-tetra-*O*-methyl-D-glucofuranoside, **7**: methyl 2,3,5,6-tetra-*O*-methyl-D-glucofuranoside, **8**: methyl 1,3,4,6-tetra-*O*-methyl-β-D-fructopyranoside (Gee and Walker; with permission)[2]

Single sugars are separated into their anomeric compounds; hence, in general, more peaks are obtained per individual product in the gas chromatogram. Figure 1 demonstrates the separation of per-*O*-methylated glucose and fructose into their different anomeric forms[2]. The same effects also occur with disaccharides[3].

For practical purposes, the methylation of free sugars is not suitable since more than one step is necessary for complete methylation. The method of methylation is useful, however, for the elucidation of the structures of polysaccharides. After hydrolysis of the per-*O*-methylated polysaccharide the partially methylated monosaccharides have to be separated and determined quantitatively. The preferred method here is gas chromatography.

The separation of completely methylated sugar alcohols of mono- and oligosaccharides have also been studied. They are obtained by reduction of the corresponding sugars with sodium borohydride and methylated according to the method of Hakomori[4-8]. An alternative method is the procedure of Ciucanu and Kerek which uses dimethyl sulfoxide, iodomethane and solid sodium hydroxide[9,10]. In Table 1 the relative retention times of the sugar alcohols of the mono- and some oligosaccharides are listed (see also Figure 2).

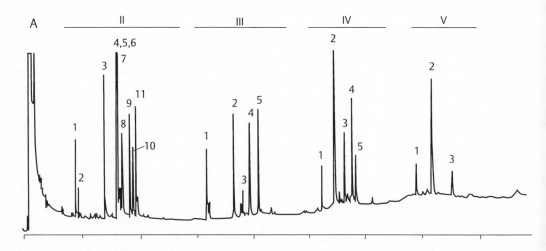

Figure 2.   Gas Chromatogram of the Separation of Permethylated Oligosaccharide Alditols
Quartz capillary: L: 12 m, i.d.: 0.25 mm; *stationary phase*: DB-1; *temperature*: 150–330 °C, 5 °C/min
Alditols of: disaccharides (zone II): **1**: xylobiose, **2**: sucrose, **3**: trehalose, **4**: cellobiose, **5**: maltose, **6**: lactose, **7**: laminaribiose, **8**: sophorose, **9**: isomaltose, **10**: melibiose, **11**: gentiobiose; trisaccharides (zone III): **1**: xylotriose, **2**: raffinose, **3**: maltotriose, **4**: laminaritriose, **5**: isomaltotriose; tetrasaccharide (zone IV): **1**: xylotetraose, **2**: stachyose, **3**: maltotetraose, **4**: isomaltotetraose, **5**: laminaritetraose; pentasaccharides (zone V): **1**: xylopentaose, **2**: isomaltopentaose, **3**: laminaripentaose (Wei-Tong Wang, Matsuura and Sweeley[5]; with permission)

Table 1. Relative Retention Times and Retention Indices of Per-*O*-methylated Sugar Alcohols

Compound	A	B	C[a]	D	E	F
meso-Erythritol	0.20	-	-	0.26	0.26	-
Adonitol (Ribitol)	-	-	-	0.70	0.60	-
Arabitol	0.47	0.48	-	0.95	1.05	-
Xylitol	0.49	0.49	-	1.00	1.00	-
Rhamnitol	0.62	0.68	-	1.05	1.25	-
Fucitol	-	-	-	1.21	1.47	-
Mannitol	0.98	0.97	-	2.80	2.84	-
Sorbitol	1.00	1.00	-	2.82	2.46	-
Galactitol	1.18	0.95	-	3.41	3.88	-
Sophoritol	-	-	2329	-	-	-
Laminaribiitol	-	-	2306	-	-	0.95
Cellobiitol	-	-	2292	-	-	0.92
Maltitol	-	-	2295	-	-	1.00
Gentiobiitol	-	-	2420	-	-	-
Isomatitol	-	-	2380	-	-	1.48
Xylobiitol	-	-	2024	-	-	-
Lactitiol	-	-	2297	-	-	1.20
Melibiitol	-	-	2400	-	-	1.82
Isomaltotriitol	-	-	3227	-	-	-
Maltotriitol	-	-	3095	-	-	-
Laminaritriitol	-	-	3171	-	-	-
Xylotriitol	-	-	2869	-	-	-
Maltotetraitol	-	-	3914	-	-	-
Isomaltotetraitol	-	-	4017	-	-	-

[a]Retention indices instead of relative retention times.

Index	Stationary Phase	Temperature (°C)	Dimension of the Column length (cm)	i.d. (cm)	Reference Substance	Literature
A	20% Apiezon L on Celite 545	180 Isothermal	100 Steel	0.4	Permethyl-sorbitol	Ovodov, Evtushenko[4]
B	15% Polyethylene glycoladipate on Chromosorb W	152 Isothermal	100 Steel	0.4	Permethyl-sorbitol	Ovodov, Evtushenko[4]
C	Quartz-capillary chem. modified with DB 1	150–330 5 °C/min	1 200	0.04	Permethyl-sorbitol	Wei-Tong Wang, Matsuura, Sweeley[5]
D	1% OV-17 on Gas-Chrom Q	130 Isothermal	180 Alumina	0.45	Permethyl-xylitol	Whyte[7,8]
E	3% QF-1	110 Isothermal	180	0.45	Permethyl-xylitol	Whyte[7,8]
F	1.5% XE-60 1.5% EGS	195 Isothermal	180	0.45	Permethyl-maltitol	Whyte[7,8]

Methylation Procedures

Method of Hakomori[4,6]
Dimethyl sulfoxide reacts with sodium hydride yielding methylsulfinyl carbanions (I).

$$CH_3S(=O)CH_3 + NaH \longrightarrow CH_3S(=O)CH_2^- + H_2 \qquad (I)$$

This system is strongly reactive towards compounds with free hydroxy groups giving the corresponding anions (II), which can be methylated by treatment with methyl iodide(III).

$$CH_3S(=O)CH_2^- + ROH \longrightarrow RO^- + CH_3S(=O)CH_3 \qquad (II)$$

$$RO^- + CH_3I \longrightarrow ROCH_3 + I^- \qquad (III)$$

Dissolve 50–100 mg of the carbohydrate in 5 mL of dry dimethyl sulfoxide; under a stream of nitrogen add this solution to 5 mL of a solution of methylsulfinyl carbanion in dimethyl sulfoxide (prepared by dissolving 0.1 g sodium hydride in 5 mL dry dimethyl sulfoxide at 65 °C and stirring for 30 min). After 10 min stirring at 20 °C add 0.5 mL of methyl iodide; stir again for 20 min, add 20 mL of chloroform, and extract the organic phase four times with 10 mL water. Dry the organic phase with anhydrous sodium sulfate.

Potassium *tert.*-butoxide has also been recommended instead of sodium hydride. This reagent tends to give less background noise in the GC[11].

Ciucanu and Kerek Procedure[9]
To 0.2 mg (0.3–2.0 mg) of the sample add 0.5 (1.0) mL of dimethyl sulfoxide, 0.5 (1.0) mL of iodomethane and about 25 (50) mg of NaOH powder; stir the mixture with a magnetic bar for 10 min and quench by adding 2 (4) mL of water and 1 (2) mL of chloroform. Remove the water phase and wash the chloroform phase 5 times with 2 (4) mL water; dry the organic phase with anhydrous sodium sulfate. The yields are superior to those obtained by the procedure of Hakomori.

This classic permethylation method has gained importance in the field of GC–MS for elucidating the structures of oligosaccharides of glycoproteins and glycosphingolipids up to molecular weight of 2200 Dalton (up to 11 monosaccharide units) (GC conditions: fused silica column 10 m x 0.25 mm i.d.; coated with cross-linked SE-54; temperature: 70–370 °C, 10 °C/min)[12].

2.1.4.2. Separation of the Mono- and Oligosaccharides Using Their *O*-Trimethylsilyl Ethers
Conversion of the polyhydroxy compounds into their *O*-trimethylsilyl ethers, compounds of strongly enhanced volatility, allows successful separation by GC. Simple procedures for carrying out these derivatization methods are of great importance in the separation of organic hydroxy compounds such as carbohydrates and related compounds. The first papers concerning this reaction were published by Hedgley and Overend[13], and Bayer and Witsch[14]. A simple method for quantitative derivatization has been developed[15] and this has become important in practice. The background paper for the analysis of the trimethylsilyl ethers of mono- and oligosaccharides by gas chromatography has been published by Sweeley and co-workers[16].

The conversion of sugars into their O-trimethylsilyl ethers is carried out, at present, with the following reagents:

-   Hexamethyldisilazane (HMDS)–trimethylchlorosilane (TMSCl)[15,16].
-   Bis(trimethylsilyl)acetamide (BSA)[17].
-   N-Methyl-N-trimethylsilylacetamide (MSA)[18].
-   Bis(trimethylsilyl)trifluoroacetamide (BSTFA)[19].
-   N-Methyl-N-trimethylsilyltrifluoroacetamide (MSTFA)[20].
-   N-Trimethylsilylimidazole (TSIM)[21].
-   N-Methyl-N-trimethylsilylheptafluorobutyramide (MSHFBA)[22].

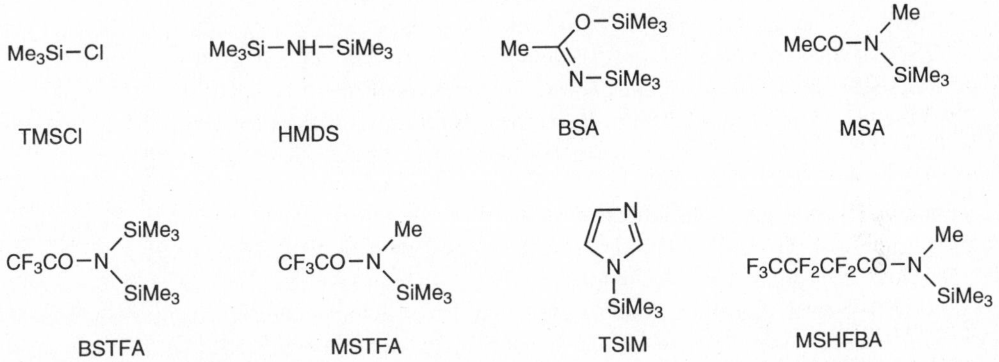

Chemical Structures of Silylation Agents

The oldest and until now most applied method for converting carbohydrates into their O-trimethylsilyl ethers is that of Sweeley et al.[16] using the following general procedure.

Procedure

Dry the sample in a closeable reaction tube containing approx. 10 mg sugar by several additions and evaporation in vacuo of anhydrous 2-propanol; dissolve the residue in dry pyridine (stored over solid KOH). Add 0.2 mL of hexamethyldisilazane (HMDS) and 0.1 mL of trimethylchlorosilane (TMSCl) and shake intensively for 30 min; after standing for 5 minutes at room temperature inject an aliquot (e.g., with a Hamilton syringe) into the gas chromatogram.

Note: The reaction with monosaccharides is complete within a short time. Ketoses react slower than aldoses, and oligosaccharides need more time; for example, maltose needs 30 min for complete reaction[23]. An incomplete reaction can be recognized by the appearance of more peaks in the gas chromatogram for one sugar than is expected[24].

It is important that the sugars are completely dissolved; for this purpose it is sometimes necessary, such as in the case of oligosaccharides or inositol, to carry out the silylation at an elevated temperature[25,26]. A disadvantage of this reaction is the formation of a white precipitate, which can block the syringe.

The other silylation reagents mentioned above have been used for the silylation of sugars. In several cases the reaction is incomplete at room temperature. Extensive studies have been performed by Hadorn and Zuercher[27,28] to find optimal conditions for the quantitative silylation of sugars. The reaction with fructose is particularly difficult, since beside the expected peaks for the α- and β-forms at least one or two side peaks appear; one of them has the same retention time as α-glucose on the commonly used OV-17 phase.

The following conditions have been found to be optimal for the silylation reagents mentioned above with regard to quantitative reaction of an absolute sugar concentration of approximately 10 mg.

- Hexamethyldisilazane (HMDS).

1 mL pyridine, 0.2 mL HMDS, 0.1 mL TMSCl; 30 min; 50 °C[16,29].

- *N*-Methyl-*N*-trimethylsilylacetamide (MSA).

1 mL pyridine, 0.4 mL MSA, 0.5 mL TMSCl; 60 min; 50 °C[30].

- *N*-Methyl-*N*-trimethylsilyltrifluoroacetamide (MSTFA).

0.5 mL pyridine, 0.5 mL MSTFA, 0.1 mL TMSCl; at room temperature or 60 min; 50 °C[27].

- *N*-Methyl-*N*-trimethylsilylheptafluorobutyramide (MSHFBA).

0.5 mL pyridine, 0.5 mL MSHFBA, 0.1 mL TMSCl; 60 min; 50 °C[27,31].

- Bis(trimethylsilyl)acetamide (BSA) combined with trimethylsilylimidazole (TSIM)[28].

(a) 0.5 mL TSIM, 0.3 Ml BSA, 0.05 mL TMSCl at room temperature (overnight).

(b) 0.5 mL ethyl methyl ketone, 0.5 mL BSA; shaking till all substances are dissolved; addition of 1 mL mixture (1 mL pyridine and 0.25 mL TSIM); heat under reflux at 100 °C overnight.

One further disadvantage of the GC separations of TMS sugars is the breakdown of the flame-ionization detector (FID). The combustion of these substances in the hydrogen flame yields silicon dioxide, which deposits on both electrodes and causes a strong reduction in the sensitivity of the FID. It is possible to prevent this effect either by addition of fluorine-containing compounds to the silylation mixture or by the application of fluorine-containing silylation agents such as MSHFBA; this reagent gave the best results and between 50–100 injections can be carried out without the need to clean the detector; in the case of MSTFA or BSTFA this is not possible[31].

When sugars in solution need to be separated by GC as their *O*-TMS ethers it must be recognized that they exist in their anomeric forms (α- and β-furanoside, α- and β-pyranoside and the open carbonyl form), which are in equilibrium. For monosaccharides up to five peaks can be obtained[24]. The adjustment of the equilibrium is accelerated in pyridine or *N,N*-dimethylformamide by the addition of 2-hydroxypyridine as a catalyst[32].

When crystalline α-D-glucose is dissolved in water or when α-D-glucose is liberated by the enzymatic hydrolysis of sucrose and the following mutarotation is interrupted by deep-freezing and lyophilization, the silylation of the dry residue after 15 seconds shows only the peak of α-D-glucopyranose in the GC chromatogram. After approximately 3 minutes, a second peak of β-D-glucopyranose appears and the peak of α-D-glucopyranose diminishes; at equilibrium, both peaks have the same area. For D-fructose in solution at its equilibrium, up to five peaks of the *O*-TMS ethers have been obtained which belong to the totally silylated pyranoside and furanoside forms and to a totally silylated ketal form (identified by IR spectroscopy and mass spectrometry[24]).

For D-sorbose one pyranoside, one furanoside and the open-chain structure can be detected, and for D-galactose the two pyranoside and one furanoside forms have been found at the equilibrium of mutarotation[24].

Stationary Phases

Nonpolar Phases

Nonpolar phases are used for the separation of mixtures of sugars with different molecular weights as their O-trimethylsilyl ethers which also have different volatilities. The separation of compounds with equal molecular weights such as diasterereoisomeric forms of aldoses are not satisfactory and for total resolutions the techniques of capillary gas chromatography are necessary (Figure 3).

Such phases are invariably silicon compounds which have been proved to be particularly temperature stable. The most common commercial products are:

– Methyl silicon rubber, trade term: SE-30[34–39], OV-1[40].
– Methyl silicon oil, trade term: OV-101[34,39,41].
– Phenylmethyl silicones, trade term: SE-52[16,32,39,42–44], OV-17[38,40], OV-25[40], SE-54[45].
– Fluoro silicon oil, trade term: QF-1[35].
– Methylvinyl silicon rubber, trade term: UCCW-982[47].
– Nitrile silicon rubber, trade term: XE-60[33,41].

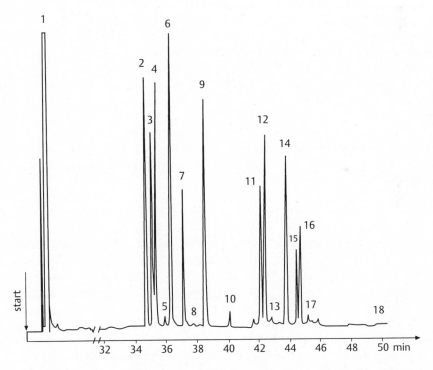

Figure 3. Gas Chromatogram of a Mixture of Pentoses and Hexoses as O-Trimethylsilyl Ethers
Capillary column: L: 16 m, i.d.: 0.17 mm; *stationary phase*: OV-101; *temperature*: 130–195 °C, 3.5 °C/min
1: solvent, **2**: lyxose, **3**: α-arabinose, **4**: rhamnose, **5**: arabinose, **6**: ribose, **7**: ribose, **8**: arabinose, **9**: α-xylose, **10**: β-xylose, **11**: mannose, **12**: fructose, **13**: tagatose, **14**: galactose, **15**: tagatose, **16**: α-glucose, **17**: mannose, **18**: β-glucose (Tesarik[41]; with permission)

Polar Phases

Polar phases are applied for separations within a small range of molecular weights. The following compounds have been approved:

– Ethylene glycol polysuccinate[16,24,48].
– Carbowax 20M[49].
– Tetracyanoethylpentaerythrol[50].

Complex sugar mixtures are formed by hydrolysis of a great number of polysaccharides, for example, thickening agents tragacanth and gum arabic, etc., yielding different neutral monosaccharides and uronic acids. The separation of such mixtures by GC as their O-trimethylsilyl ethers on packed columns gives very complex gas chromatograms where it is often impossible to allocate single peaks to the corresponding sugars as a result of overlapping. A complete resolution is only possible by capillary gas chromatography.

In the following tables (Tables 2 and 3) the relative retention times for a large number of mono- and oligosaccharides on different stationary phases are listed; these have been described in the literature.

Methanolysis

The hydrolysis of polysaccharides with aqueous mineral acids leads to the formation of side products (e.g., furan derivatives) which appear as auxiliary peaks in the gas chromatogram. This effect causes considerable losses.

A milder procedure is the cleavage of polysaccharides with alcohol–acid mixtures, where the corresponding glycosides are formed. The most frequently applied technique is the reaction with methanolic hydrochloric acid[51,52], so called methanolysis using the following procedure.

Procedure

Place 1–20 mg of the polysaccharide into a thick-walled test tube (e.g., Sovirel vials with Teflon screw caps), add 1 mL of methanolic hydrochloric acid (2.0 mol/L or 0.5 mol/L), seal the tube and heat it for 4 h at 100 °C; after cooling add 40 µL of pyridine and evaporate the solution in a rotatory evaporator under vacuo at 40 °C to dryness.

Table 4 lists the stability of individual sugar compounds towards treatment with methanolic hydrochloric acid at different concentrations, temperatures and times[53].

For polysaccharides, which are used as thickening agents, cleavage with 2 M methanolic acid at 100 °C for 4 hours is recommended. Polysaccharides consisting of neutral sugar units are cleaved almost quantitatively, whilst uronic acid containing products up to two-thirds. 3,6-Anhydro-L(D)-galactose, the units of agar and carrageenan, is destroyed by this procedure. For their determination, methanolysis must be carried out with 0.5 M methanolic hydrochloric acid[51].

After methanolysis the hydrochloric acid has to be removed. Evaporating with a rotatory evaporator leads to losses of 8–26% for monosaccharides. Addition of pyridine before evaporating diminishes the losses. The optimal way to prevent this effect is neutralization with silver carbonate.

Table 2. Relative Retention Times of *O*-Trimethylsilyl Ethers of Monosaccharides (equil = equilibrium after mutarotation; m = major, s = small)

Compound	A	B	C	D[a]	E	F	G
Erythrose	0.10	0.29	-	-	0.13	-	-
	0.12	0.35	-	-	-	-	-
	0.14	-	-	-	-	-	-
Arabinose	0.28(s)	0.97	-	0.31	0.19	0.284	0.236
(equil)	0.33(m)	1.31(m)	-	-	0.22(m)	0.294	0.273
	0.38	1.10(s)	-	-	0.26	0.332	0.345
	-	-	-	-	0.27	0.399	-
D-Xylose	0.31(s)	1.64	-	0.45	0.30	0.462	0.272
(equil)	0.43	2.11(m)	-	0.57	0.37(m)	0.560	0.345
	0.54(m)	-	-	-	-	-	0.425
	-	-	-	-	-	-	0.563
D-Ribose	0.27	1.22(m)	-	0.36(m)	0.23	0.325	0.330
(equil)	0.32(m)	1.33	-	0.37	0.24	0.337	0.378
	0.35	1.48(s)	-	-	0.25(m)	0.350	0.421
	-	-	-	-	0.15	0.395	-
D-Lyxose	0.26(m)	0.94(m)	-	0.31(m)	0.21(m)	0.282	0.257
(equil)	0.33	1.26	-	0.38	0.26	-	0.345
	-	1.42	-	-	-	-	0.421
D-Allose	0.76	-	0.78	0.72	-	-	-
(equil)	0.81(m)	-	0.91(m)	0.76(m)	-	-	-
	0.91	-	1.20(s)	0.82	-	-	-
D-Altrose	0.65(m)	-	0.63(m)	0.64(m)	-	-	-
(equil)	0.68	-	0.75	0.67	-	-	-
	0.94(s)	-	1.16(m)	0.82	-	-	-
D-Galactose	0.76(s)	-	0.91(s)	0.72(m)	0.46	0.752	0.743
(equil)	0.88	-	1.03	0.80	0.55	0.893	0.912
	1.08	-	1.38(m)	0.90(m)	0.67	1.060	1.212
	-	-	-	-	0.71(m)	-	-
D-Glucose	1.00	-	1.00	0.86	0.58	1.000	1.000
(equil)	1.57	-	1.94	1.13	1.00	1.520	1.834
D-Gulose	0.66	-	0.84	0.65	-	-	-
(equil)	0.74(m)	-	0.95	0.71(m)	-	-	-
	0.95(s)	-	2.11(s)	1.00	-	-	-
D-Mannose	0.70(m)	-	0.62(m)	0.66	0.42(m)	0.701	0.657
(equil)	1.08	-	1.31	0.89	0.69	1.070	1.147
D-Talose	0.86(m)	-	1.06(s)	0.74	-	-	-
(equil)	1.00(s)	-	1.22(m)	0.78(m)	-	-	-
	1.13	-	1.51	0.91	-	-	-
2-Deoxyribose (equil)	0.16	0.49	-	-	0.20	-	-

Table 2.   (Continued)

Compound	A	B	C	D[a]	E	F	G
L-Rhamnose	0.30	-	0.20	0.33	0.21(m)	-	0.255
(equil)	-	-	-	0.42(m)	0.28	-	0.321
	-	-	-	-	-	-	0.365
L-Fucose	0.33	-	0.25	0.35	0.20	-	-
(equil)	0.38(m)	-	-	0.39	0.26	-	-
	0.45	-	-	0.45(m)	0.30(s)	-	-
	-	-	-	-	0.32(m)	-	-
2-Deoxygalactose	0.42	-	0.59	-	-	-	-
(equil)	0.45	-	-	-	-	-	-
	0.53(m)	-	-	-	-	-	-
2-Deoxyglucose	0.16(s)	-	-	-	0.26	-	-
(equil)	0.25	-	-	-	0.40	-	-
	0.46	-	-	-	-	-	-
	0.64(m)	-	-	-	-	-	-
D-Fructose	-	-	-	-	0.41	0.704	0.706
(equil)	-	-	-	-	0.42(m)	0.721	0.823
	-	-	-	-	0.44	0.752	1.082
	-	-	-	-	0.51(s)	1.02	1.372
L-Sorbose	-	-	-	-	0.34	0.521	0.586
(equil)	-	-	-	-	0.44(m)	0.668	0.851
	-	-	-	-	0.51	0.891	1.291
	-	-	-	-	0.67	-	-

[a]Only the significant peaks.

Index	Stationary Phase	Temperature (°C)	Dimension of the Column length (cm)	i.d. (cm)	Reference Substance	Literature
A	3% SE-52 on acid washed silan. Chromosorb W (80–100 mesh)	140 Isothermal	183 Steel	0.6	α-Gluco-pyranose	Sweeley, Bentley, Makita, Wells[16]
B	15% EGS on acid washed silan. Chromosorb W (80–100 mesh)	140 Isothermal	244 Steel	0.6	O-Methyl-arabino-pyranoside	Sweeley, Bentley, Makita, Wells[16]
C	15% EGS on acid washed silan. Chromosorb W (80–100 mesh)	150 Isothermal	244 Steel	0.6	α-Gluco-pyranose	Sweeley, Bentley, Makita, Wells[16]
D	3–3.8% SE-30 on Diatoport S	140–200 0.5 °C/min	250 Glass	0.32	Mannitol	Bhatti, Chamber, Clamp[36]
E	1.2% XE-60 on Chromosorb W-DMCS	150	300-600	0.32	β-Gluco-pyranose	Ellis[33]
F	OV-101: Glass capillary	160	1 600		α-Gluco-pyranose	Tesarik[41]
G	XE-60: Glass capillary		2 800		α-Gluco-pyranose	Tesarik[41]

Table 3.   Relative Retention Times of O-Trimethylsilyl Ethers of Oligosaccharides

Compound	A	B	C	D	E	F
Sucrose	10.4	1.00	1.00	1.00	1.00	1.000
Lactose	10.5	0.99	1.00	1.09	1.10	1.000
	-	1.58	1.50(m)	1.54	1.44	1.445(m)
Turanose	10.0(s)	-	1.28	1.28(m)	1.21(m)	1.236^a
	12.7	-	-	1.56	1.47	-
Maltose	11.7	1.17	1.12	1.19	1.18	1.119
(equil)	13.1(m)	1.36	1.30(m)	1.33	1.26(m)	1.273(m)
αα'-Trehalose	9.0	1.37	1.34	1.38	-	1.330
	13.5	-	-	-	-	-
Cellobiose	11.9(s)	1.18	1.15	1.22	1.21	1.145
	16.6	1.77	1.67	1.70	1.57	1.579
Melibiose	15.1	-	2.02	2.16	1.98	1.905
	19.0	-	2.26	2.33	2.08	2.050
	20.0	-	-	-	-	-
Gentiobiose	22.6	2.56	2.02	2.36	2.14	2.240
	-	2.68	2.50(m)	2.76(m)	2.40(m)	2.255(m)
Mannobiose	-	1.22	1.21	1.18(m)	1.30(m)	-
	-	1.64	1.59	1.52	1.40	-
Palatinose	-	1.37	1.25	1.39	1.34	1.383^a(m)
	-	1.49	1.35	1.48(m)	1.40	-
	-	-	1.43(m)	1.79	1.70	1.630
Laminaribiose	-	1.57	1.56	1.64	1.53	1.514
	-	1.77	1.76	1.80	1.62	1.670(m)
Sophorose	-	1.59	1.59	1.66	1.57	-
	-	1.86	1.85(m)	1.99(m)	1.82(m)	-
Isomaltose	-	1.96	2.02	2.37	2.14	1.977
	-	2.36	2.48	2.77	2.42(m)	2.346
ββ-Trehalose	-	-	1.77	1.90	1.70	-
Kojibiose	-	-	1.38	1.40	1.31	1.330
	-	-	1.69	1.82	1.65	1.639
Lactulose	-	-	1.00	0.94	0.97	0.927
	-	-	-	-	-	0.959
Primverose	-	-	1.49	1.92	1.68	-
	-	-	-	1.96	1.78	-
Raffinose	99.0	1.00	-	-	-	-
Melizitose	120.5	-	-	-	-	-
Planteose	135.5	-	-	-	-	-
Gentianose	138.0	-	-	-	-	-
Isomaltofriose	-	1.87	-	-	-	-
Mannotriose	-	1.22	-	-	-	-
Fucolactose	-	0.82	-	-	-	-
Stachyose	-	1.00	-	-	-	-
Isomaltotetraose	-	1.67	-	-	-	-

^{a}Peaks not resolved.

Table 3. (Continued)

Index	Stationary Phase	Temperature (°C)	Dimension of the Column length (cm)	i.d. (cm)	Reference Substance	Literature
A	3% SE-52 on acid washed silan. Chromosorb W (80–100 mesh)	210 Isothermal	183 Steel	0.6	α-Glucose abs. time 1.1 min	Sweeley, Bentley, Makita, Wells[16]
B	3–3.8% SE-30 on Diatoport S	200[a] 300[b] Isothermal	250[a] 100[b] Glass	0.32	Sucrose[a] Raffinose[c] Stachyose[d]	Bhatti, Chambers, Clamp[36]
C	3% OV-1 on Chromosorb W (80–100 mesh)	228 Isothermal	270 Steel	0.32 (OD)	Sucrose	Haverkamp, Kamerling, Vliegenhart[40]
D	3% OV-17 on Chromosorb W (80–100 mesh)	228 Isothermal	270 Steel	0.32 (OD)	Sucrose	Haverkamp, Kamerling, Vliegenhart[40]
E	3% OV-25 on Chromosorb W (80–100 mesh)	228 Isothermal	270 Steel	0.32 (OD)	Sucrose	Haverkamp, Kamerling, Vliegenhart[40]
F	SE-54 on Quartz-capillary	240 Isostherm	3 000	0.026	Sucrose	Nikolov, Reilly[46]

[a]Disaccharides. [b]Tri-, Tetra-, Pentasaccharides. [c]Trisaccharides. [d]Tetrasaccharides.

Table 4. Dependence of the Loss of Sugars Upon Concentration of HCl and Reaction Time During Methanolysis

Compound	Concentration of Methanolic HCl (M)	Time (h)	Temperature (°C)	Loss (%)
Neutral sugars	1	24	100	0
	2	24	85	0
	4	24	100	33–65
	6	24	100	45–80
Uronic acid	1	24	100	9
	2	24	85	3
	4	24	100	45
	6	24	100	64
Acetamido hexoses	4	24	100	27
	6	24	100	35
N-Acetylneuraminic acid	4	24	100	96
	6	24	100	100

For trimethylsilylation of methyl glycosides the following technique has been reported[51]: Dissolve the residue in 0.5 mL of dry pyridine (stored over solid KOH) and add 0.25 mL BSA, 0.05 mL TMSCl and 0.25 mL 1,1,2-trichlorotrifluoroethane (TCFE). Close the vessel and, after shaking, heat for 1 hour at 60 °C. After cooling, inject an aliquot into the gas chromatograph.

Table 5 lists the relative retention times for the individual methyl glycosides of the monosaccharides. In this case four peaks are also obtained for a single sugar according to the different anomeric structures of the methyl glycosides.

Table 5.   Relative Retention Times of O-Trimethylsilyl Ethers of O-Methyl Glycosides

Compound	A	B	C[a]	Compound	A	B	C[a]
Erythrose	-	-	-	Altrose	-	-	0.42
Arabinose	0.29	-	0.22		-	-	0.49
	0.30	-	0.23		-	-	0.57
	0.35	-	-		-	-	0.66
Xylose	0.43	0.36	0.37	Galactose	0.68	0.57	0.61
	0.47	0.39	0.37		0.74	0.60	0.68
Ribose	-	-	0.24		0.75	0.65	0.74
	-	-	0.25		0.80	-	-
Lyxose	-	-	0.24	Glucose	0.85	0.67	0.79
	-	-	0.28		0.90	0.71	0.86
Rhamnose	0.33	-	0.26	Mannose	0.67	0.535	0.59
	0.34	-	-		0.70	-	0.65
Fucose	0.34	0.295	0.25	Gulose	-	-	0.37
	0.37	0.30	0.28(m)		-	-	0.51
	0.40	0.34	0.31		-	-	0.55
	0.41	-	-		-	-	0.66
Allose	-	-	0.53	Talose	-	-	0.65
	-	-	0.55(m)		-	-	0.70
	-	-	0.60				
	-	-	0.63				

[a]Only the significant peaks.

Index	Stationary Phase	Temperature (°C)	Dimension of the Column length (cm)	i.d. (cm)	Reference Substance	Literature
A	SE-30, 0.05% coating Glass Capillary	120–190 4 °C/min	3 500	0.03	D-Sorbitol	Preuss, Thier[51]
B	SE-30, 3% on Chromosorb WAW-DMCS (100-120 m)	110–200 2 °C/min	200 Glass	0.3	myo-Inositol	Yokota, Mori[52]
C	SE-30, 3–3.8% on Diatoport S	140–200 0.5 °C/min	250 Glass	0.32	Mannitol	Bhatti, Chambers, Clamp[36]

Modification of Sugars before Trimethylsilylation

Since several peaks are observed in the gas chromatogram of a single compound, this can make evaluation of mixtures difficult. To improve this situation, the sugars can be converted to other derivatives before the trimethylsilation which yield a reduction in the number of peaks in the gas chromatogram.

Oximation

Reaction of reducing sugars with hydroxylamine produces the open-chain sugar oximes, whose TMS derivatives exist in the *cis(Z)*- and *trans(E)*-form. For each of the reducing sugars two peaks appear in the gas chromatogram although some stationary phases have identical retention times.

The following modifications of the procedure exist.

(a) In Pyridine

Dissolve 10 mg of the dry sugar mixture in 1 mL of dry pyridine, add ~15 mg of hydroxylamine hydrochloride and heat for 30–60 min in a closed vessel at 70–80 °C. Under these conditions the conversion to the open-chain sugar oximes is quantitative. This solution can be used directly for silylation[16,54].

According to another procedure these high temperatures are not necessary; treatment for two hours at 30 °C in an ultrasound bath or mechanical shaking at room temperature is enough for quantitative conversion[55].

In some procedures the pyridine is evaporated before addition of the silylation reagent[56–58].

(b) In Methanol

Dissolve the dry sugar mixture (~ 35 mg) in 2 mL of absolute methanol in pressure resistant, lockable vessels. Add 60 mg of water-free sodium acetate and 50 mg of hydroxylamine hydrochloride, seal the vessel and heat for 60 min at 80 °C. After cooling, evaporate 1 mL of the solution in a rotatory evaporator to dryness and silylate the dry residue[31].

For the silylation the following silylation reagents have been recommended: HMDS–TMSCl[16,56], BSTFA–TMSCl[55], MSHFBA–TMSCl[31], HMDS–TFA[57,58].

In the following tables (Tables 6 and 7) the relative retention times of the common sugars on different stationary phases are listed.

For disaccharides, the appearance of a few very small peaks in addition to the two main peaks have been reported which have been assumed to be cyclic oximes[57].

Quantitative Determination in Food

The application of special stationary phases enables the appearance of only one peak for each of the food-relevant mono- and oligosaccharides in the gas chromatogram which facilitates their quantitative determinations. For this purpose the following method has been described[22].

1.   Standard mixture of glucose, fructose, sucrose, lactose, maltose and raffinose in water–methanol 1:1 v/v containing 10 mg of each compound per 1 mL.

2.   Reaction vessel: thick-walled reaction tube with screw cap and connection piece to a rotatory evaporator.

3.   Procedure.

(a) Evaporation of the Sugar Solution

Pipette 0.2 mL of the standard sugar solution into the reaction tube and evaporate the liquid by means of the rotatory evaporator under vacuo at 70 °C to a syrupy consistency; add ~0.5 mL of 2-propanol and evaporate under vacuo to dryness.

For food samples, pipette an aliquot of the aqueous or methanolic or ethanolic extract into the tube and carry out the same procedure.

In some cases the extract has to be clarified to remove proteins and polysaccharides (e.g., by the method Carrez, see page 106)

Table 6. Relative Retention Times of *O*-Trimethylsilyl Ethers of Monosaccharide Oximes

Compound	A	B	C	D	E
Glyceraldehyde	0.06	0.09	0.06	-	0.78
	-	0.08	-	-	1.00
D-Erythrose	0.16	0.23	0.15	-	2.28
		0.20	-	-	2.78
D-Threose	-	-	-	-	2.95
	-	-	-	-	3.00
L,D-Arabinose	0.43	0.53	0.40[a]	0.23	-
D-Xylose	0.42	0.55	0.40	-	-
D-Ribose	0.48	0.60	0.46[a]	0.26	-
D-Lyxose	-	-	-	-	-
L-Rhamnose	0.58	0.68	0.58[a]	-	-
D-Fucose	0.58[a]	0.69	0.57	-	-
	-	-	0.65	-	-
D-Glucose	1.34	1.55	1.32	0.73	-
	-	-	1.50	0.77	-
D-Mannose	1.30[a]	1.45	1.31	-	-
	-	-	1.51	-	-
D-Allose	-	-	-	-	-
D-Altrose	-	-	-	-	-
D-Idose	-	-	-	-	-
D-Gulose	-	-	-	-	-
D-Talose	-	-	-	-	-
D-Galactose	1.27	1.43	1.26	0.69	-
	1.41	1.57	1.54	0.78	-
Fructose	-	-	-	0.56	-
	-	-	-	0.59	-

[a]Shoulder before the major peak.

Index	Stationary Phase	Temperature (°C)	Dimension of the Column length (cm) i.d. (cm)		Reference Substance	Literature
A	0.5% OV-1 on Chromosorb G (100–120 mesh)	160 Isothermal	200 Steel	0.2	Sorbitol: abs. time 13.2 min	Petersson[55]
B	0.5% OV-17 on Chromosorb G (100–120 mesh)	160 Isothermal	200 Steel	0.2	Sorbitol: abs. time 6.6 min	Petersson[55]
C	5% QF-1 on Chromosorb G (100–120 mesh)	120 Isothermal	200 Steel	0.2	Sorbitol: abs. time 6.6 min	Petersson[55]
D	1% SE-30 on Gas-Chrom Q (100–120 mesh)	170 Isothermal	460 Steel	0.32	*myo*-Inositol	Mason, Slover[56]
E	OV-17 on steel capillary	140 Isothermal	4 500	0.02	Dihydroxy-acetone	Anderle, König-stein, Kovacik[54]

Table 7.   Relative Retention Times of O-Trimethylsilyl Ethers of the Oximes of Oligosaccharides

Compound	A	B	C	Compound	A	B	C
Sucrose	-	-	1.000	Kojibiose	1.56	1.36	1.218
Trehalose	1.00	1.00	-		1.72	1.50	1.224
Maltose	1.45	1.26	1.209	Lactulose	1.15	0.94	-
	-	-	1.224	Neolactose	1.22	0.99	-
Cellobiose	1.24	1.09	-		1.35	1.08	-
	1.35	1.16	-	Maltulose	-	-	1.177
Lactose	1.16	0.97	-		-	-	1.188
Gentiobiose	2.01	1.82	1.171	Palatinose	-	-	1.256
	2.16	-	1.264		-	-	1.272
	-	-	1.293	Isomaltose	-	-	1.224
Melibiose	2.20	1.95	1.256		-	-	1.305
	2.46	2.12	1.300		-	-	1.340
	-	-	1.34	Raffinose	-	-	1.800
Turanose	1.47	1.23	1.206	Melezitose	-	-	1.918
	-	-	1.214	1-Kestose	-	-	1.824
Nigerose	1.21	1.05	1.143				
	1.48	1.23	1.204				
	1.57	1.35	1.233				

Index	Stationary Phase	Temperature (°C)	Dimension of the Column length (cm)	i.d. (cm)	Reference Substance	Literature
A	1.5% SE-52 on Chromosorb WAW DMCS (60–80 mesh)	215 Isothermal	200 Steel	0.3	Trehalose: abs. time 14.5 min	Toba, Adachi[58]
B	1.5% OV-17 on Shimalite W (80–100 mesh)	215	200 Steel	0.3	Trehalose: abs. time 17.05 min	Toba, Adachi[58]
C	OV-101 Quartz capillary	180–280 3 °C/min 280–290 2 °C/min	2 500	0.023	Sucrose	Mateo, Bosch, Pastor, Jimenez[57]

(b) Preparation of the Sugar Oximes and Their O-Silyl Ethers

Dissolve 1.25 g of hydroxylamine hydrochloride and 0.3 g of internal standard (phenyl-β-glycoside or octadecane) in 50.0 g of accurately weighed dry pyridine (1 g of the solution contains 25 mg hydroxylamine hydrochloride and 6.0 mg internal standard). Weigh the dry reaction tube, closed by a screw cap with a septum, then inject 0.3 mL of the oxime reagent and internal standard with a syringe through the septum and weigh the tube again. Heat the closed reaction tube for 30 min at 80 °C and then cool to room temperature. Inject 0.3 mL MSHFBA and 0.1 mL TMSCl with a syringe through the septum; heat the closed reaction tube 30 min at 80 °C.

For the gas chromatographic analysis the septum is pierced with a syringe and an aliquot taken and injected into the gas chromatograph.

Gas chromatographic separation:

Column: stainless steel, 250 cm x 3.2 mm (i.d.); 5% OV-17 on Varaport 30; 80–100 mesh.

Temperature: 130–300 °C; 6 °C/min; injector and FID detector 310 °C.

Calculation: from the calibration chromatogram, the specific peak area factors for each sugar relative to the internal standards are calculated.

For the evaluation of the gas chromatograms of the samples, the peak areas of each sugar are multiplied with the corresponding factors and the concentrations are then calculated from the concentration of the internal standard.

The accuracy ranges between 2–5% rel. Figure 4 shows the gas chromatogram of a standard mixture. This method, the conversion of the common sugars glucose and fructose into their oximes, enables the simultaneous quantitative determination of sugars and organic hydroxy acids (citric acid, malic acid, etc.) in fruits such as apples and citrus fruits[59,60].

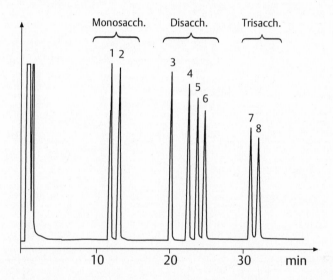

Figure 4.   Gas Chromatogram of Trimethylsilyl Derivatives of a Model Mixture of Sugar Oximes Prepared by Derivatization with a Simplified Method
Steel column: L: 2.5 m, i.d.: 3.2 mm; *stationary phase*: 5% OV-17 on Varaport-30; *temperature*: 130–300 °C, 6 °C/min
1: fructose, **2**: glucose, **3**: internal stand, **4**: sucrose, **5**: lactose, **6**: maltose, **7**: raffinose, **8**: melezitose (Zürcher, Hadorn and Strack[22]; with permission)

Methoximation

The reaction occurs as for oximation, but using *O*-methoxylamine hydrochloride[55]. All the procedures reported use dry pyridine as the reaction medium[61–65].

Procedure

Dissolve the dry sample (~15 mg) in 0.5 mL of dry pyridine in a thick-walled reaction tube with a screw cap, add 12 mg of dry *O*-methoxylamine hydrochloride and heat the closed vessel for 2 h at 80 °C. After cooling add the silylation reagent to this mixture; the following reagents are recommended: BSTFA–TMSCl[55,61,62,64], HMDS–TMSCl[63], BSTFA, BSTFA–TSIM[65].

Instead of O-methoxylamine, the reaction can be carried out with O-benzylhydroxylamine[66,67]. Methoximation is preferred especially in GC–MS studies since the interpretation of the mass spectra of the sugar methoximes is easier than those of the sugar oximes. The O-TMS methoxime compounds are more stable then the O-TMS oximes. In Table 8 the relative retention times for the TMS methoximes of common mono- and oligosaccharides are listed.

Table 8. Relative Retention Times of Trimethylsilyl Ethers of Sugar Methoximes

Compound	A	B	C	D	E
Glyceraldehyde	-	-	-	-	-
D-Erythrose	0.325	0.35	-		
D-Threose	-	-			
D-Arabinose	-	-	-	0.39	0.86
D-Xylose	0.970	0.99ᵃ	0.25	-	0.84
D-Ribose	1.000	1.00	-		0.90
2-Deoxyribose	-	-	-	0.31	
L-Rhamnose	-		-	0.46	
	-		-	0.48	
L-Fucose	1.36	1.22	-	-	1.00
	-	-	-	-	1.03
2-Deoxyglucose	-		-	0.57	
D-Glucose	3.05ᵃ	2.60ᵃ	0.80	0.84	1.44
	-	-	-	0.89	1.47
D-Galactose	2.96ᵃ	2.46ᵃ	-	0.82	1.43
	-	-	-	0.88	1.46
D-Mannose	-	-	0.77		1.42
	-	-	-		1.45
D-Fructose	2.79ᵃ	2.30	0.73	0.77	1.38
	-	-	-	0.80	1.14

ᵃ2 peaks; not resolved.

Index	Stationary Phase	Temperature (°C)	Dimension of the Column length (cm) i.d. (cm)		Reference Substance	Literature
A	3% SE-30 on Supelcoport (100–120 mesh)	170 Isothermal	200	0.2	D-Ribose	Laine, Sweeley[62]
B	3% OV-17 on Supelcoport (100–120 mesh)	170 Isothermal	200	0.2	D-Ribose	Laine, Sweeley[62]
C	0.5% OV-1 on Chromosorb	175 Isothermal	200 Steel	0.2	Sorbitol: abs. time 13.2 min	Petersson[55]
D	OV-101 Glass capillary	175 Isothermal	5 000	0.03	Sorbitol	Zegota[63]
E	SE-30 Glass capillary	150–200 2 °C/min	2 500	0.03	Fucose, 1ˢᵗ peak	Storset, Stokke, Jellum[64]

Diethyl Dithioacetals

In acidic media ethanethiol reacts with aldoses yielding the corresponding dithioacetals. The stereochemistry at C-1 disappears and only the open-chain structure of the sugar remains giving one peak per compound[68].

Procedure

Evaporate to dryness 100 µL of an aqueous sample of the aldoses (2–2.000 µg) and *myo*-inositol, as internal standard, in a thick-walled reaction tube. Add 20 µL of a mixture of ethanethiol and trifluoroacetic acid 2:1 v/v and close the vessel. Dissolve the dry residue by carefully shaking and keep it at 25 °C for 10 min; then, add 50 µL dry pyridine, 100 µL HMDS and 50 µL TMSCl. Close the vessel again and incubate the mixture for 30 min at 50 °C; remove the turbid matter by centrifugation and inject an aliquot of the clear supernatant into the GC.

The separation must be carried out using a sodium chloride treated capillary column, since the resolution on a packed column is too poor. With a 50-m SCOT column (i.d. 0.28 mm) coated with SF-96 at 225 °C (isothermal) the complete separation of the common units of most polysaccharides (aldoses, alduronic acids, *N*-acetylaldosamines) can be achieved. A difficult resolution of the pentoses especially the pair xylose–ribose has been observed. Figure 5 shows a gas chromatogram of a mixture of *O*-TMS-diethyl dithioacetals. With prolonged reaction times and enhanced acid concentration, monoacetals and thioglycosides are formed as side products.

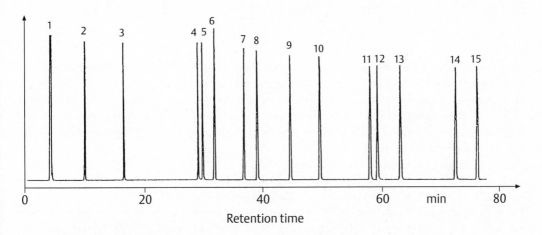

Figure 5.   Gas Chromatogram of *O*-Trimethylsilyl-diethyl Dithioacetals of Different Aldoses
Column: SCOT SF-96, L: 50 m, i.d.: 0.28 mm; *temperature*: 225 °C
1: solvent, **2**: glyceraldehyde, **3**: erythrose, **4**: D-ribose + D-xylose, **5**: L-arabinose, **6**: inositol, **7**: L-rhamnose, **8**: L-fucose, **9**: D-galacturonic acid, **10**: D-galacturonic acid, **11**: D-glucose, **12**: D-mannose, **13**: D-galactose, **14**: *N*-acetyl-D-glucosamine, **15**: *N*-acetyl-D-galactosamine (Honda, Yamauchi and Kakehi[68];with permission)

Aldonitriles

The reaction of aldoses with hydroxylamine leads to the corresponding aldoximes which yield nitriles of aldonic acids upon further dehydration. This can be performed using *O*-hydroxylamine sulfonic acid in the presence of triethylamine. The aldonic acid nitriles exist in the open-chain structure and their *O*-TMS ethers afford only one peak in the gas chromatogram[69] (see reaction pathway below).

Reaction Pathway for the Formation of Aldonitriles (Example: D-Glucose)

Procedure

Evaporate the aqueous or methanolic solution of the sugars (50–1.000 µg per single compound), in a thick-walled reaction tube with a PTFE screw cap (Sovirel vials), completely to dryness. Add 50–100 µL of a molar methanolic solution of *O*-hydroxylamine sulfonic acid which contains an equimolar amount of triethylamine; close the tube and keep it for 30 min at room temperature. Then, after evaporation of the solvent add 50 µL BSTFA and 50 µL pyridine, and heat the mixture at 80 °C for 30 min. An aliquot is injected into the GC.

For the separation fused silica columns coated with SE-54 or CP-Sil 19B (50% phenylmethylpoly-siloxane) are suitable. A good separation of the common aldoses can be achieved as shown in Figure 6. In Table 9 the relative retention times for both stationary phases are listed.

The reaction conditions must follow the described procedure exactly, otherwise side reactions occur. An excess of triethylamine and pH greater than 5 or temperatures over 40 °C lead to methanolysis of the *O*-hydroxylamine sulfonic acid yielding *O*-TMS oximes instead of aldonitriles. The oxime reagent must always be prepared fresh as it decomposes after a few hours.

Methylated Sugars

The separation and quantitative determination of methylated sugars is a very important operation for the elucidation of the structures of polysaccharides by the so-called "methylation analysis". In this case, more than one structure exists per single compound giving rise to more than one peak in the gas chromatogram. For heteropolysaccharides with many different units the gas chromatograms become very complicated; an exact evaluation is only possible by capillary GC. The procedures of silylation are the same as those for unsubstituted sugars[70–73].

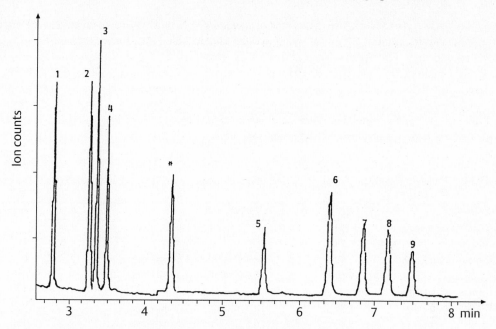

Figure 6. Chromatogram of the *O*-Trimethylsilyl Ethers of Aldonitriles
Column: Quartz capillary, L: 25 m, i.d.: 0.32 mm; *stationary phase*: CP-Sil-19 CB; *temperature*: 180 °C; *detection*: total ionization current/mass spectrometer
1: deoxyribose, **2**: arabinose, **3**: xylose, **4**: ribose, **5**: 2-deoxyglucose, **6**: galactose, **7**: mannose, **8**: idose, **9**: glucose (Rubino[69]; with permission)

Table 9. Relative Retention Times of *O*-Trimethylsilyl Ethers of Aldonic Acid Nitriles

Compound	A	B	Compound	A	B
D-Arabinose	0.755	0.751	D-Mannose	1.706	1.575
D-Xylose	0.756	0.774	D-Talose	1.862	1.631
D-Ribose	0.790	0.804	D-Allose	1.759	1.609
D-Lyxose	0.769	0.769	D-Altrose	1.838	1.610
2-Deoxy-D-ribose	0.537	0.644	D-Idose	1.792	1.647
D-Glucose	1.876	1.724	D-Gulose	1.752	1.631
D-Galactose	1.700	1.475	D-Glucuronic acid	2.098	2.109

Index	Stationary Phase	Temperature (°C)	Dimension of the Column length (cm)	i.d. (cm)	Reference Substance	Literature
A	SE-54 Quartz-capillary column	180 Isothermal	2 500	0.025	Hexachlorobenzene: abs. time 4.525 min	Rubino[69]
B	CP-Sil-19CB Quartz-capillary column	180 Isothermal	2 500	0.032	Hexachlorobenzene: abs. time 4.537 min	Rubino[69]

The separation of methylated TMS monosaccharides has been used especially for the elucidation of the structures of the sugar residues of naturally occurring physiologically active glycosides. These glycosides are first methylated, then hydrolyzed by a mixture of 3.5 mL of pure acetic acid and 1 mL of conc. hydrochloric acid and 5.5 mL of water, and the liberated methylated monosaccharides are then converted into their TMS ethers. Their structures are elucidated by GC–MS (separation conditions: fused silica capillary column: 50 m x 0.25 mm i.d.; coated with SE-54; temperature: 100–250 °C; 3 °C/min)[73].

In Table 10 the relative retention times of different methylated glucoses and mannoses are listed. Generally two peaks are obtained per single compound[71,72].

Table 10. Relative Retention Times of O-Trimethylsilyl Ethers of Distinctly Methylated Glucoses and Mannoses

Compound	A	B	Compound	A	B
2-O-Methyl-D-glucose	5.98	-	2,3,4,6-Tetra-O-methyl-D-glucose	1.00	-
	9.26	-		0.91	-
3-O-Methyl-D-glucose	4.80	-	2,4-Di-O-methyl-D-mannose	-	0.92
	9.50	-	2,6-Di-O-methyl-D-mannose	-	1.27
4-O-Methyl-D-glucose	-	-		-	1.67
6-O-Methyl-D-glucose	7.81	-	3,4-Di-O-methyl-D-mannose	-	0.67
	9.67	-	3,5-Di-O-methyl-D-mannose	-	1.41
2,3-Di-O-methyl-D-glucose	2.63	-		-	1.70
	3.44	-	5,6-Di-O-methyl-D-mannose	-	1.45
2,4-Di-O-methyl-D-glucose	-	-		-	1.92
2,6-Di-O-methyl-D-glucose	4.33	-	2,3,4-Tri-O-methyl-D-mannose	-	0.66
3,4-Di-O-methyl-D-glucose	-	-	2,5,6-Tri-O-methyl-D-mannose	-	0.97
3,6-Di-O-methyl-D-glucose	3.40	-	3,4,6-Tri-O-methyl-D-mannose	-	0.42
	3.99	-	3,5,6-Tri-O-methyl-D-mannose	-	1.05
4,6-Di-O-methyl-D-glucose	3.91	-		-	1.23
	4.86	-	2,3,4,6-Tetra-O-methyl-D-mannose	-	0.95
2,3,6-Tri-O-methyl-D-glucose	1.82	-		-	1.04
	1.59	-	2,3,5,6-Tetra-O-methyl-D-mannose	-	1.12

Index	Stationary Phase	Temperature (°C)	Dimension of the Column		Reference Substance	Literature
			length (cm)	i.d. (cm)		
A	8.7% Polyethylene glycol succinate on Kieselgur (60–100 mesh)	125 Isothermal	120 Glass	0.4	2,3,4,6-Tetra-O-methyl-α-D-glucose	Haworth, Roberts, Sagar[72]
B	2% Neopentylglycol succinate Polyester on Chromosorb W (80–100 mesh)	136 Isothermal	370	0.6	2,3,4,6-Tetra-O-methyl α-D-methylmannoside	Jones[71]

Sugar Alcohols

Sugar alcohols have open-chain structures and only one peak per compound appears in the GC. The differences between the relative retention times of the stereoisomeric compounds with the same numbers of carbons are often extremely small and their resolution is only possible by capillary GC. Monosaccharides and reducing oligosaccharides can be readily converted quantitatively into their corresponding sugar alcohols by reaction with sodium or potassium borohydride in aqueous solution. Aldoses yield only one compound but ketoses yield equimolar parts of the two epimeric sugar alcohols.

Procedure

To the dry sample or to an aqueous solution, pH 8.0–9.0, add slightly more than an equimolar amount of a 10% solution of sodium or potassium borohydride in water or in water–methanol 1:1 v/v. Stir for 2 h at room temperature and destroy the excess reagent with a strong acidic cationic exchanger (e.g., Dowex 50X8; 80–100 mesh); by this procedure the cations are removed from the solution.

Table 11. Relative Retention Times of O-Trimethylsilyl Ethers of Sugar Alcohols (C$_4$–C$_6$: A–D), and Their Kovats Indices (E)

Compound	A	B	C	D	E	
Glycerol	-	-	0.07	0.28	1258.1	± 1.0
Erythritol	0.16	0.49[a]	0.24	0.56	1512.5	± 0.5
Threitol	-	-	-	0.55	1501.7	± 0.7
Arabitol	0.46	2.13[a]	0.58	1.01	1758.7	± 0.2
Xylitol	0.42	2.06[a]	-	0.98	17445	± 0.2
Ribitol	0.46	2.08[a]	-	1.02	17670	± 0.2
Allitol	1.24	1.39[b]	-	-	-	-
Talitol	1.30	1.59[b]	-	-	-	-
Galactitol (Dulcitol)	1.28	1.38[b]	-	1.55	2004.8	± 0.1
Mannitol	1.21	1.31[b]	1.00	1.54	1993.6	± 0.2
Sorbitol	1.24	1.42[b]	1.00	1.56	2000.1	± 0.1
Rhamnitol	-	-	0.60	-	1817.0	± 0.2
Fucitol	-	-	-	1.18	1837.6	± 0.2
2-Deoxyribitol	-	-	0.36	-	1615.3	± 0.4
2-Deoxysorbitol	-	-	0.77	-	1859.7	± 0.2
3-Deoxysorbitol	-	-	0.79	-	-	-
3-Deoxymannitol	-	-	0.77	-	-	-

Index	Stationary Phase	Temperature (°C)	Dimension of the Column length (cm) i.d. (cm)		Reference Substance	Literature
A	3% SE-52 on Chromosorb WDMCS (80–100 mesh)	140 Isothermal	183 Glass	0.63	α-D-Gluco-pyranoside	Sweeley, Bentley, Makita, Wells[16]
B	15% EGS (Polyethylene glycol succinate) on Chromosorb W (80–100 mesh)	[a]140 [b]150	244 Glass	0.63	[a]Methyl-α-D-arabinopyranoside [b]α-D-Gluco-pyranoside	Sweeley, Bentley, Makita, Wells[16]
C	3% JXR on Gas-Chrom Q (100–120 mesh)	120–180 2 °C/min	183 Steel	0.32	Sorbitol	El-Dasch, Hodge[43]
D	SE-30 Glass capillary column	120–200 2 °C/min	2 500	0.03	Fucose 1st peak	Storset, Stokke, Jellum[64]
E	OV-101 Glass capillary column	180–230 1.7 °C/min	5 000	0.03	-	Adam[75]

Table 12.   Relative Retention Times of O-Trimethylsilyl Ethers of Disaccharide Sugar Alcohols

Compound	A	B	C
Maltitol	0.76	0.68	1.9
Cellobiitol	0.64	0.59	1.7
Lactitol	0.62	0.54	-
Gentiobiitol	0.85	0.89	-
Melibiitol	1.00	1.00	2.7
Laminaribitol	0.67	0.63	1.9

Index	Stationary Phase	Temperature (°C)	Dimension of the Column length (cm)	i.d. (cm)	Reference Substance	Literature
A	2.2% SE-30 on Gas-Chrom S (100–200 mesh)	263 Isothermal	200 Glass	0.35	Melibiitol	Kärkkäinen[76]
B	1% OV-22 on Supelcoport (80–100 mesh)	218 Isothermal	200 Glass	0.35	Melibiitol	Kärkkäinen[76]
C	3% SE-30 on Gas-Chrom	210	150 Glass	0.4	Sucrose	Percival[77]

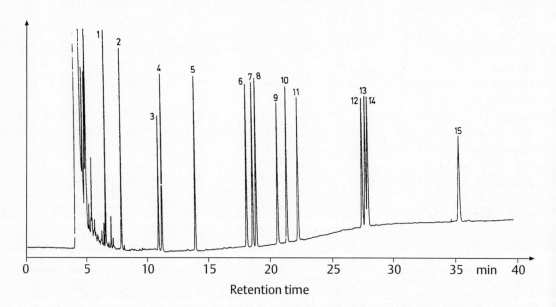

Figure 7.   Gas Chromatogram of the Separation of the Trimethylsilyl Ethers of Monosaccharide Alditols
Column: Glass capillary, L: 50 m, i.d.: 0.3 mm; *stationary phase*: OV-101; *temperature*: 180–230 °C, 1.7 °C/min
1: glycerol, **2**: butane-1,2,4-triol, **3**: threitol, **4**: erytrol, **5**: 2-deoxytibitol, **6**: xylitol, **7**: arabitol, **8**: ribitol, **9**: rhamnitol, **10**: fucitol, **11**: 2-deoxyglucitol, **12**: mannitol, **13**: sorbitol (glucitol), **14**: galactitol, **15**: *myo*-inositol (Adam[75]; with permission)

Evaporate the free boric acid by tenfold evaporation with methanol as the boric acid methyl ester. After, dry over P_2O_5 silylate according to the common procedure[74]. The silylation is carried out in the same manner to that as described for sugars.

In Tables 11 and 12 the relative retention times and Kovats indices of sugar alcohols of mono- and disaccharides are listed[16,43,64,75–77]. Figure 7 shows the separation of some TMS sugar alcohols.

Sugar Acids

Oxidation of sugars yields hydroxycarbonic acids which can be classified into the following groups:

Alduronic Acids

Alduronic acids are compounds in which the primary alcohol group of aldoses has been oxidized (for pentoses at C-5, for hexoses at C-6). These compounds are widespread and are units of many polysaccharides and glycosides.

Aldonic Acids

Aldonic acids are formed by oxidation at the aldehyde semiacetal group at C-1. Branched acids which are formed by alkaline rearrangement of aldoses (e.g., saccharic acids) also belong to this group.

Aldaric Acids

Aldaric acids are sugar acids with the general chemical formula $HOOC–(CHOH)_n–COOH$. They are formed by strong oxidation of aldoses (e.g., with HNO_3) All these compounds can be converted into their *O*-trimethylsilyl sugar acid trimethylsilyl esters (the carboxyl groups are trimethylsilylated in the same manner as the hydroxyl groups) which can be separated by GC. This also applies to sugar acid lactones which are formed by dehydration of free sugar acids in acid media.

Alduronic Acids

The samples must be present as salts and the lactones are converted into these compounds by the following procedure[55].

Dissolve 0.1–10 mg of the sample in 2–5 mL distilled water and add 0.05 M NaOH until pH 8.5 is reached; hereafter this value is adjusted in the autotitrator. Stir for 4 h at room temperature and then evaporate the solution to dryness in vacuo.

When the compounds are base labile, adjust these solutions by adding 0.01 M HCl to pH 7.0 before evaporation. Before trimethylsilylation it is sometimes convenient to convert such compounds into their oximes or methoximes, these reactions are carried out in absolute pyridine according to the procedures on pages 214–218.

Trimethylsilylation

After the oximation reaction in dry pyridine, add approximately 0.4 mL of BSTFA and 0.2 mL of TMSCl and keep this mixture for 1 h at 60 °C or 2 h at room temperature. Evaporate the pyridine and dissolve the residue in water-free diethyl ether containing 0.5% BSTFA or inject the whole reaction mixture directly into the gas chromatograph[36,37,55,79].

In Table 13 the relative retention times of the oximes of common uronic acids for some of the common stationary phases are listed.

Methanolysis of polysaccharides affords uronic acid methyl ester–methyl glycosides. After trimethylsilylation up to four peaks appear for each of these compounds in the gas chromatogram (see Table 14)[51].

Table 13. Relative Retention Times of *O*-Trimethylsilyl Ethers of the Oximes of Uronic Acid Trimethylsilyl Esters (A, B, C) and of the Nonaltered Uronic Acid Compounds (D)

Compound	A	B	C	D
Glucuronic acid	1.47	2.04	1.57	0.82
	-	-	1.74	-
Mannuronic acid	1.47	1.96	1.57	-
	1.65	-	1.84	-
Guluronic acid	1.62	2.30	1.73	-
	1.90	2.53	2.20	-
Galacturonic acid	1.56	2.20	1.60	1.07
	1.74	2.41	1.91	1.27
xylo-5-Hexulosonic acid	1.42.	2.00	1.55	-
lyxo-5-Hexulosonic acid	1.20	1.46	1.31	-

Index	Stationary Phase	Temperature (°C)	Dimension of the Column length (cm) i.d. (cm)		Reference Substance	Literature
A	0.5% OV-1 on Chromosorb G (100–120 mesh)	160	200 Steel	0.2	D-Sorbitol: abs. time 13.2 min	Petersson[55]
B	0.5% OV-17 on Chromosorb G (100–120 mesh)	160	200 Steel	0.2	D-Sorbitol: abs. time 6.6 min	Petersson[55]
C	3% QF-1 on Chromosorb G (100–120 mesh)	160	200 Steel	0.2	D-Sorbitol: abs. time 4.3 min	Petersson[55]
D	3% SE-30 on Diatoport S	140–200 0.5 °C/min	250 Glass	0.32	D-Mannitol	Bhatti, Chambers, Clamp[36]

Table 14. Relative Retention Times of *O*-Trimethylsilyl Ethers of Uronic Acid Methyl Ester–1-*O*-Methyl Glycosides

Compound	A	B	C
Glucuronic acid methyl ester–*O*-methyl glycosides	2.9	3.0	0.57
	3.1	4.8	0.88
	4.7	-	0.89
	4.9	-	-
D-Galacturonic acid methyl ester–*O*-methyl glycosides	2.6	2.5	0.59
	3.1	3.0	0.65
	4.1	3.9	0.78
	4.3	4.0	0.78
D-Mannuronic acid methyl ester–*O*-methyl glycosides	2.7	2.6	0.52
	2.9	3.0	0.60
	3.2	3.5	0.79
	3.5	-	0.71
L-Guluronic acid methyl ester–*O*-methyl glycosides	-	-	0.39
	-	-	0.52
	-	-	0.61
	-	-	0.64

Table 14.   (Continued)

Index	Stationary Phase	Temperature (°C)	Dimension of the Column length (cm) i.d. (cm)		Reference Substance	Literature
A	2% SE-52 on Chromosorb GAW DMCS (80–100 mesh)	190 Isothermal	200	0.3	*meso*-Erythritol	Raunhardt, Schmidt, Neukom[78]
B	1% SE-30 on Chromosorb GAW DMCS (80–100 mesh)	170 Isothermal	200	0.3	*meso*-Erythritol	Raunhardt, Schmidt, Neukom[78]
C	SE-30 Glass capillary	120–190 4 °C/min	3 500	0.03	D-Sorbitol	Preuss, Thier[51]

Aldonic Acids, Deoxyaldonic Acids and Aldaric Acids

The acyclic free acids exist in aqueous solution in equilibrium with their cyclic lactones; the presence of acids causes a shift towards the latter group of compounds.

In the presence of alkali these lactones are saponified and the salts of the free acids are obtained. These acids have acyclic structures and yield only one peak per compound in the gas chromatogram (for procedure see Alduronic Acids, *vide supra*)[55,79]. For four stationary phases with different polarity the relative retention times are listed in Table 15. With polar phases the differences in the retention times are found to be larger than with nonpolar phases as a result of superpolar interactions with the ester groups. In general, the retention time of the individual compounds depends on the number of trimethylsilyl groups.

The corresponding aldonic acid lactones are formed from the acids by evaporation with volatile acids (e.g., HCl). After trimethylsilation (BSTFA–TMSCl) more than one peak is obtained due to the different structures of the lactones. Complete separations of the *O*-TMS ethers of the free acids as well as of their corresponding lactones are obtained on capillary columns (e.g., 60 m SE-30; 25 m, OV-101[37,168]).

Amino Sugars

These compounds, especially as their *N*-acetyl derivatives, are widespread units of mucopolysaccharides, lipopolysaccharides, glycoproteins, peptidoglycans of bacteria and antibiotics, and oligosaccharides of milk and urine. They can be separated by GC as their *O*-trimethylsilyl derivatives, but some problems exist since the silylation reagents react at different rates with the hydroxyl and amino groups. This property sometimes causes incomplete silylation leading to large multiplicities of peaks per compound in the gas chromatogram.

With the HMDS–TMSCl reagent in pyridine only the hydroxyl groups are trimethylsilylated at room temperature; even at 70 °C, the amino groups are only attacked to a minor extent[35,80].

Better results are obtained with BSA and BSTFA. With the mixture BSA–TMSCl–HMDS in pyridine (1:0.5:1:10 v/v) complete silylation is achieved at room temperature in 30 minutes. With BSA and BSTFA alone, the silylation is incomplete even at elevated temperatures[35].

In pyridine, which has not been distilled and stored over solid KOH, application of BSA alone leads to complete silylation of the hexosamines (giving pentaTMS derivatives) as the HCl of hexosamine hydrochlorides serves as a catalyst for the reaction[80].

Table 15. Relative Retention Times of O-Trimethylsilyl Ethers of Aldonic Acid Trimethylsilyl Esters, Deoxy-aldonic Acid Trimethylsilyl Esters and Aldaric Acid Trimethylsilyl Esters

Compound	A	B	C	D
Glyceric acid	0.052	0.092	0.086	0.087
Erythonic acid	0.151	0.218	0.212	0.214
Threonic acid	0.156	0.251	0.237	0.256
Ribonic acid	0.438	0.569	0.565	0.567
Arabonic acid	0.463	0.646	0.632	0.661
Xylonic acid	0.428	0.617	0.570	0.621
Lyxonic acid	0.457	0.599	0.621	0.641
Allonic acid	1.115	1.276	1.410	1.380
Altonic acid	1.247	1.531	1.568	1.637
Gluconic acid	1.286	1.569	1.639	1.713
Mannonic acid	1.113	1.254	1.385	1.433
Gulonic acid	1.087	1.256	1.309	1.332
Idonic acid	1.368	1.743	1.812	1.930
Galactonic acid	1.256	1.561	1.570	1.671
Talonic acid	1.235	1.507	1.656	1.656
2-Deoxytetronic acid	0.084	0.146	0.145	0.149
3-Deoxytetronic acid	0.078	0.136	0.129	0.135
4-Deoxyerythronic acid	0.058	0.094	0.091	0.086
4-Deoxythreonic acid	0.061	0.099	0.092	0.094
2-Deoxy-*erythro*-pentonic acid	0.252	0.379	0.395	0.409
2-Deoxy-*threo*-pentonic acid	0.251	0.373	0.392	0.402
3-Deoxy-*erythro*-pentonic acid	0.236	0.380	0.375	0.404
3-Deoxy-*threo*-pentonic acid	0.252	0.419	0.418	0.455
2-Deoxy-*arabino*-hexonic acid	0.659	0.912	1.007	1.031
2-Deoxy-*lyxo*-hexonic acid	0.673	0.935	1.035	1.080
3-Deoxy-*ribo*-hexonic acid	0.673	0.946	1.007	1.107
3-Deoxy-*arabino*-hexonic acid	0.676	0.971	1.004	1.080
3-Deoxy-*xylo*-hexonic acid	0.659	0.995	1.026	1.116
3-Deoxy-*lyxo*-hexonic acid	0.644	0.894	0.973	1.070
6-Deoxymannonic acid	0.528	0.629	0.687	0.685
6-Deoxyglactonic acid	0.615	0.809	0.801	0.805
Erythratic acid	0.174	0.313	0.354	-
Threaric acid	0.212	0.432	0.471	-
Ribaric acid	0.532	0.873	1.037	-
Arabaric acid	0.522	0.864	0.945	-
Xylaric acid	0.545	0.978	1.034	-
Allaric acid	1.242	1.669	1.904	-
Talaric acid	1.419	2.098	2.195	-
Glucaric acid	1.295	1.814	1.920	-
Mannaric acid	1.074	1.346	1.401	-
Idaric acid	1.602	2.518	2.589	-
Galactaric acid	1.500	2.272	2.347	-
2-Deoxy-*arabino*-hexaric acid	0.777	1.277	1.565	-
2-Deoxy-*lyxo*-hexaric acid	0.847	1.467	1.751	-
3-Deoxy-*ribo*-hexaric acid	0.732	1.290	1.550	-
3-Deoxy-*arabino*-hexaric acid	0.767	1.425	1.675	-
3-Deoxy-*xylo*-hexaric acid	0.761	1.317	1.639	-
3-Deoxy-*lyxo*-hexaric acid	0.789	1.470	1.713	-

Table 15.   (Continued)

Index	Stationary Phase	Temperature (°C)	Dimension of the Column length (cm)	i.d. (cm)	Reference Substance	Literature
A	0.5% OV-1 on Chromosorb G (100–120 mesh)	160 Isothermal	200 Steel	0.2	D-Sorbitol: abs. time 11.4 min	Petersson[55,79]
B	0.5% OV-17 on Chromosorb G (100–120 mesh)	160 Isothermal	200 Steel	0.2	D-Sorbitol: abs. time 5.6 min	Petersson[55,79]
C	3% DC QF-1 on Chromosorb Q (100–120 mesh)	120 Isothermal	200 Steel	0.2	D-Sorbitol: abs. time 15.2 min	Petersson[55,79]
D	1% GE XE-60 on Gaschrom Q	120 Isothermal	200 Steel	0.2	D-Sorbitol: abs. time 11.8 min	Petersson[55,79]

Different sialic acid derivatives (*N*-acetylneuraminic acid, *N*-acetyl-4-*O*-acetylneuraminic acid, *N*-glycolylneuraminic acid, etc.) can also be separated as their *O*-TMS derivatives. The following silylation reagents are suitable:

1.   HMDS–TMSCl–pyridine 2:1:10 v/v, giving complete silylation of the sialic acid at room temperature in 30 minutes.

2.   TSIM–BSA–TMSCl 3:2:2 v/v[81].

Conversion of the Amino Sugars to the Corresponding Anhydro Sugars
When glycoconjugates are cleaved by methanolysis the corresponding methyl glycosides are formed. The *N*-acetylamino sugars are deacetylated by this operation. When the free amino sugars are treated with nitrous acid, the corresponding anhydro sugars are formed (e.g., from free hexosamines the 2,5-anhydrohexoses, 2-amino-2-deoxyglucose yields 2,5-anhydro-D-mannose, 2-amino-2-deoxygalactose yields 2,5-anhydro-D-talose). These anhydro sugars are distinctly less polar and the retention times of the *O*-TMS derivatives are much shorter compared to the 2-acetamido-2-deoxy sugar. When these anhydro sugars are reduced with sodium borohydride to the anhydrohexitols, the corresponding TMS derivatives give only one peak per compound in the gas chromatogram, which can be identified easily[82].

Procedures[82]

(a) Methanolysis
Heat the dry sample with 500 μL 0.5 M methanolic hydrochloric acid containing 1.5% H_2O for 18 h at 80 °C in a closed reaction tube, then, after cooling, neutralize with silver carbonate. After centrifugation and removal of the supernatant, wash the precipitate twice with methanol and evaporate the whole supernatant to dryness under a nitrogen atmosphere.

(b) Deamination (microvariation of the procedure)
Dissolve the dry residue in a mixture of 50μL distilled water and 50 μL of distilled water–acetic acid 12:1. Add 20 μL of a 5.5 M aqueous sodium nitrite solution and shake the mixture for 30 min at room temperature. Remove the sodium ions with 100 μL of an aqueous suspension of a cationic exchanger [Dowex 50 X2(H^+)] and evaporate the clear solution to dryness.

(c) Reduction
Dissolve the dry residue in 50 μL distilled water and add 50 μL of a 0.22 M aqueous solution of sodium borohydride. Allow the reaction to proceed for 1 h at room temperature, then destroy the

excess reagent by addition of 20 µL pure acetic acid. Remove the free boric acid as its methyl ester by evaporation with methanol 3–5 times. If deaminated sialic acid is present, heat the dry residue again with 0.5 M methanolic hydrochloric acid for 1 h at 80 °C to convert the substance into its methyl ester.

(d) Trimethylsilylation
Add the mixture HMDS–TMSCl–pyridine 2:1:10 v/v to the dry residue and keep the mixture for 30 min–1 h at room temperature.

Table 16 shows the relative retention times of some of the free and substituted amino sugars; the corresponding alcohols and the anhydro sugar alcohols are also listed.

An alternative procedure for the GC determination of amino sugars is conversion to their TMS dithioacetals after hydrolysis, and deamination as described by Honda et al[68]. This method offers the possibility to determine amino sugars simultaneously in the presence of neutral sugars and uronic acids, it is rapid and the coefficient of variations is less than 3%[83]. This method is especially suitable for the determination of the units of amino sugar containing polysaccharides.

Procedure[83]
To a dry residue of the hydrolysate of the polysaccharide (20–200 µg) in a reaction tube add 3-*O*-methyl-D-glucose as the internal standard. Evaporate the mixture to dryness in vacuo and add 100 µL of barium nitrite (0.1 M) and 75 µL cold sulfuric acid (0.1 M), keep the mixture for 1 h in an ice bath and then evaporate again in the same manner. Add 20 µL of a mixture of ethanethiol–trifluoroacetic acid (2:1 v/v), tightly close the reaction tube and keep it for 10 min at 25 °C. Then add 50 µL dry pyridine, 100 µL HMDS and 50 µL TMSCl, and incubate the mixture occasionally shaking at 50 °C for 30 min. After centrifugation inject an aliquot into the GC.
Separation conditions:
Column: SCOT, coated with SF-96; sodium chloride treated; L: 50 m, i.d.: 0.28 mm.
Temperature: 225 °C, isothermal.
The peaks of the *O*-TMS anhydro sugars dithioacetals appear in the range of the *O*-TMS pentose dithioacetals.

2.1.4.3. Acetylation
This method also has a widespread application and involves the conversion of sugars and related compounds into their acetyl derivatives by treatment with acetic anhydride, according to the reaction scheme below:

$$R(OH)_n + n(CH_3CO)_2O \longrightarrow R(OCOCH_3)_n + n\ CH_3COOH$$

The volatilities of these compounds are less than those of the TMS ethers and for this reason gas chromatographic separations are successful only in the field of monosaccharides and their derivatives. For pure monosaccharides, separation by GC as their acetyl derivatives is limited to special cases such as, for example:[84]
–   Separation of monosaccharides and the corresponding sugar alcohols[85].
–   Separation of D-glucose from erythritol in a publication focusing on the chemical structure of oat glucans[86].
–   Studies of the reaction mechanism of acetoxonium rearrangements[84,87].

Table 16. Relative Retention Times of *O*-Trimethylsilyl Ethers of the Amino Sugars, Their Alcohols and Their Anhydro Sugar Alcohols

Compound	A	B	C	D	E
2-Amino-2-deoxy-D-glucose	0.88	1.20			
(Tetrasilyl derivative)	1.12	1.38			
2-Amino-2-deoxy-D-glucose	1.10	-	-		
(Pentasilyl derivative)	1.32				
2-Amino-2-deoxy-D-galactose	0.77	1.11	-		
(Tetrasilyl derivative)		1.22			
2-Amino-2-deoxy-D-galactose	0.97	-	-		
(Pentasilyl derivative)	1.32				
1-*O*-Methyl-2-amino-2-deoxy-D-glucoside (Tetrasilyl derivative)	0.70	-	-		
1-*O*-Methyl-2-amino-2-deoxy-D-glucoside (Pentasilyl derivative)	1.00	-	-		
1-*O*-Methyl-2-amino-2-deoxy-D-galactoside (Tetrasilyl derivative)	0.59	-	-		
1-*O*-Methyl-2-amino-2-deoxy-D-galactoside (Pentasilyl derivative)	0.82	-	-		
2-Amino-2-deoxy-D-sorbitol (Tetrasilyl derivative)	1.15	-	-		
2-Amino-2-deoxy-D-sorbitol (Pentasilyl derivative)	1.53	-	-		
2-Amino-2-deoxy-D-galactitol (Tetrasilyl derivative)	1.15	-	-		
2-Amino-2-deoxy-D-galactitol (Pentasilyl derivative)	1.49	-	-		
2-Acetamido-2-deoxy-D-glucose	2.25	8.3	-		
(Tetrasilyl derivative)		10.5			
2-Acetamido-2-deoxy-D-galactose	2.04	-	-		
(Tetrasilyl derivative)					
2,5-Anhydro-D-mannitol	-	-	0.430	0.492	0.163
2,5-Anhydro-D-talitol	-	-	0.545	0.506	0.193
Deaminated sialic acid	-	-	2.15	2.25	2.17

Index	Stationary Phase	Temperature (°C)	Dimension of the Column length (cm)	i.d. (cm)	Reference Substance	Literature
A	2.2% SE-30 on Gas Chrom S (100–120 mesh)	187 Isothermal	200 Glass	0.35	α-D-Gluco-pyranoside	Kärkkainen[76]
B	3% QF-1 on Gas Chrom Q (80–100 mesh)	140	200 Glass	0.35	α-Gluco-pyranoside	Kärkkainen[76]
C	2.2% SE-30 on Gas Chrom Q (110–120 mesh)	140[a], 190[b] Isothermal	200 Glass	0.2	*myo*-Inositol	Mononen[82]
D	1% OV-101 on Gas Chrom Q (100–120 mesh)	140[a], 190[b] Isothermal	200 Glass	0.2	*myo*-Inositol	Mononen[82]
E	3% QF-1 on Gas Chrom Q	140[a], 190[b] Isothermal	200 Glass	0.2	*myo*-Inositol	Mononen[82]

[a]Anhydro compounds. [b]Sialic acid.

–   Studies concerning the equilibrium of aldohexoses and their 1,6-anhydrides[88].

Recently, the acetylation of methyl glycosides has been found to be suitable for the determination of the structures of glycosphingolipids, which had been cleaved by methanolysis. The peracetylated methyl glycosides show improved GC separation on capillary columns compared to the TMS derivatives and, furthermore, they are stable towards humidity. The fatty acid component of these compounds can be determined in the same chromatograms. Due to small amounts of these substances being generally present in biological materials, the following micromethod has been developed[89].

Procedure

Dissolve the samples (10 pmol–10 nmol) of the neutral glycosphingolipids into chloroform–methanol (1:1 v/v) and transfer the solution into 2.5 cm x 2 mm capillary tubes with one end sealed. Remove the solvent quickly under high vacuum, add 0.75 M dry methanolic hydrochloric acid (25 µL) and methyl acetate (5 µL), and seal the other end of the tube. Heat the tube for 30 min at 110 °C. After opening one end of the tube carefully and removing the methanolic hydrochloric acid under high vacuum, add 5 µL of freshly prepared pyridine–acetic anhydride, and reseal the tube. After keeping for 1.5 h at room temperature, open the tube and inject the mixture directly into the gas chromatograph.
Separation conditions:
Column: DB-1 capillary column; L: 60 m.
Temperature: 150 °C, isothermal 15 min; then 150–300 °C, 4 °C/min.

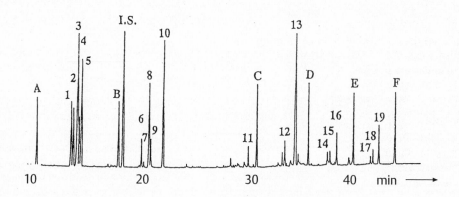

Figure 8.   Gas Chromatogram Showing the Separation of the Peracetylated Methanolysis Products from GM$_1$ Ganglioside
Column: capillary, L: 60 m; *stationary phase*: DB-1; *temperature*: 150–300 °C, 15 min isothermal, then 4 °C/min
1: methyl D-galactofuranoside, **2**: methyl α-D-galactopyranoside, **3**: methyl β-D-galactopyranoside, **4**: methyl α-D-glucopyranoside, **5**: methyl β-D-glucopyranoside, **6**: methyl 2-acetamido-2-deoxy-α-D-galactofuranoside, **7**: methyl 2-acetamido-2-deoxy-β-D-galactofuranoside, **8**: methyl 2-acetamido-2-deoxy-α-D-galactopyranoside, **9**: methyl 2-acetamido-2-deoxy-β-D-galactopyranoside, **10**: methyl palmitate (C$_{18:0}$), **11**: methyl eicosanate (C$_{20:2}$), **12**: methyl docosanate (C$_{22:0}$), **13**: methyl (methyl-5-acetamido-3,5-dideoxy-D-glycero-D-galacto-α-nonulopyranosid)onate, **14**: *O,O,N*-triacetylsphing-4-enine (d$_{18:1}$), **15**: 3-*O*-methyl-*O,N*-diacetylsphing-4-enine, **16**: 5-*O*-methyl-*O,N*-diacetylsphing-3-enine, **17**: *O,O,N*-triacetyleicosasphing-4-enine (d$_{20:1}$), **18**: *O*-methyl-*O,N*-diacetyleicosasphing-4-enine, **19**: *O*-methyl-*O,N*-diacetyleicosasphing-3-enine; **A, B**: hydrocarbons C$_{18-20}$; **C, D, E, F**: hydrocarbons C$_{24-30}$ (Wiesner and Sweeley[89]; with permission)

Figure 8 shows the gas chromatogram of the acetylated methanolysis products of the GM_1 ganglioside. The direct acetylation of free sugars has also been described recently. This method is suitable for those cases where the reduction to the corresponding alditols is incomplete[90]. A mixture of acetic anhydride and 1-methylimidazole is used as the acetylating agent.

Sugar Alcohols
Reduction of monosaccharides leads to the corresponding sugar alcohols, which can be separated as their acetyl derivatives with high efficiency. Only one peak per compound is obtained. The reduction is carried out today exclusively by reaction with sodium borohydride.

This method is applied especially for the determination of the chemical structure of polysaccharides. These compounds are converted into their *O*-methyl derivatives in the first step and they are then cleaved, by acid hydrolysis, to their partially methylated monosaccharides, which are reduced to their corresponding partially methylated sugar alcohols. The latter compounds are then acetylated. The following gas chromatographic separation is usually coupled with a mass spectrometer to obtain individual mass spectra for the exact determination of the structure of a single, separated, partially methylated sugar alcohol.

Reduction with Sodium Borohydride
Several procedures exist. For the classical operation dissolve 20 mg of the monosaccharide mixture in 0.5 mL of aqueous ammonium hydroxide (1 M) and add 10–20 mg of sodium borohydride. Stir for 1 h at room temperature and destroy the excess of sodium borohydride by addition of pure acetic acid. Evaporate the solution under vacuo to dryness and remove the boric acid as boric acid methyl ester by repeated evaporation with methanol. The last step is important as the acetylation reaction is disturbed by the formation of sugar boric acid complexes[91,92].

In another variation, the destruction of borohydride is carried out by addition of cation exchangers (H^+-form), e.g. Dowex 50 $H^{+[93]}$, Amberlite CG-120 $H^{+[94]}$. By this procedure sodium ions are also removed from the solution.

For reduction of disaccharides an aqueous methanol medium is used (see pages 223ff). Especially for the reduction of oligosaccharides with borohydrides the pH of the solution must be maintained around 9.0, since at higher values a remarkable de Bruyn–van Ekenstein transformation occurs. For example, the reduction of maltotriose at pH 12.1 yields, after hydrolysis and reduction, up to 3.5% mannitol (at elevated temperature of 50 °C) in addition to glucitol. The reduction at pH 9.1 yields only around 1% mannitol using the same treatment[95].

In another procedure the sodium borohydride reduction is carried out in dimethyl sulfoxide.
Reagent: Solution of 2 g sodium borohydride in 100 mL dry dimethyl sulfoxide (dissolved at 100 °C). Add 1 mL of the reagent to 0.1 mL of a solution of the monosaccharide mixture in aqueous ammonium hydroxide (1 M) and keep the mixture at 40 °C for 90 min. Add 0.1 mL of acetic acid for the destruction of the sodium borohydride. This solution is ready for acetylation using 1-methylimidazole as the catalyst. In this case, removal of boric acid is not necessary[96,97].

Acetylation
Acetylation is carried out by reaction with acetic anhydride. For accelerating the reaction the following catalysts are used: anhydrous sodium acetate, water-free pyridine and 1-methylimidazole.

Procedure

(a) With Anhydrous Sodium Acetate

After destroying the excess of sodium borohydride and removing the boric acid by evaporation with methanol, add 100 mg of anhydrous sodium acetate and 1 mL of acetic anhydride to the dry residue of the alditols, and heat the mixture in a closed vessel for 2 h at 100–120 °C. Then, remove the excess of acetic acid anhydride by evaporation in vacuo, extract the residue with 2 mL of dichloromethane, wash the organic phase several times with distilled water, and dry it. Then inject an aliquot into the GC[91,98].

(b) With Pyridine

Suspend the dry residue of the borohydride reduction, after removing the excess of boric acid, in 1 mL of pyridine and add 1 mL of acetic anhydride. Heat the mixture in a closed vessel at 100 °C for 2 h. After cooling the mixture, add 1 mL of toluene, evaporate the mixture to dryness, dissolve the residue in an adequate amount of chloroform, wash the organic phase twice with 5 mL of water, and evaporate it again to dryness. Dissolve this residue again in chloroform and inject an aliquot into the GC. The time of heating of the pyridine–acetic anhydride mixture varies from either 30 min at 60 °C, 60 min at 100 °C, or 60 min at 70 °C[99–102].

The removal of the pyridine–acetic anhydride mixture is important, since it causes tailing in the gas chormatogram. It can be performed by evaporation after addition of toluene[99].

(c) With 1-Methylimidazole

It has been found, that 1-methylimidazole is a very efficient catalyst for acetylation[103]. When using this substance for the acetylation of alditols it is not necessary to remove the boric acid after the sodium borohydride reduction in dimethyl sulfoxide, and the reaction can be carried out at room temperature. For this purpose add 0.2 mL of 1-methylimidazole and 2 mL of acetic anhydride to the reaction mixture from the reduction (~ 1.2 mL) and keep it for 10 min at room temperature with intense stirring. Then, add distilled water to destroy the excess of acetic anhydride. After cooling the mixture add 1 mL of dichloromethane and stir thoroughly again. After separation, dry the organic phase and inject an aliquot into the GC[96,97].

In the case of the reaction with pyridine, as well as with 1-methylimidazole, the reaction mixture can turn a brown color and artifact peaks can appear in the gas chromatogram[104]. This effect can be prevented by acetylation with water-free sodium acetate.

Gas Chromatographic Separation of the Nonsubstituted Alditol Acetates

Among the several stationary phases the most suitable are the following: ECNSS-M[92,93], OV-225[105], neopentylsebacic acid polyester[106], OV-275[98], QF-1/BDS (trifluoropropylmethylsiloxane/butandiol polysuccinate)[100], a mixture of polyethylene glycol adipate, XF-1150 and polyethylene glycol succinate[91], Silar-10C[96].

Table 17 lists the relative retention times of the alditol acetates for some of these stationary phases. For packed columns, strong polar phases, such as ECNSS-M, yield good separations of the single alditol acetates; for capillary columns, medium polar phases can also be used (Figure 9).

The alditol acetate GC method has been established for the quantitative determination of the units of polysaccharides obtained after acid hydrolysis. The results have been proved to be adequate compared to the modern anion exchange HPLC method using an amperometric detector. Disadvantages of the GC method are: (a) fructose-containing polysaccharides can not be determined, since upon reduction with borohydride, glucitol and mannitol are formed which can lead to misinterpretation of the gas chromatogram, (b) uronic acids can not be determined in contrast to HPLC and, (c) the sample preparation before the separation is more complicated than for the HPLC method (see Polysaccharides, Chapter 3)[107]. In the case of branched alditols, acetylation of the tertiary hydroxyl groups is slow and

Table 17.   Relative Retention Times of *O*-Alditol Acetates

Compound	A	B	C	D
Glycerol	0.13	-	0.184	-
D-Erythritol	0.42	-	0.431	0.28
D-Threitol	0.50	-	-	-
D-Ribitol (Adonitol)	0.83	-	0.784	0.53
L-Arabitol	0.87	1.00	0.820	0.56
D-Xylitol	1.00	1.24	1.000	0.66
L-Rhamnitol	0.66	-	0.596	0.41
L-Fucitol	0.69	0.76	0.625	0.43
2-Desoxyribitol	0.60	-	0.589	0.39
D-Sorbitol	1.41	2.43	1.366	0.90
D-Mannitol	1.26	2.11	1.117	0.79
D-Galactitol (Dulcitol)	1.33	2.27	1,249	0.84
D-Allitol	1.20	-	1.11	0.75
D-Altritol	1.27	-	1.192	-
D-Iditol	1.55	-	1.584	-
2-Desoxysorbitol	-	-	1.016	0.68
2-Desoxygalactitol	-	-	1.033	-
myo-Inositol	1.57	-	1.589	1.00

Index	Stationary Phase	Temperature (°C)	Dimension of the Column length (cm)	i.d. (cm)	Reference Substance	Literature
A	10% ECNSS-M on Chromo-sorb (60–80 mesh), washed, silanized	170–220 0.8 °C/min	184 Glass	0.32	Xylitol	Oades[93]
B	1% OV-225 on Chromosorb G-HP (80–100 mesh)	170–230 1 °C/min	150	0.3	Arabitol: abs. time 2.5 min	Metz, Ebert, Weicker[105]
C	OV-275 Glass capillary	165–215 2 °C/min	2 500	0.025	Xylitol	Klok, Cox, de Leeuw, Schenk[98]
D	Silar-10C Scot-Glass capillary column	190–230 4 °C/min	2 850	0.05	*myo*-Inositol	Henry, Blakeney, Harris, Stone[96]

incomplete. The best acetylation can be achieved with 1-methylimidazole and dimethyl sulfoxide[97], giving a mixture of the tetraacetate (16%) and pentaacetate (84%)[108].

Gas Chromatographic Separation of Methylated Alditols

These separations are important for elucidation of the structures of polysaccharides with the classical methylation analysis, which is carried out by the following methods, namely:

(a)   Methylation of the polysaccharide.

(b)   Hydrolysis.

(c)   Reduction of the *O*-methyl monosaccharide to the corresponding *O*-methyl alditols by borohydride.

(d)   Acetylation of the *O*-methylated alditols.

(e)   Separation of the acetylated-*O*-methyl alditols and identification of their structures by mass spectroscopy (see Polysaccharides, Chapter 3).

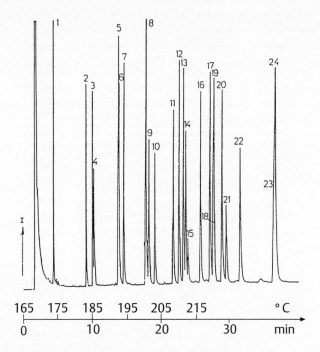

Figure 9.   Gas Chromatogram Showing the Separation of Alditol Acetates
Capillary column: L: 25 m, i.d.: 0.25 mm; *stationary phase*: OV-275
1: glycerol, **2**: 2,3,4,5-tetra-*O*-methylsorbitol, **3**: erythritol, **4**: digitoxitol, **5**: 2-deoxyribitol, **6**: rhamnitol, **7**: fucitol, **8**: 6-deoxysorbitol, **9**: ribitol, **10**: arabitol, **11**: 1,4-anhydromannitol, **12**: 1,5-anhydromannitol, **13**: xylitol, **14**: 2-deoxysorbitol, **15**: 2-deoxygalactitol, **16**: allitol, **17**: mannitol, **18**: 3-*O*-methylsorbitol, **19**: altritol, **20**: galactitol, **21**: 4-*O*-methylsorbitol, **22**: sorbitol (glucitol), **23**: iditol, **24**: *myo*-inositol (Klok, Cox, de Leeuw and Schenk[98]; with permission)

The acetylation of the *O*-methyl alditols is carried out in the same manner to that of the mono-substituted compounds.

The separations of the partially *O*-methylated acetyl alditols on packed columns with stationary phases of high polarity but low stability, such as ECNSS-M or OV-255, are incomplete. Better results are obtained with capillary columns using the following phases:
(a)   SP-1000 (glass capillary column)[109].
(b)   OV-275, BP-75 (WCOT quartz column)[98,110].
(c)   SP-2100 (WCOT column)[111].
(d)   SP-2330 (WCOT column)[112].

With weak polar phases such as SP-2100 the separation of the partially *O*-methylated acetylated pentitols can be achieved, but it is less suitable for the corresponding hexitols. The best results are obtained with polar phases OV-275 (bound on the wall of the capillary = BP-75) and SP-2330. In Tables 18 and 19 the relative retention times of partially *O*-methylated alditol acetates are listed.

Table 18. Relative Retention Times of Partially Methylated Alditol Acetates Obtained from the Corresponding Monoaldoses, Index A

Positions of the Methyl Groups	Rham-nose	Fucose	Ribose	Arabi-nose	Xylose	Man-nose	Galac-tose	Glucose
unsubstituted	0.596	0.625	0.784	0.820	1.000	1.170	1.249	1.366
1-	0.370	0.386	0.518	0.534	0.666	0.865	0.918	0.983
2-	0.511	0.541	0.640	0.685	0.777	1.035	1.098	1.123
3-	0.589	0.623	0.623	0.714	0.768	1.166	1.247	1.188
4-	0.554	0.611	0.640	0.707	0.777	1.166	1.247	1.277
5-	0.445	0.505	0.518	0.550	0.666	1.035	1.098	1.142
6-						0.865	0.918	0.995
1,2-	0.274	0.278	0.357	0.379	0.443	0.671	0.715	0.743
1,3-	0.296	0.300	0.355	0.368	0.430	0.746	0.798	0.789
1,4-	0.337	0.347	0.397	0.425	0.492	0.857	0.875	0.931
1,5-	0.276	0.304	0.333	0.339	0.411	0.758	0.799	0.824
1,6-	-	-	-	-	-	0.586	0.631	0.689
2,3-	0.398	0.459	0.424	0.493	0.545	0.883	0.989	0.961
2,4-	0.390	0.425	0.408	0.498	0.492	0.922	0.994	0.914
2,5-	-	-	0.397	0.433	0.492	0.893	0.974	0.951
2,6-						0.758	0.799	0.829
3,4-	0.383	0.475	0.424	0.509	0.545	0.951	1.049	0.963
3,5-	-	-	0.335	0.384	0.430	0.922	0.994	0.980
3,6-						0.857	0.875	0.871
4,5-	-	-	0.357	0.401	0.443	0.883	0.989	0.949
4,6-						0.746	0.798	0.844
5,6-						0.671	0.715	0.761
1,2,3-	-	-	0.195	0.232	0.256	0.498	0.579	0.543
1,2,4-	-	-	0.210	0.252	0.254	0.565	0.602	-
1,2,5-	-	-	0.216	0.232	0.266	0.560	0.629	0.616
1,2,6-						0.446	0.469	0.494
1,3,4-	-	-	0.218	0.250	0.277	0.590	0.661	0.585
1,3,5-	-	-	0.182	0.194	0.231	0.577	0.612	0.625
1,3,6-						0.493	0.501	0.524
1,4,5-	-	-	0.216	0.227	0.266	0.623	0.678	0.667
1,4,6-						0.493	0.501	0.573
1,5,6-						0.446	0.469	0.494
2,3,4-	0.234	0.297	0.216	0.305	0.314	0.657	0.776	0.654
2,3,5-	-	0.279	0.218	0.255	0.277	0.690	0.756	0.673
2,3,6-						0.623	0.678	0.694
2,4,5-	-	-	0.210	0.258	0.254	0.690	0.756	-
2,4,6-						0.577	0.612	0.568
2,5,6-						0.560	0.629	0.610
3,4,5-	-	-	0.195	0.239	0.256	0.657	0.776	0.679
3,4,6-						0.590	0.661	0.593
3,5,6-						0.565	0.602	0.602
4,5,6-						0.498	0.579	0.543
1,3,4,5-						0.387	0.447	0.388
1,3,4,6-						0.315	0.361	0.317
2,3,4,6-						0.387	0.447	0.392
2,3,5,6-						0.396	0.432	-

Table 19.   Relative Retention Times of Partially Methylated Alditol Acetates, Index B

Positions of the Methyl Groups	Rham-nose	Fucose	Arabi-nose	Xylose	Mannose	Galac-tose	Glucose
unsubstituted	1.420	1.503	1.839	2.143	2.396	2.500	2.600
2-	1.241	1.304	1.567	1.729	2.229	2.332	2.370
3-	1.392	1.469	1.623	1.715	2.376	2.480	2.439
4-	1.320	1.450	1.604	1.729	2.376	2.480	2.439
6-	-	-	-	-	1.960	2.125	
2,3-	0980	1.103	1.170	1.274	1.991	2.171	2.146
2,4-	-	1.052	1.184	1.169	2.050	2.195	-
2,6-	-	-	-	-	1.760	1.848	1.904
3,4-	0.945	1.149	1.199	1.274	2.091	2.246	-
3,6-	-	-	-	-	1.934	1.983	1.968
4,6-	-	-	-	-	1.746	1.848	1.934
2,5-	-	-	1.041	-	-	-	-
3,5-	-	-	0.944	-	-	-	-
2,3,4-	0.583	0.732	0.748	0.779	1.545	1.781	-
2,3,5-	-	-	0.628	-	-	-	-
2,3,6-	-	-	-	-	1.491	1.587	1.637
2,4,6-	-	-	-	-	1.387	1.484	1.406
3,4,6-	-	-	-	-	1.409	1.566	1.437
2,3,4,6-	-	-	-	-	0.984	1.120	1.000

Index	Stationary Phase	Temperature (°C)	Dimension of the Column length (cm)	i.d. (cm)	Reference Substance	Literature
A	OV-275 Glass capillary column	165–215 2 °C/min	2 500	0.025	Xylitol pentaacetate	Klok, Cox de Leeuw, Schenk[98]
B	SP-2330 Quartz capillary column	160–210 2 °C/min 210–240 5 °C/min	3 000	0.025	1,5-Diacetyl 2,3,4,6-Tetra-O-methylsorbitol	Shea, Carpita[112]

Amino Sugar Alcohols

Reduction of reducing amino sugars, which are present as units of many mucopolysaccharides and glycoproteins, with sodium borohydride affords the corresponding amino sugar alcohols, which can be acetylated in the same manner as the sugar alcohols themselves.

A disadvantage is that these acetylated amino sugar alcohols have long retention times on columns whose stationary phases are suitable for the separation of sugar alcohols.

This fact can be overcome by converting the amino sugar alcohols into the corresponding dimethylamino compounds with formaldehyde and sodium cyanoborohydride according to the following reaction scheme[99].

$$CH_2OH[CHOH]_n-CH(NH_2)-CH_2OH \xrightarrow[\text{[BH}_3\text{CN]}^-]{CH_2=O} CH_2OH[CHOH]_n-CH[N(CH_3)_2]-CH_2OH$$

Procedure

To an aqueous solution of 0.1–2 mg of the mixture of different monosaccharides add 3 mg of sodium borohydride and keep the mixture for 2 h at room temperature. Adjust the pH of the solution to 6.0 by adding 1 M acetic acid, add 0.5 mL of a 0.7% aqueous solution of formaldehyde and a further 3 mg of sodium cyanoborohydride, and keep the mixture for 3 h at room temperature. Destroy the excess of cyanoborohydride with a few drops of concd. hydrochloric acid and evaporate the solution to dryness. Remove the boric acid as its methyl ester by repeated evaporation with methanol, then carry out the acetylation in pyridine–acetic anhydride, as described on page 234, of the dimethylaminoalditols.

Another possibility for shortening the retention times of the acetylated amino sugars is by conversion into their anhydro sugars by reaction with nitrous acid which are then further reduced by sodium borohydride into the corresponding anhydroalditols. The procedures are described on pages 229, 230. The acetylation is carried out in the usual manner using acetic anhydride and 1-methylimidazole[96].

Table 20. Relative Retention Times of *O*-Acetylated Amino Sugar Alcohols

Compound	A	B	C	D
Allosaminitol	0.83	-	-	-
Glucosaminitol	1.00	2.98	2.39	-
Talosaminitol	1.03	-	-	-
Altrosaminitol	1.05	-	-	-
Idosaminitol	1.13	-	-	-
Galactosaminitol	1.22	3.39	-	-
Gulosaminitol	1.31	-	-	-
Mannosaminitol	1.36	3.30	2.66	-
N-Dimethylglucosaminitol	-	-	-	1.16
N-Dimethylgalactosaminitol	-	-	-	1.21
N-Dimethylmannosaminitol	-	-	-	1.36
2,5-Anhydromannitol	-	0.54	-	-
2,5-Anhydrotalitol	-	0.74	-	-

Index	Stationary Phase	Temperature (°C)	Dimension of the Column length (cm)	i.d. (cm)	Reference Substance	Literature
A	Poly A-103 (Appl. Science Lab.) Glass capillary, coated	175, 50 min Isothermal 175–210, 0.5 °C/min	2 500	0.025	*O*-Acetyl-glucos-aminitol	Kontrohr, Kocsis[94]
B	Silar-10C SCOT Glass capillary	230, 4 min Isothermal 230–250, 4 °C/min	2 850	0.05	*O*-Acetyl-*myo*-inositol	Henry, Blakeney, Harris, Stone[96]
C	SP-2340 Glass capillary	185, 10 min 185–220, 30 min 220, 5 min 235, 15 min 235, 30 min	3 000	0.03	*O*-Acetyl-allitol	Kraus, Shinnick, Marlett[113]
D	2% EG SS-X on Chromo-sorb WAW-DMCS (60–80 mesh)	195 Isothermal	200 Glass	0.3	*O*-Acetyl-xylitol	Kiho, Ukai, Hara[99]

For the separation of O-acetyl aminoalditols and O-acetyl anhydroalditols the following stationary phases are recommended: Poly-A-103[94], OV-225[101], Silar-10C[96], SP-2340[113]. Table 20 shows the separation on these stationary phases and the relative retention times of some of the O-acetylated amino sugar compounds.

Aldonitrile Acetates

Aldonitrile acetates are derivatives of monoaldoses which also give only one peak per compound in the gas chromatogram. They are formed by reaction with hydroxylamine to the corresponding oximes which are dehydrated and acetylated with acetic anhydride to the corresponding aldonitrile acetates according to the following scheme[114].

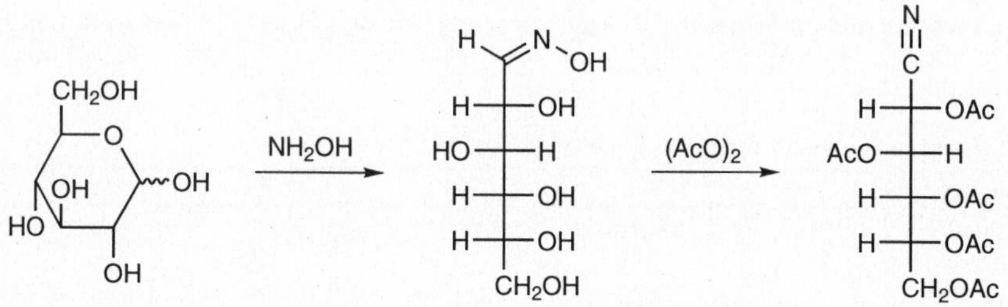

Reaction Pathway for the Formation of Aldonitrile Acetates (Example: D-Glucose)

Procedures

Several procedures exist. The most important are described below.

(a) Reaction with Pyridine[115–120]

Dissolve the dry mixture of monosaccharides (~ ca. 10 mg; dried over P_2O_5) in dry pyridine; add 10–15 mg of dry hydroxylamine hydrochloride and heat the mixture in a sealed reaction tube for 30 min at 90 °C. Then cool, add 1.5 mL of acetic anhydride and heat the mixture again for 30 min at 90 °C. Cool, evaporate the mixture of pyridine and acetic anhydride at 50 °C, dissolve the residue in 1 mL of chloroform, and inject an aliquot (~ 2 µL) into the GC[118].

For this procedure several variations exist regarding the pyridine–acetic anhydride ratio and the purification of the aldonitrile acetates[115,116,119]. For example, the sugar mixture can be dissolved in 0.5 mL of pyridine and, after addition of hydroxylamine, the mixture is heated for only 20 min at 60–65 °C. Then 0.5 mL of acetic anhydride is added and the mixture is again heated at 60–65 °C for 20 min. After cooling, 1 mL of water and 2 mL of chloroform are added, the organic phase is washed twice with water, and 2,2-dimethoxypropane is then added to remove traces of water. After standing for 15–20 min, the mixture is evaporated to dryness again and the residue dissolved in an aliquot of chloroform[119,121].

(b) Reaction with Methanol

Dissolve the dry mixture of the sugars (~ 5 mg) in 1 mL of methanol, add 5 mg of hydroxylamine hydrochloride and 12.5 mg of anhydrous sodium acetate, and heat the mixture in a sealed reaction tube (preferably a vial with a Teflon screw cap) at 60 °C for 30 min. Evaporate the methanol with a nitrogen stream to dryness, add 0.3 mL of acetic anhydride and heat for a further 60 min at 100 °C. Cool and inject an aliquot into the GC[122].

(c) Reaction with 1-Methylimidazole

This compound is a very effective catalyst for the acetylation of alcohols, glycols, phenols and sugars[123] and it has also been used for the formation of aldonitrile acetates of monoaldoses.

Dissolve the mixture of sugars (approx. 5 mg of each substance) in 0.5 mL of 1-methylimidazole containing hydroxylamine hydrochloride (50 mg/mL) and heat the mixture in a vial with a screw cap at 50 °C for 5 min. After cooling, add 0.3 mL of acetic anhydride and then, after 5 min, add 1 mL of chloroform. Wash the organic phase three times with water, and remove the traces of water with anhydrous sodium sulfate to give a sample which is ready for analysis by GC[124].

This variation shortens the formation of the aldonitrile acetates from 40–100 min in pyridine to 10 min. In another variation, aqueous solutions of sugars are used without evaporation of the water before the reaction itself[125]. Another effective catalyst for acetylation is 4-dimethylaminopyridine[126].

For GC separation strong polar phases are used. Suitable materials for packed columns are: Neopentyl glycol polysuccinate[117,118,127], neopentyl glycol polysebacinate[128], diethylene glycol polyadipate[129], OV-225[119]. For separation on glass capillaries, SE-30[122] and OV-1[126] have been used.

Figure 10 shows the separation of aldonitrile acetates of common monoaldoses which are units of polysaccharides.

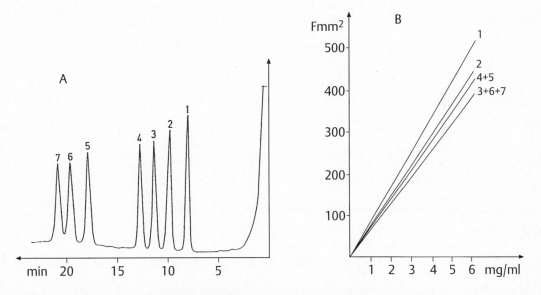

Figure 10. Gas Chromatogram Showing the Aldonitrile Acetates of Units of Polysaccharides
Column: L: 2.0 m, i.d.: 0.2 cm; *stationary phase*: 3% neopentyl glycol polysuccinate; *temperature*: 190–230 °C/2 °C min; *concentration*: 3 mg/mL; *injection*: 4 µL
1: rhamnose, **2**: fucose, **3**: arabinose, **4**: xylose, **5**: mannose, **6**: glucose, **7**: galactose (Mergenthaler and Scherz[118]; with permission)

Ketoses react with the system hydroxylamine–acetic anhydride to give compounds which do not appear in the gas chromatograms under these conditions. Amino sugars (e.g., 2-amino-2-deoxyaldoses) also yield suitable aldonitrile acetates, although the yields are not constant and the retention times are very long. An advantage is to convert these compounds into their corresponding anhydro sugars whose aldonitrile acetates have similar retention times to those of the aldopentoses and aldohexoses[119,130].

For improved isolation of amino sugars from a mixture of neutral sugars, the solution of both is passed through a cation exchanger (e.g., Biorad AG 50WX8; H^+). The amino sugars are fixed on the resin and can be eluted with hydrochloric acid (1 mol/L) and deaminated with sodium nitrite[130].

Uronic Acids

Uronic acids, important units of many polysaccharides, do not afford volatile derivatives after reaction with hydroxylamine and acetic anhydride. They can be analyzed with this reaction using the following procedure[131]:

(a)   Cleavage of the polysaccharides with methanolic hydrochloric acid (methanolysis) to form the 1-O-methyl glycosides and 1-O-methyluronic acid methyl esters.

(b)   Reduction of the 1-O-methyluronic acid methyl esters to the corresponding 1-O-methyl glycosides with sodium borohydride.

(c)   Cleavage of the 1-O-methyl glycosides to the free sugars and derivatization to their aldonitrile acetates.

Figure 11 shows the gas chromatogram of the aldonitrile acetates of the units of tragacanth, obtained by methanolysis with and without reduction with borohydride. The enhancement of the galactose peak after the methanolysis and borohydride procedure corresponds to the galacturonic acid part of the polysaccharide.

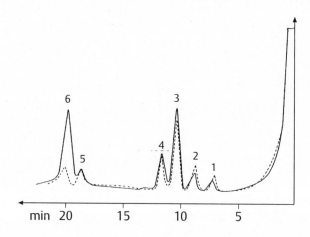

Figure 11.   Gas Chromatogram Illustrating the Separation of the Hydrolysis Products of Tragacanth Obtained by Methanolysis with (—) and without (---) Reduction with Borohydride
Column: L: 2 m, i.d.: 3.2 mm; *stationary phase*: 3% neopentylglycol polysuccinate; *temperature*: 190–230 °C, 2 °C/min
1: rhamnose, **2**: fucose, **3**: arabinose, **4**: xylose, **5**: glucose, **6**: galactose (Mergenthaler and Scherz[131]; with permission)

In Table 21 the relative retention times of the aldonitrile acetates of the common monoaldoses and amino sugars (as anhydro sugars) are listed.

Table 21.   Relative Retention Times of *O*-Aldonitrile Acetates

Compound	A	B	C
DL-Glyceraldehyde	0.25	-	-
D-Erythrose	1.00	-	-
D-Ribose	2.25	1.00	-
D- or L-Arabinose	2.35	1.15	0.58
D-Xylose	2.50	1.34	0.65
D-Lyxose	2.35	1.08	-
L-Rhamnose	2.15	0.80	0.42
L-Fucose	2.35	0.98	0.51
D-Allose	3.40	-	-
D-Mannose	3.55	1.88	0.91
D-Talose	3.55	-	-
D-Glucose	3.65	2.14	1.00
D-Galactose	3.75	2.23	1.07
L-Idose	3.90	-	-
2-Amino-2-deoxygalactose (as 2,5-Anhydrotalose)	-	1.60	-
2-Amino-2-deoxyglucose (as 2,5-Anhydromannose)	-	1.25	-

Index	Stationary Phase	Temperature (°C)	Dimension of the Column length (cm)	i.d. (cm)	Reference Substance	Literature
A	2% OV-17 on Chromosorb WHP (80–100 mesh)	130–300 5 °C/min	123 Glass	0.2	D-Erythrose: abs. time 4.15 min	Seymour, Chen, Bishop[121]
B	3% OV-225 on Gas Chrom Q (100–120 mesh)	170–180 1 °C/min 180–200 2 °C/min	200 Glass	0.2	D-Ribose	Turner, Cherniak[119]
C	3% Neopentyl glycol polysuccinate on Chromosorb WAW (100–120 mesh)	190–230 2 °C/min	200 Glass	0.2	D-Glucose	Mergenthaler, Scherz[118]

This derivatization is suitable for the quantitative determination of monoaldoses and is used especially in the analysis of thickening agents[117,132,133]. The conversion of pentoses to the corresponding aldonitrile acetates is quantitative; in the case of hexoses (e.g., glucose, galactose), side products are formed (e.g., hydroxyhexosylamine acetates) which do not appear in the gas chromatogram under the described conditions. For the latter compounds, the reaction conditions have to be kept constant for the samples and for the standards. It is further recommended to use all glass equipment (glass columns and glass inlets) to prevent decomposition of the side products on the walls of the metal columns.

In Table 22 the results of a quantitative analysis of the neutral monoaldoses of two thickening agents by the aldonitrile acetate method are compared with those of the TMS method; they show good agreement[115].

Table 22. Comparison of the Results of the Quantitative Analysis of the Neutral Monosaccharides in Guar and Gum Arabic with the Aldonitrile Acetate and the TMS Method

| | % Monosaccharide | |
	Aldonitrile Acetate	TMS Ether
Guar		
Mannose	61.82	62.10
Galactose	34.31	34.58
Gum arabic		
Rhamnose	12.20	12.29
Arabinose	16.71	16.83
Galactose	47.05	46.99

O-Methoxime Acetates

Reaction of monosaccharides and reducing oligosaccharides with O-methylhydroxylamine hydrochloride yields the corresponding open-chain methoximes, which have to be transformed into volatile compounds for separation by GC. A disadvantage of the O-TMS compounds is that the differences in relative retention times of the single sugars are very small, although sufficient separations can be obtained with long capillary columns and nonpolar phases. With the O-acetyl compounds better results are obtained and, furthermore, neutral monosaccharides and amino sugars can be determined in one step, which occurs often in the hydrolysis of glycoproteins[65,126,129,134].

Procedure[129,134]

Reagent: 300 mg of O-methylhydroxylamine is dissolved in 1.0 mL of methanol, 1.78 mL of pyridine and 0.22 mL of 1-dimethylamino-2-propanol.

Dissolve the dry sample (1–9 mg total amount of neutral monosaccharides and amino sugars) in 0.4 mL of the reagent, in a vial with a Teflon screw cap, and heat at 70–80 °C for 20 min. After cooling, remove the solvent with a stream of dry nitrogen. Carry out the acetylation with 0.25 mL of dry pyridine and 0.75 mL of acetic anhydride at 70–80 °C for 30 min. Then, concentrate the mixture in vacuo, add 1 mL of dichloromethane, and wash the organic phase once with 1 mL of 1 M aqueous hydrochloric acid and three times with water. Dry the organic phase with anhydrous sodium sulfate, then dilute it with an adequate amount of a mixture of heptane–dichloromethane, and inject an aliquot into the GC[129,134]. Figure 12 shows the separation of some compounds of both groups.

For the separation of monosaccharides and amino sugars, capillary columns coated with Carbowax 20M was found to be suitable. One single compound yields two peaks corresponding to the E- and Z-forms of the methoximes. The ratios of the amounts of both isomers were found to be constant. The relative retention times of common monosaccharides and amino sugars are listed in Table 23. With this method the sugar units in glycoproteins and dietary fibres can be determined quantitatively.

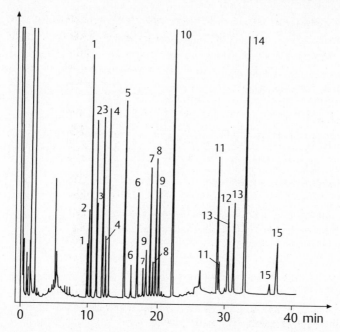

Figure 12. Gas Chromatogram Showing the Separation of Neutral and Amino Sugars as their *O*-Methoxime Acetates
Column: fused silica: L: 25 m; i.d.: 0.3 mm; stationary phase: Carbowax 20 M; temperature: 80–230 °C **1**: rhamnose, **2**: fucose, **3**: arabinose, **4**: xylose, **5**: xylitol, **6**: 3-*O*-Me-Glc, **7**: mannose, **8**: galactose, **9**: glucose, **10**: *myo*-inositol, **11**: glucosamine, **12**: galactosamine, **13**: mannosamine, **14**: NMe-Glc, **15**: muramic acid (Neeser and Schweizer[134]; with permission)

Table 23. Relative Retention Times of *O*-Acetyl Methoximes of Neutral Monosaccharides and Amino Sugars

Compound	A	Compound	A
Arabinose	0.513	Galactose	0.833
	0.553		0.915
Xylose	0.572	Mannose	0.812
	0.589		0.857
Ribose	-	Glucosamine (HCl & *N*-Acetyl)	0.877
Rhamnose	0.455		0.885
	0.477	Galactosamine (HCl & *N*-Acetyl)	0.924
Fucose	0.462	Mannosamine (HCl & *N*-Acetyl)	0.924
	0.507		0.950
Glucose	0.875		
	0.893		

Index	Stationary Phase	Temperature (°C)	Dimension of the Column length (cm)	i.d. (cm)	Reference Substance	Literature
A	Carbowax 20M capillary column	80–180 20 °C/min 180–210 2 °C/min	2 500	0.03	*myo*-Inositol (sugar) *N*-Methylglucosamine (amino sugar)	Neeser, Schweizer[134]

2.1.4.4. Trifluoroacetylation

The *O*-trifluoroacetyl esters of mono- and oligosaccharides are the most volatile sugar derivatives allowing gas chromatographic separations up to tetrasaccharides without difficulties. Another advantage is the possibility of using an electron capture detector suitable for extending the detection limit to the nano- and picomolar range.

Derivatization Methods

In most cases trifluoroacetic anhydride (TFA) is the derivatization reagent. Procedures exist with and without catalysts[135,136].

Reaction with Trifluoroacetic Anhydride

I. Reaction of the Sugars with TFA in the Absence of Catalysts

This modification requires higher temperatures and reaction times.

Procedure

Heat 0.5 mg of the sample, in a thick-walled vial with a Teflon screw cap, with 200 µL of dichloromethane and 100 µL TFA at 110 °C for 20 min. Cool the vial and remove the excess reagent with a stream of dry nitrogen and dissolve the dry residue in 200 µL of the same solvent or in benzene. An aliquot is used for analysis by GC[135].

Instead of dichloromethane other solvents, for example, ethyl acetate have been used[137–139]. In ethyl acetate the reaction between sugar alcohols and TFA was found to be quantitative at room temperature after 30 min[139]. The reaction with TFA alone leads to shifts of the isomerization equilibria for single sugars. For solid galactose, for example, four peaks are obtained with TFA–dichloromethane, which are the different pyranose and furanose forms of the sugar; this has been evaluated by mass spectrometry studies[140]. In contrast, it is well-known that galactose exists only in one form in the solid state.

II. Reaction of the Sugars with TFA in the Presence of Catalysts

In the presence of a catalyst the reaction occurs quantitatively under much gentler conditions. The following compounds have been used as catalysts: anhydrous sodium trifluoroacetate, pyridine, dimethylformamide and 1-methylimidazole.

Procedures

(a) With Sodium Trifluoroacetate

Analogous to the system anhydrous sodium acetate–acetic anhydride, this system accelerates the reaction between TFA and polyols with acetonitrile[141] or formamide[142] as solvents. The following describes that using formamide as solvent[142].

Dissolve 1 mg of the dry sample and 0.3 mg of anhydrous sodium trifluoroacetate in 50 µL of formamide at room temperature. Add 200 µL TFA whilst cooling in an ice bath and after 10 min at room temperature the reaction is complete. Then 1–2 µL of the reaction mixture is used for analysis by GC.

(b) With Pyridine

Only the addition of a small amount is necessary as described in the following procedure for micro-amounts[140,143].

Dry approximately 100 µg of the sample overnight in vacuo over P_2O_5. Then add 65 µL of dichloromethane, 30 µL of TFA and 5 µL of pyridine. After a reaction time of 3 h at 20 °C, inject an aliquot into the GC[140].

With this procedure the equilibria of the isomers, as present in the sample, are not altered. The pyridine trifluoroacetate, which is formed during the reaction, decomposes upon injection into the GC to pyridine and trifluoroacetic acid. The pyridine elutes as a broad peak. A higher pyridine content in the reaction mixture interferes with the GC separation.

(c) With Dimethylformamide

Using this solvent the reaction between sugar or sugar alcohols with TFA is complete within a few minutes[148]. The following describes the micromethod procedure.

Dissolve the dry sample in a microvial in 50–100 µL of dry dimethylformamide and add 30–60 µL of TFA. After 5 minutes inject an aliquot into the GC (1–3 µL).

(d) With 1-Methylimidazole

This compound enables complete trifluoroacetylation in the presence of borate ions[144], which are present after the reduction of reducing sugars to the corresponding alcohols (similar to acetylation, see page 234). With methods a–c, the borate ions must be removed. The following describes the micromethod procedure[144].

Dissolve the dry sample (max. 2 mg) in 40 µL of pyridine and 10 µL of 1-methylimidazole in a microvial with a Teflon screw cap. Heat at 75 °C for 20 min, add 80 µL of TFA and heat the closed vessel again for 10 min at 75 °C. After cooling, evaporate the reaction mixture with a stream of dry nitrogen and dissolve the residue in 100 µL of dichloromethane.

Reaction with N-Methylbis(trifluoroacetamide)

An alternative to the reaction with TFA is the derivatization with compounds on which the trifluoroacetyl group is weakly bound and easy transferable. The most suitable of these substances is N-methylbis(trifluoroacetamide) (MBTFA). With this compound the trifluoroacetylation of polyhydroxy compounds such as sugars and sugar alcohols is quantitative under mild conditions and with short reaction times[145–147].

$$Me-N\begin{array}{l}{}^{COCF_3}\\{}_{COCF_3}\end{array}$$

N-methylbis(trifluoroacetamide)

Procedure

Dissolve 2–5 mg of the dry sample, placed in a vial with a Teflon screw cap, in 0.5 mL of dry pyridine and add 0.5 mL of MBTFA. Heat the closed vessel at 65 °C for 1 h and then inject an aliquot into the GC[145,146].

For a micromodification dissolve 200 µg of the dry sample, placed in a closeable microvial, in 20 µL of pyridine at 75 °C. Then, add 40 µL of MBTFA and heat the closed vessel for 10 min at 75 °C. After cooling for 15 min inject an aliquot (~ 1 µL) into the GC[147].

For the gas chromatographic separation all glass systems are needed, otherwise a degradation of the volatile trifluoroacetates of the sugars may occur on metal surfaces. The following stationary phases are used: OV-17[140,145,148], Dexsil 300[140], Dexsil 410[147], GE-XF 1105[137,139,148], OV-210[145,149], FS-1265

(Fluoralkyl polysiloxane)[140], CNSI (Cyanethylmethyl silicon)[150]. In Tables 24–26, the relative retention times of the O-trifluoroacetyl derivatives of the common mono- and oligosaccharides as well as sugar alcohols are listed for several stationary phases.

Table 24.   Relative Retention Times of O-Trifluoroacetyl Monosaccharides (A,B,C) and Their O-Methyl Glycosides (D)

Compound	A	B	C	D[a]
D(L)-Arabinose	-	-	0.78[b]	0.465
				0.470
				0.535
D-Xylose	-	-	0.75[b]	0.446
				0.467
				0.505
D-Ribose	-	-	-	0.500
				0.538
				0.553
				0.669
L-Fucose	0.39	-	-	0.441
	0.41	-	-	0.444
	0.64	-	-	0.573
L-Rhamnose	-	-	-	0.376
				0.503
D-Glucose	1.00	0.398	1.00	1.000
	1.22	0.504	1.21	1.069
D-Galactose	0.99	-	-	0.889
	1.23	-	-	0.919
				0.941
				1.099
D-Mannose	0.96	-	-	0.967
	1.39	-	-	1.148
D-Fructose	-	0.400	0.97[b]	-
N-Acetylglucosamine	1.76	-	-	1.417
	2.11	-	-	1.651
	2.41	-	-	1.661
	-	-	-	1.725
N-Acetylgalactosamine	1.79	-	-	1.513
	1.97	-	-	1.569
	2.08	-	-	1.618
	2.38	-	-	1.715

[a]Methyl glycoside. [b]Mean peak.

Index	Stationary Phase	Temperature (°C)	Dimension of the Column length (cm) i.d. (cm)		Reference Substance	Literature
A	2% XF-1105 on Gas-Chrom P (80–100 mesh)	80 2 °C/min	180 Glass	0.2	α-D-glucose	Tomana, Niedermeier, Spivey[137]
B	3% Dexsil 410 on Chromosorb W-HP	complex[a]	180 Glass	0.2	β-D-phenyl-glucopyranoside	Englmaier[144]
C	10% OV-17 on Chromosorb W (HP)	75–225 1 °C/min	180 Glass	0.4	α-D-glucose abs. time 4.62 min	Sullivan, Schewe[145]
D	OV-210 capillary column	120–210	700	0.5	α-D-glucose abs. time 37.2 min	Wrann, Todd[149]

[a]1.5 min 100 °C; 3.5 min 3.5 °C/min; 6.0 min 6.0 °C/min; 5.0 min 15 °C/min; 3.7 min 35 °C/min; 6.3 min 310 °C (final temperature).

Table 25.   Relative Retention Times of *O*-Trifluoroacetyl Oligosaccharides

Compound	A	B	C	Compound	A	B	C
Sucrose	1.017	1.00	-	Melezitose	-	-	1.131
Maltose	1.045	1.20	-	Panose	-	-	1.181
	1.070	1.36	-		-	-	1.215
Lactose	-	1.61	-	Cellotriose	-	-	1.423
	-	2.18	-		-	-	1.488
Trehalose	-	1.54	-	1-Kestose	-	-	1.043
Cellobiose	-	1.22	-	Isopanose	-	-	1.372
	-	1.56	-		-	-	1.643
Raffinose	1.165	-	1.00	Xylotriose	-	-	1.988
Fructotriose	1.151	-	-		-	-	2.026
Maltotriose	-	-	1.326	Stachyose	1.245	-	-
	-	-	1.342	Fructotetraose	1.228	-	-
Isomaltotriose	-	-	1.056				
	-	-	1.306				

Index	Stationary Phase	Temperature (°C)	Dimension of the Column length (cm)	i.d. (cm)	Reference Substance	Literature
A	3% Dexsil 410 on Chromosorb W-HP (80–100 mesh)	complex[a]	180 Glass	0.2	Phenyl-β-D-glucoside abs. time 14.28 min	Engelmaier[147]
B	3% SE-52 on Chromosorb W (silanized)	160 Isothermal	100 Steel	0.3	Sucrose abs. time 6 min	Ueno, Kurihara, Nakajima[142]
C	DB-5 (liquid phase) Capillary column	180, 5 min 180–200 5 °C/min 200, Isoth.	3 000 Quartz	0.03	Raffinose abs. time 7.63 min	Selosse, Reilly[146]

[a]1.5 min 100 °C; 3.5 min 3.5 °C/min; 6.0 min 6.0 °C/min; 5.0 min 15 °C/min; 3.7 min 35 °C/min; 6.3 min 310 °C (final temperature).

Table 26.   Relative Retention Times of *O*-Trifluoroacetyl Sugar Alcohols

Compound	A	B	C	D	E
Erythritol	0.35	0.134	-	0.50	-
Threitol	0.45	-	-	0.59	-
Ribitol (Adonitol)	0.82	0.251	0.70	0.93	-
Arabitol	1.00	0.285	0.85	1.00	0.731
Xylitol	1.17	-	1.00	1.07	-
Sorbitol	2.25	0.445	1.92	1.50	0.906
Dulcitol (Galactitol)	2.48	0.490	2.12	1.74	0.946
Mannitol	1.69	0.392	1.44	1.47	0.876
Talitol (Altritol)	1.90	-	-	1.54	-
Allitol	1.49	-	-	1.82	-
Iditol	2.25	-	-	-	-
Rhamnitol	-	-	0.42	0.71	-
Fucitol	-	-	0.57	0.82	0.624
2-Desoxyribitol	-	-	0.55	-	-
2-Desoxysorbitol	-	-	1.23	-	-
2-Desoxygalactitol	-	-	1.57	-	-
Glucosaminitol	-	-	-	-	1.349
Galactosaminitol	-	-	-	-	1.432

Table 26.   (Continued)

Index	Stationary Phase	Temperature (°C)	Dimension of the Column length (cm)	i.d. (cm)	Reference Substance	Literature
A	2% Cyanoethylmethyl silicon (CNSI) on Anachrom	140 Isothermal	150	0.4	L-Arabitol abs. time 4.3 min	Matsui, Okada, Imanari, Tamura[150]
B	3% Dexsil 410 on Chromosorb	complex[a]	180	0.2	Phenyl-β-D-glucopyrano-side	Englmaier[147]
C	2% XF-1105 Gas-Chrom P	140	180	0.4	Xylitol	Imanari, Arakawa Tamura[139]
D	FS-1265 Capillary column	150	1 600	0.01	Arabitol	Shapira[143]
E	OV-210 Capillary column	120–210 1 °C/min	700	0.05	α-D-glucose	Wrann, Todd[149]

[a]1.5 min 100 °C; 3.5 min 3.5 °C/min; 6.0 min 6.0 °C/min; 5.0 min 15 °C/min; 3.7 min 35 °C/min; 6.3 min 310 °C (final temperature).

Figure 13 shows the gas chromatogram of some trifluoroacetyl trisaccharides.

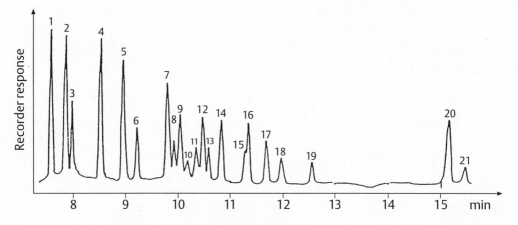

Figure 13.   Gas Chromatogram Illustrating the Separation of Thirteen Trisaccharides as *O*-Trifluoroacetates *Column*: Quartz capillary, L: 30 m, i.d.: 0.26 mm; *stationary phase*: DB-5; *temperature*: 5 min 180 °C, then 5 °C/min till 200 °C, then further isothermal
1: raffinose, **2**: 1-kestose, **3**: isomaltotriose 1, **4**: melezitose, **5**: panose 1, **6**: panose 2, **7**: 3-*O*-α-isomaltosyl-D-glucose 1, **8**: isomaltotriose 2, **9**: maltotriose 1, **10**: maltotriose 2, **11**: 4-*O*-β-laminaribiosyl-D-glucose 1, **12**: isopanose 1, **13**: 3-*O*-α-isomaltosyl-D-glucose 2, **14**: cellotriose 1 + 4-*O*-β-laminaribiosyl-D-glucose 2, **15**: laminaritriose 1, **16**: laminaritriose 2 + cellotriose 2, **17**: 3-*O*-β-cellobiosyl-D-glucose 1, **18**: 3-*O*-β-cellobiosyl-D-glucose 2, **19**: isopanose 2, **20**: xylotriose 1, **21**: xylotriose 2 (Selosse and Reilly[146]; with permission)

O-Alkyl Sugar Oxime Trifluoroacetates

O-Alkyloximation is an effective method for protection of carbonyl groups in mono- and oligosaccharides. Subsequent *O*-trifluoroacetylation leads to volatile compounds which can easily be separated by GC. *O*-methylhydroxylamine, *O*-1-butylhydroxylamine[151–153] and *O*-1-benzylhydroxylamine[67], all as their hydrochlorides, are the most suitable reagents for this purpose.

Procedure

In a thick-walled vial fitted with a septum cap (Vol: ~ 1 mL) dissolve 1 mg of the sample, 3 mg of O-methylhydroxylamine hydrochloride or 5 mg O-butylhydroxylamine and 6 mg of sodium acetate in 0.1 mL of water. Heat the mixture for 1 h at 60 °C in the closed vessel. Then, remove the water with a stream of dry air at 60 °C, add 0.1 mL of methanol, evaporate it in the same manner, and remove the last traces of water by the same procedure with benzene. Close the septum, inject 30 µL of TFA and 15 µL of ethyl acetate with a 50-µL syringe through the septum. After keeping for 12 h in a refrigerator or for 2 h at room temperature the reaction is complete. Inject an aliquot into the GC. The latter must be equipped with all glass systems otherwise decomposition of the compounds occurs.

For the gas chromatographic separation OV-225 is used as stationary phase (see Figure 14), except for the case of derivatives formed with O-1-benzylhydroxylamine when DB-1701 is used. The retention times found with both stationary phases are shown in Table 27.

Table 27. Retention Times (in min) of O-Trifluoroacetyl Monosaccharide O-Methoximes (A1), O-Trifluoroacetyl Monosaccharide O-Butoximes (A2) and O-Trifluoroacetyl Monosaccharide O-Benzyloximes (B)

Compound	A (1)	A (2)	B	Compound	A (1)	A (2)	B
D-Erythrose	20.28	23.38	7.63	D-Allose	25.17	28.61	10.15
	21.13	25.61	8.87		26.81	31.91	13.85
D-Threose	20.74	23.79	7.85	D-Altrose	25.88	29.59	10.76
	22.42	26.98	9.77		27.61	33.12	14.88
D-Erythrulose	21.84	25.34	-	D-Talose	26.81	31.31	12.23
	22.66	26.88	-		28.42	34.59	15.91
D-Ribose	22.83	26.48	8.97	D-Gulose	27.05	31.03	11.40
	24.22	28.82	11.22		28.90	34.93	15.27
D-Arabinose	23.23	26.85	9.24	D-Idose	27.77	32.37	12.71
	25.48	30.61	11.22		30.21	37.09	17.02
D-Lyxose	24.71	28.52	9.87	D-Fructose	-	31.96	14.40
	25.67	30.95	12.23			34.55	18.00
D-Xylose	25.03	29.01	10.33	L-Sorbose	-	33.66	15.10
	26.41	31.67	12.80			35.52	17.80
D-Glucose	27.46	31.60	11.74	D-Psicose	-	32.08	12.90
	28.90	35.13	16.50			33.08	13.90
D-Galactose	27.05	31.47	12.35				16.50
	30.21	37.09	17.34	D-Tagatose	-	35.68	15.00
D-Mannose	26.51	30.65	12.06			36.76	16.70
	28.00	33.96	15.89				19.10

Index	Stationary Phase	Temperature (°C)	Dimension of the Column length (cm)	i.d. (cm)	Reference Substance	Literature
A	OV-225 Capillary column	70, 2 min 70–180 5 °C/min then 180, Isoth.	5 000 Glass	-	none	Decker, Schweer[152,153]
B	DB-1701 Capillary column	130–180 180, 15 min 180–280 20 °C/min	3 000	0.025	none	Andrews[67]

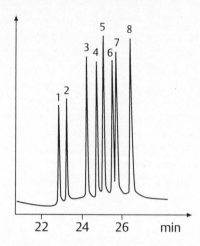

Figure 14.   Gas Chromatogram Showing the Separation of Methoximes of the *O*-Trifluoracetyl Derivatives of Aldopentoses
Column: capillary, L: 50 m; *stationary phase*: OV-225; *temperature*: 70–180 °C, 5 °C/min
1, 3: D-ribose, **2, 6**: D-arabinose, **4, 7**: D-lyxose, **5, 8**: D-xylose (Decker and Schweer[152]; with permission)

2.1.4.5. Butane Borates

The reaction of vicinal diol compounds with 1-butane boric acid [$C_4H_9B(OH)_2$] leads to cyclic butane boric esters which are sufficiently volatile for gas chromatographic separations. The yields and the shapes of the peaks are strongly dependent on the individual compounds; it depends primarily on whether the 1-butane boric acid esters still have a free OH group or not. The first case causes complications in the GC separation and a subsequent trimethylsilylation step is recommended[154–156].

Procedure

Dissolve approximately 4 mg of the dry sugars or sugar alcohols in 2 mL of dry pryridine. To 0.5 mL of this solution add 5 mg of 1-butane boric acid and heat the mixture for 10 min at 100 °C. Inject an aliquot into the GC. For the possible trimethylsilylation afterwards, cool the mixture to room temperature, and add 0.2 mL of HMDS and 0.1 mL TMSCl (trimethylsilylation, see pages 205ff).

Both reactions are very sensitive to humidity and must be carried out with the absolute exclusion of water. Stationary phases which are used: ECNSS-M, OV-17. The 1-butane boric esters of sugar alcohols appear as single peaks with different retention times. Those of arabitol and xylitol are broad and have longer retention times than the hexitols due to a free hydroxyl group. Trimethylsilylation afterwards leads to sharper peaks but also to the formation of side products[156].

The 1-butane boric acid esters of fucose, arabinose and xylose show only one peak per compound, and hexoses more than one with broad shapes.

In Table 28 the relative retention times of the 1-butane boric acid esters of the common monosaccharides and sugar alcohols are listed.

The reaction with 1-butane boric acid and following trimethylsilylation is a good method for quantitative determinations of sugars in mixtures, as demonstrated in the case of glucose–fructose where only one peak per compound is obtained in yields of 97–100%.

Table 28. Relative Retention Times of 1-Butane Boric Acid Esters of Sugars and Sugar Alcohols (A) and Their TMS Compounds (B, C)[a]

Compound	A	B	C	Compound	A	B	C
Erythrol	0.73	0.44(w)	0.39	D-Ribose	1.40	0.60(m)	-
	-	0.47(w)	0.42(m)		-	0.90	-
	-	0.69(w)	0.52	L-Arabinose	1.24	1.24	0.75
	-	0.78(m)	0.56	D-Xylose	1.39	1.36	0.80
L-Arabitol	2.03[b]	0.61(w)	0.54(w)	L-Fucose	1.07	1.07	0.70
	-	0.95	0.66(w)	L-Rhamnose	1.10(w)	0.45(m)	-
	-	1.11(m)	0.70		1.44(m)	1.43	-
	-	-	0.78(m)	D-Glucose	-	1.56	0.90
D-Xylitol	2.09[b]	1.06	0.57(w)	D-Mannose	-	1.34	0.62(w)
	-	-	0.75(m)		-	-	0.78
D-Mannitol	1.76	1.22	0.67(w)		-	-	0.86(m)
	-	1.27	0.77(w)	D-Galactose	-	1.28	0.68(w)
	-	1.74(m)	0.80		-	-	0.78(w)
	-	-	0.84		-	-	0.85(m)
	-	-	0.86(m)	D-Fructose	-	1.21	0.61(w)
D-Sorbitol	1.97	1.93	0.80(w)		-	-	0.81(m)
	-	-	0.84				
	-	-	1.05(m)				
D-Galactitol	2.10	1.18(m)	0.81(m)				
	-	2.07	0.82(w)				
	-	-	1.10				

[a] w = weak, m = main. [b] Broad peak.

Index	Stationary Phase	Temperature (°C)	Dimension of the Column length (cm)	i.d. (cm)	Reference Substance	Literature
A	1% ECNSS-M on Gas-Chrom Q (80–100 mesh)	115–190 5 °C/min	213 Glass	0.64	Methyl-palmitate	Wood, Siddiqui, Weisz[156]
B	1% ECNSS-M on Gas-Chrom Q (80–100mesh)	115–190 5 °C/min	213 Glass	0.64	Methyl-palmitate	Wood, Siddiqui, Weisz[156]
C	3% OV-225 on Gas-Chrom Q (80–100 mesh)	100–240 5 °C/min	152 Glass	0.64	Methyl-arachidate	Wood, Siddiqui, Weisz[156]

2.1.4.6. Gas Chromatographic Separation of the Enantiomeric Forms of the Monosaccharides

Tetroses, pentoses and hexoses are built up from the basic substance glyceraldehyde and they exist in two enantiomeric forms, those which derive from D-glyceraldehyde (D-configurations) and those from L-glyceraldehyde (L-configurations) (see Introduction, pages 1, 2).

Most of the natural monosaccharides belong to the D-series. In some cases the same compound also exists in the L-configuration in substances of natural origin as in the following examples:

– L-Galactose, the unit in the polysaccharide of linseed.
– 3,6-Anhydro-L-galactose, the unit of agar.
– L-Arabinose, the unit of several plant gums (e.g., gum arabic).
– D-Arabinose as glycosides.
– D- and L-Galactose as two different units of the polysaccharide of the mucus of snails.

By determination of the optical rotation as well as by treatment with specific enzymes, the D- or L-configuration of a sugar can be determined.

Gas chromatography enables the separation of both enantiomers of the individual monosaccharides using the following methods:

1.   Conversion of the sugars into volatile compounds and separation on chiral stationary phases.
2.   Conversion of the enantiomeric sugars with chiral reagents into diastereoisomeric derivatives and separation on nonchiral phases.

Method 1. The chiral phase XE-60–valine-*S*-α-phenylethylamide is used, which is prepared according to the reaction scheme below.

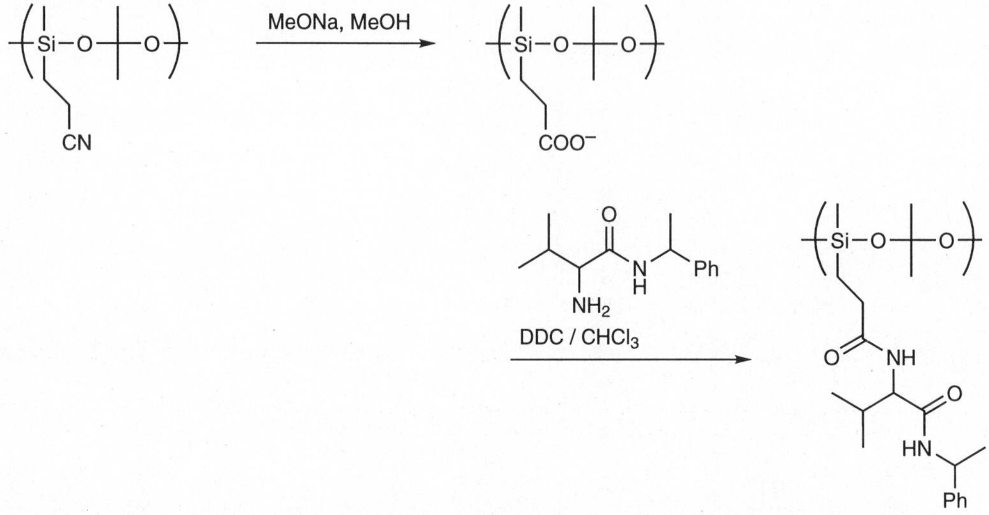

Reaction Scheme for the Preparation of the Chiral Phase XE-60–Valine-*S*-α-phenylethylamide

In practice, a solution of XE-60 in methanol is heated with sodium methylate at 120 °C for 5 h. This operation converts the nitrile groups of the silicon compound into carboxylate groups which are coupled in a further step with L-valine-*S*(or *R*)-α-phenylethylamide using dicyclohexylcarbodiimide (DCC) in chloroform[157,158].

The sugars are converted as free molecules or as *O*-methyl glycosides into their trifluoroacetyl derivatives according to the procedures on pages 246–248. The *O*-methyl glycosides are prepared by treatment of the monosaccharides, in thick-walled vials with Teflon screw caps, with 1.5 mol/L methanolic hydrochloric acid at 100 °C for 15 min[159].

Figure 15 shows the separation of the D- and L-enantiomers of the different trifluoroacetyl 1-*O*-methyl glucosides on a 40-m glass capillary with the above stationary phase. The differences in the retention times of the D- and L-forms can be explained by assuming diastereoisomeric association complexes between the polar trifluoroacetyl groups and the stationary phase; the separation factors are between 1.02–1.07 and, in most cases, the D-compounds have larger retention times than the L-compounds. The D- and L-enantiomers of the *O*-trifluoroacetyl sugar alcohols can be separated in the same manner[160]. With the corresponding *O*-trimethylsilyl derivatives such separations are not possible.

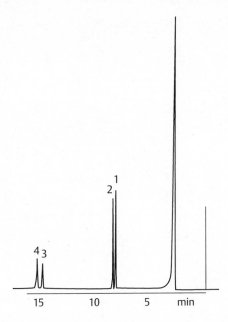

Figure 15. Gas Chromatogram Illustrating the Separation of the Enantiomeric Glucose Isomers as *O*-Trifluoroacetylated Methyl Glucosides
Column: glass capillary, L: 40 m; *stationary phase*: XE-60–valine-*S*-α-phenylethylamide; *temperature*: 100 °C (isothermal)
1: L-glucose-α-pyranosides, **2**: D-glucose-α-pyranosides, **3**: L-glucose-β-pyranosides, **4**: D-glucose-β-pyranosides (König, Benecke and Sievers[157]; with permission)

Method 2. The enantiomeric monosaccharides are converted into their corresponding glycosides, oximes, amides or dithioacetals using chiral reagents. As examples of the derivatization of sugars with optically active reagents, the reactions with (–)-2-butanol and (–)-*O*-menthylhydroxylamine are presented.

Procedures

Reaction with (–)-2-Butanol
After flushing with nitrogen, heat 0.5 mg of the dry monosaccharides in 0.5 mL 1 M (–)-2-butanolic hydrochloric acid in a closed vial at 80 °C for 8 h. Cool and neutralize the solution with silver carbonate. After centrifugation, evaporate the solution at 45 °C in vacuo and dry the residue in vacuo over P_2O_5. Carry out the trimethylsilylation with a mixture of hexamethyldisilazane–trimethylchlorosilane–pyridine (1:1:5 v/v; room temperature, 30 min)[161,162].
In nearly all cases, separation of the D- and L-enantiomers of the monosaccharides has been achieved using a 25-m glass capillary column coated with SE-30. A disadvantage is the contamination of the (–)-2-butanol with small amounts of (+)-2-butanol, which causes additional weak peaks of the corresponding TMS-(+)-2-butyl glycosides. This effect is not possible with chiral phases.

In Table 29 the relative retention times of the D- and L-TMS-(–)-2-butyl glycosides of some monosaccharides are listed.

Table 29. Relative Retention Times of O-Trimethylsilyl-(–)-2-butyl Glycosides of Different D- and L-Mono-saccharides

Compound	A (D-)	A (L-)	Compound	A (D-)	A (L-)
Arabinose	0.70	0.69	Fucose	0.72	0.74
	0.71	0.70		0.79	0.78
	0.73	0.73		0.82	0.87
	0.80	0.82		0.89	0.88
Ribose	0.75	0.77	Glucose	1.36	1.32
	0.76	0.78		1.53	1.53
Xylose	0.88	0.87	Mannose	1.15	1.12
	1.02	1.03		1.34	1.32
Rhamnose	0.72	0.71	Galactose	1.18	1.21
	0.82	0.81		1.28	1.25
				1.34	1.38

Index	Stationary Phase	Temperature (°C)	Dimension of the Column length (cm)	i.d. (cm)	Reference Substance	Literature
A	SE-30 Glass capillary column	135–200 1 °C/min	2 500	0.03	α-D-methylgalacto-pyranoside abs. time 31 min	Gerwig, Kamerling, Vliegenhart[161]

Reaction with (–)-Menthylhydroxylamine

Heat 0.5 mg of the sugars with 4 mg O-(–)-menthylhydroxylamine and 3 mg of sodium acetate in 100 µL water. Convert these oximes into the trifluoroacetyl derivatives according to the procedure on page 249.

The separation of the D- and L-enantiomers of the monosaccharides can be achieved on a 50-m glass capillary coated with OV-225. For each compound two peaks appear due to the E- and Z-isomers of the oximes[152,163] (see Figure 16).

In Table 30 the absolute retention times of the O-trifluoroacetyl D- and -L-monosaccharide menthyloximes are listed.

Table 30. Relative Retention Times of O-Trifluoroacetyl-(–)-menthyloximes (Always Z- and E-Isomers) of Different D- and L-Monosaccharides

Compound	A (D-)	A (L-)	Compound	A (D-)	A (L-)
Glyceraldehyde	11.94	11.71	Fucose	15.31	14.64
	15.24	15.14		23.36	22.89
Ribose	18.55	17.50	Glucose	25.54	25.78
	25.01	24.84		40.30	40.28
Arabinose	18.15	19.01	Mannose	25.03	26.15
	29.53	29.94		37.05	37.63
Xylose	22.10	22.18	Galactose	27.95	26.63
	31.44	31.17		45.60	44.63

Index	Stationary phase	Temperature (°C)	Dimension of the Column length (cm)	i.d. (cm)	Reference Substance	Literature
A	OV-225 Capillary column	180, Isothermal	5 000	-	-	Schweer[163,164]

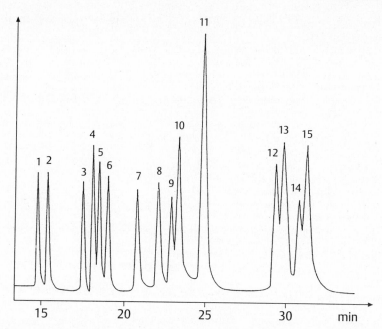

Figure 16.   Gas Chromatogram Showing the Separation of the Enantiomers of Fucose, Ribose, Arabinose and Lyxose as (–)-Menthyloximes Pertrifluoroacetates
Column: capillary, L: 50 m; *stationary phase*: OV-225; *temperature*: 180 °C (isothermal)
1,9: L-fucose, **2,10**: D-fucose, **3,11***: L-ribose, **4,12**: D-arabinose, **5,11***: D-ribose, **6,13**: L-arabinose, **7,14**: D-lyxose, **8,15**: L-lyxose. *Overlapping (Schweer[163]; with permission)

Further reagents also have been described for this purpose, these are the following: (–)-*O*-bornyloxime[164], L-α-methylbenzylamine[165], (+)-2-octanol[166] and (+)-1-phenylethanethiol[167]. Trifluoroacetylation leads to better separations than trimethylsilylation.

References

1   McInnes, A. G.; Ball, D. H.; Cooper, F. P.; Bishop, J. *J. Chromatogr.* **1958**, *1*, 556.
2   Gee, M.; Walker, H. G. *Anal. Chem.* **1962**, *34*, 650.
3   Bishop, C. T. *Methods Biochem. Anal.* **1962**, *10*, 1.
4   Ovodov, Y. S.; Evtushenko, E. V. *J. Chromatogr.* **1967**, *31*, 527.
5   Wei-Tong Wang; Matsuura, F.; Sweeley, C. C. *Anal. Biochem.* **1983**, *134*, 398.
6   Hakomori, S. *J. Biochem. (Tokyo)* **1964**, *55*, 205.
7   Whyte, J. *J. Chromatogr.* **1973**, *87*, 163.
8   Whyte, J. *Can. J. Chem.* **1973**, *51*, 3198.
9   Ciucanu, I.; Kerek, F. *Carbohydr. Res.* **1984**, *131*, 209.
10 Yang, Z. C.; Cashman, J. R. *J. Chromatogr.* **1992**, *596*, 79.
11 Finne, J.; Krusius, T.; Rauvala, H. *Carbohydr. Res.* **1980**, *80*, 336.
12 Karlsson, H.; Carlstedt, I.; Hansson, G. *Anal. Biochem.* **1989**, *182*, 438.
13 Hedgley, E. J.; Overend, W. G. *Chem. Ind.* **1960**, 378.

14   Bayer, E.; Witsch, H. G. In *Gas Chromatography*; Bayer, E., Ed.; Elsevier: New York, 1961; Witsch, H. G. Dissertation, Karlsruhe 1961.
15   Makita, M.; Wells, W. W. *Anal. Biochem.* **1963**, *5*, 523.
16   Sweeley, C. C.; Bentley, R.; Makita, M.; Wells, W. W. *J. Am. Chem. Soc.* **1963**, *85*, 2497.
17   Klebe, J. F.; Finkbeiner, H. White, D. M. *J. Am. Chem. Soc.* **1966**, *88*, 3390.
18   Birkhofer, L.; Donike, M. *J. Chromatogr.* **1967**, *26*, 270.
19   Stalling, D. L.; Gehrke, C. W.; Zumwalt, R. W. *Biochem. Biophys. Res. Commun.* **1968**, *31*, 616.
20   Donike, M. *J. Chromatogr.* **1969**, *42*, 103.
21   Pierce, A. E. In *Silylation of Organic Compounds*; Pierce, A. E., Ed.; Chemical Co., Rockford, Ill. USA, 1968.
22   Zürcher, K.; Hadorn, H.; Strack, Ch. *Dtsch. Lebensm. Rdsch.* **1975**, *71*, 393.
23   Bhatti, T.; Clamp, J. R. *Clin. Chim. Acta* **1968**, *22*, 563.
24   Curtius, H. C.; Völlmin, J. A.; Müller, M. *Fresenius' Z. Anal. Chem.* **1968**, *243*, 341.
25   Cayle, T.; Viebrock, F.; Schiaffino, J. *Cereal Chem.* **1968**, *45*, 154.
26   Lee, Y. C.; Ballou, C. E. *J. Chromatogr.* .**1965**, *18*, 147.
27   Zürcher, K.; Hadorn, H. *Dtsch. Lebensm. Rdsch.* **1974**, *70*, 425.
28   Zürcher, K.; Hadorn, H. *Dtsch. Lebensm. Rdsch.* **1975**, *71*, 68.
29   Schmolck, W.; Mergenthaler, K. *Z. Lebensm. Unters. Forsch.* **1973**, *152*, 263.
30   Müller, B.; Göke, G. *Dtsch. Lebensm. Rdsch.* **1972**, *68*, 222.
31   Zürcher, K.; Hadorn, H.; Strack, Ch. *Mitt. Geb. Lebensm Hyg.* **1975**, *66*, 92.
32   Reid, P. E.; Donaldson, B.; Secret, A.; Bradford, B. *J. Chromatogr.* **1970**, *47*, 199.
33   Ellis, W. C. *J. Chromatogr.* **1969**, *41*, 325, 335.
34   Gerwig, G. J.; Kamerling, J. P.; Vliegenthart, J. P.G. *Carbohydr. Res.* **1978**, *62*, 349; **1979**, *77*, 1.
35   Kärkkäinen, J. Vihko, R. *Carbohydr. Res.* **1969**, *10*, 113.
36   Bhatti, T. B.; Chambers, R. E.; Clamp, J. R. *Biochim. Biophys. Acta* **1970**, *222*, 339.
37   Szafranek, J.; Pfaffenberger, C. O.; Hornig, E. C. *J. Chromatogr.* **1974**, *88*, 149.
38   Harvey, J.; Horning, J. *J. Chromatogr.* **1973**, *76*, 51.
39   Freemann, B. H.; Stephen, A. M.; van Bijl, P. *J. Chromatogr.* **1972**, *73*, 29.
40   Haverkamp, J.; Kamerling, J. P.; Vliegenthart, J. F. G. *J. Chromatogr.* **1971**, *59*, 281.
41   Tesarik, K.; *J. Chromatogr.* **1972**, *65*, 295.
42   Vidaurreta, L.; Fournier, L. *Anal. Chim. Acta* **1970**, *52*, 507.
43   El-Dash, A. A.; Hodge, J. E. *Carbohydr. Res.* **1971**, *18*, 259.
44   Nozawa, Y.; Hiraguri, J.; Ito, Y. *J. Chromatogr.* **1969**, *45*, 244.
45   Nurok, D.; Reardon, T. J. *Carbohydr. Res.* **1977**, *56*, 165.
46   Nikolov, Z.; Reilly, P. *J. Chromatogr.* **1983**, *254*, 157.
47   Rumpf, G. *J. Chromatogr.* **1969**, *43*, 247.
48   Gheorghiu, Th.; Oette, K. *J. Chromatogr.* **1970**, *48*, 430.
49   Sawardeker, J.; Sloneker, J. *Anal. Chem.* **1965**, *37*, 947.
50   Farshtchi, D.; Moss, C. W. *J. Chromatogr.* **1969**, *42*, 108.
51   Preuss, A.; Thier, H. P. *Z. Lebensm. Unters. Forsch.* **1982**, *175*, 93.
52   Yokota, M.; Mori, T. *Carbohydr. Res.* **1977**, *59*, 289.
53   Chambers, R. E.; Clamp, J. B. *Biochem. J.* **1971**, *125*, 1009.
54   Anderle, D.; Königstein, J.; Kovacik, V. *Anal. Chem.* **1977**, *49*, 137.
55   Petersson, G. *Carbohydr. Res.* **1974**, *33*, 47.
56   Mason, B.; Slover, H. T. *J. Agric. Food Chem.* **1971**, *19*, 551.
57   Mateo, R.; Bosch, F.; Pastor, A.; Jimenez, M. *J. Chromatogr.* **1987**, *410*, 319.
58   Toba, T.; Adachi, S. *J. Chromatogr.* **1977**, *135*, 411.
59   Morvai, M.; Molnar-Perl, I. *J. Chromatogr.* **1990**, *520*, 201.
60   Morvai, M.; Molnar-Perl, I. *Chromatographia* **1992**, *34*, 502.
61   Adam, S.; Jennings, W. G. *J. Chromatogr.* **1975**, *115*, 218.
62   Laine, R. A.; Sweeley, C. C. *Carbohydr. Res.* **1973**, *27*, 199.
63   Zegota, H. *J. Chromatogr.* **1980**, *192*, 446.
64   Storset, P.; Stokke, O.; Jellum, E. *J. Chromatogr.* **1978**, *145*, 351.
65   Mawhinney, T. P. *J. Chromatogr.* **1986**, *351*, 91.
66   den Drijver, L.; Holzapfel, W. C. *J. Chromatogr.* **1986**, *363*, 345.
67   Andrews, M. *Carbohydr. Res.* **1989**, *194*, 1.
68   Honda, S.; Yamauchi, N.; Kakehi, K. *J. Chromatogr.* **1979**, *169*, 287.
69   Rubino, F. M. *J. Chromatogr.* **1989**, *473*, 125.

70   Bhattacharjee, S. S.; Gorin, P. A. *Can. J. Chem.* **1969**, *47*, 1207.
71   Jones, H. G.; In *Methods in Carbohydrate Chemistry*, Vol. VI; Whistler, R. L.; BeMiller, J.N., Eds.; Academic: New York, 1972; p 25.
72   Haworth, S.; Roberts, J. G.; Sagar, B. F. *Carbohydr. Res.* **1969**, *9*, 491.
73   De Bettignies-Dutz, A.; Reznicek, G.; Kopp, B.; Jurenitsch, J. *J. Chromatogr.* **1991**, *547*, 299.
74   Disdaroglu, M.; Scherz, H.; von Sonntag, C. *Z. Naturforsch.* **1972**, *27b*, 29.
75   Adam, S. *Z. Lebensm. Unters. Forsch.* **1981**, *173*, 109.
76   Kärkkäinen, J. *Carbohydr. Res.* **1969**, *11*, 247.
77   Percival, E. *Carbohydr. Res.* **1967**, *4*, 441
78   Raunhardt, O.; Schmidt, H. W.; Neukom, H. *Helv. Chim. Acta* **1967**, *50*, 1267.
79   Petersson, G. *J. Chromatogr. Sci.* **1977**, *15*, 245.
80   Hurst, R. *Carbohydr. Res.* **1973**, *30*, 143.
81   Casals-Stenzel, J.; Buscher, H. P.; Schauer, R. *Anal. Biochem.* **1975**, *65*, 507.
82   Mononen, I. *Carbohydr. Res.* **1981**, *88*, 39.
83   Honda, S.; Kakehi, K.; Okada, K. *J. Chromatogr.* **1979**, *176*, 367.
84   Dutton, G. G. S. *Adv. Carbohydr. Chem. Biochem.* **1973**, *28*, 12.
85   Lineback, D. R. *Carbohydr. Res.* **1968**, *7*, 106.
86   Fraser, C. R.; Wilkie, K. C. B. *Phytochemistry* **1971**, *10*, 199.
87   Paulsen, H.; Garrido, E. F.; Trautwein, W. P.; Heyns, K. *Chem. Ber.* **1968**, *101*, 179, 186, 191.
88   Paulsen, H.; Herold, C. P. *Chem. Ber.* **1970**, *103*, 2450.
89   Wiesner, D. A.; Sweeley, C. C. *Anal. Biochem.* **1994**, *217*, 316.
90   Hämäleinen, M.; Theander, D.; Nordkvist, E.; Ternrud, I. E. *Carbohydr. Res.* **1990**, *207*, 167.
91   Albersheim, P.; Nevins, D.; English, P.; Karr, A. *Carbohydr. Res.* **1967**, *5*, 340.
92   Sawardeker, J.; Sloneker, J.; Jeanes, A. *Anal. Chem.* **1965**, *37*, 1602.
93   Oades, J. M. *J. Chromatogr.* **1967**, *28*, 246.
94   Kontrohr, T.; Kocsis, B. *J. Chromatogr.* **1984**, *291*, 119.
95   Sturgeon, R. J. *Carbohydr. Res.* **1992**, *227*, 375.
96   Henry, R. T.; Blakeney, A. B.; Harris, P. H.; Stone, B. A. *J. Chromatogr.* **1983**, *256*, 419.
97   Blakeney, A. B.; Harris, P. J.; Henry, R. J.; Stone, B. A. *Carbohydr. Res.* **1983**, *113*, 291.
98   Klok, J.; Cox, H.; de Leeuw, J. W.; Schenk, P. A. *J. Chromatogr.* **1982**, *253*, 55.
99   Kiho, T.; Ukai, S.; Hara, C. *J. Chromatogr.* **1986**, *369*, 415.
100   Jones, H. G.; Smith, M.; Sahasrabudhe, M. *J. AOAC* **1966**, *49*, 1183.
101   Weber, P. L.; Carlson, D. M. *Anal. Biochem.* **1982**, *121*, 140.
102   Horn, L. R.; Machlin, L. G.; Hamilton, J. G. *J. Chromatogr. Sci.* **1979**, *17*, 538.
103   Chen, C. C.; McGinnis, G. *Carbohydr. Res.* **1981**, *90*, 127.
104   Fox, A.; Morgan, S. L.; Gilbert, J. In: *Analysis of carbohydrates by GC-MS*; Biermann, Ch.; McGinnis, G.D., Eds.; CRC-Press Inc.: Boca Raton, USA, 1989; p 87ff.
105   Metz, J.; Ebert, W.; Weicker, H. *Clin. Chim. Acta* **1971**, *34*, 31.
106   Perry, M. B.; Webb, A. C. *Can. J. Biochem.* **1968**, *46*, 1163.
107   Patz, C. D.; Zimmer, E.; Dietrich, H. *Dtsch. Lebensm. Rdsch.* **1993**, *89*, 1.
108   Harris, P. J.; Sadek, M.; Brownlee, T. C.; Blakeney, A. B.; Webster, J.; Stone, B. A. *Carbohydr. Res.* **1992**, *227*, 365.
109   Lomax, J. A.; Conchie, J. *J. Chromatogr.* **1982**, *236*, 385.
110   Bacic, A.; Harris, P. J.; Hak, E. W.; Clarke, A. *J. Chromatogr.* **1984**, *315*, 373.
111   Harris, P. J.; Bacic, A.; Clarke, A. *J. Chromatogr.* **1985**, *350*, 304.
112   Shea, E.; Carpita, N. *J. Chromatogr.* **1988**, *445*, 424.
113   Kraus, R. J.; Shinnick, F. L.; Marlett, J. A. *J. Chromatogr.* **1990**, *513*, 71.
114   Szafranek, J.; Pfaffenberger, C. D.; Horning, *Anal. Lett.* **1973**, *6*, 479.
115   Varma, R.; Varma, R.; Wardi, A. *J. Chromatogr.* **1973**, *77*, 222.
116   Varma, R.; Varma, R.; Allen, W.; Wardi, A. *J. Chromatogr.* **1973**, *86*, 205.
117   Baird, J. K.; Holroyd, M.; Ellwood, D .C. *Carbohydr. Res.* **1973**, *27*, 464.
118   Mergenthaler, E.; Scherz, H. *Z. Lebensm. Unters. Forsch.* **1976**, *162*, 25.
119   Turner, S.; Cherniak, R. *Carbohydr. Res.* **1981**, *95*, 137.
120   Lehrfeld, J. *Anal. Biochem.* **1981**, *115*, 410.
121   Seymour, R. F.; Chen, C. M. E.; Bishop, S. *Carbohydr. Res.* **1979**, *73*, 19.
122   Pfaffenberger, C. D.; Szafranek, J.; Horning, M. G.; Horning, E. C. *Anal. Biochem.* **1975**, *61*, 501.
123   Wachowiak, R.; Connors, K. A. *Anal. Chem.* **1979**, *51*, 27.

124   Chen., C. C.; McGinnis, G. D. *Carbohydr. Res.* **1981**, *90*, 127.
125   McGinnis, G. *Carbohydr. Res.* **1982**, *108*, 284.
126   Guerrant, G.; Moss, C. W. *Anal. Chem.* **1984**, *56*, 635.
127   Dmitriev, B. A.; Backinovski, L. V.; Chizov, O. S.; Zolotarev, B. M.; Kochetkov, N. K. *Carbohydr. Res.* **1971**, *19*, 432.
128   Scherz, H. unpublished.
129   Mawhinney, T. P.; Feather, M. S.; Barbero, G.; Martinez, J. *Anal. Biochem.* **1980**, *101*, 112.
130   Varma, R.; Varma, R.; Allen, W.; Wardi, A. *J. Chromatogr.* **1974**, *93*, 221.
131   Mergenthaler, E.; Scherz, H. *Z. Lebensm. Unters. Forsch.* **1976**, *162*, 159.
132   Scherz, H.; Mergenthaler, E. *Z. Lebensm. Unters. Forsch.* **1980**, *170*, 280.
133   Glück, U.; Thier, H. P. *Z. Lebensm. Unters. Forsch.* **1980**, *170*, 272.
134   Neeser, J. R.; Schweizer, T. F. *Anal. Biochem.* **1984**, *142*, 58.
135   König, W.; Benecke, J. *J. Chromatogr.* **1983**, *269*, 19.
136   König, W.; Benecke, J.; Sievers, S. *J. Chromatogr.* **1981**, *217*, 71.
137   Tomana, M.; Niedermeier, W.; Spivey, C. *Anal. Biochem.* **1978**, *89*, 110.
138   Ando, S.; Yamakawa, T. *J. Biochem.* **1971**, *70*, 335.
139   Imanari, T.; Arahawa, J.; Tamura, Z. *Chem. Pharm. Bull.* **1969**, *17*, 1967.
140   König, W.; Bauer, H.; Voelter, W.; Bayer, E. *Chem. Ber.* **1973**, *106*, 1905.
141   Vilkas, M.; Miu-I-Jan.; Boussac, M.; Bonnard, C. *Tetrahedr. Lett.* **1966**, *14*, 1441.
142   Ueno, T.; Kurihara, N.; Nakajima, M. *Agric. Biol. Chem.* **1967**, *31*, 1189.
143   Shapira, J. *Nature* **1969**, *222*, 792.
144   Englmaier, J. In *Analysis of Carbohydrates by GC*; Bierman, Ch.; McGinnis, G.D., Eds.; CRC: Boca Raton, Florida, USA, 1989; p 131ff.
145   Sullivan, J. E.; Schewe, L. *J. Chromatogr. Sci.* **1977**, *15*, 196.
146   Selosse, E.; Reilly, P. *J. Chromatogr.* **1985**, *328*, 253.
147   Englmaier, P. *Carbohydr. Res.* **1985**, *144*, 177.
148   Nakamura, H.; Tamura, Z. *Chem. Pharm. Bull.* **1970**, *18*, 2314.
149   Wrann, M.; Todd, Ch. *J. Chromatogr.* **1978**, *147*, 309.
150   Matsui, M.; Okada, M.; Imanari, T.; Tamura, Z. *Chem. Pharm. Bull.* **1968**, *16*, 1383.
151   Schweer, H. *J. Chromatogr.* **1982**, *236*, 355.
152   Decker, P.; Schweer, H. *J. Chromatogr.* **1982**, *236*, 369.
153   Decker, P.; Schweer, H. *J. Chromatogr.* **1982**, *243*, 372.
154   Eisenberg, F., Jr. *Carbohydr. Res.* **1971**, *19*, 135.
155   Eisenberg, F., Jr. *Methods Enzymol.* **1972**, *28*, 168.
156   Wood, P. S.; Siddiqui, I.; Weisz, J. *Carbohydr. Res.* **1975**, *42*, 1.
157   König, W. A.; Benecke, I.; Sievers, S. *J. Chromatogr.* **1981**, *217*, 71.
158   König, W. A.; Benecke, I.; Bretting, H. *Angew. Chem., Int. Ed. Engl.* **1981**, *20*, 693.
159   König, W. A.; Stölting, K.; Kruse, K. *Chromatographia* **1977**, *10*, 444.
160   König, W. A.; Benecke, I. *J. Chromatogr.* **1983**, *269*, 19.
161   Gerwig, G.; Kamerling, J. P.; Vliegenthart, F. G. *Carbohydr. Res.* **1978**, *62*, 349.
162   Gerwig, G.; Kamerling, J. P.; Vliegenthart, F. G. *Carbohydr. Res.* **1979**, *77*, 1.
163   Schweer, H. *J. Chromatogr.* **1982**, *243*, 149.
164   Schweer, H. *J. Chromatogr.* **1983**, *259*, 164.
165   Oshima, R.; Kumanotani, J.; Watanabe, C. *J. Chromatogr.* **1983**, *259*, 159.
166   Leontein, K.; Lindberg, B.; Lönngren J. *Carbohydr. Res.* **1978**, *62*, 359.
167   Little, M. *Carbohydr. Res.* **1982**, *105*, 1.
168   Hyppänen, T.; Sjöström, E.; Vuorinen, T. *J. Chromatog.* **1983**, *261*, 320

General References

Analysis of Carbohydrates by GLC and MS; Biermann, Ch.; McGinnis, G. D., Eds.; CRC: Boca Raton, Florida, USA, 1989.

2.2. Electrophoretic Methods

When molecules contain functional groups which have an electronic charge they migrate in an electric field. Their speeds are dependent on several parameters, for example, their chemical structure and the nature of the electrolyte, especially its pH. On account of the differences in the migration rates it is possible to separate mixtures of such substances.

The method most used is zone electrophoresis on an inert support. The material is soaked with the electrolyte and, after equilibration, the mixture of substances to be separated is spotted at the maximum distance from the pole to which the charged molecules are moving. When the current is turned on, the compounds begin to move to that pole, which has an opposite charge.

The following materials have been used as supports:

- Filter paper[1,2].
- Glass fibre paper[3].
- Cellulose acetate[4].
- Thin-layer materials such as silica gel, aluminum oxide or cellulose[5,6].
- Granulated gels[7].
- Compact gels of starch, agarose or polyacrylamide[8,9].

Recently, support-free zone electrophoresis in capillaries using very high voltages has become of great importance (capillary zone electrophoresis, abbreviation: CZE).

The construction of an electrophoresis apparatus is simple in principle. The support is located in a flat or cylindrical system connected at both ends to vessels containing electrolytes in which electrodes are fixed. The support system has to be cooled sufficiently during electrophoresis since it is under voltage and the current produces considerable heat, which leads to evaporation of water from the buffer and causes distortions in the concentration of the electrolytes. For high-voltage electrophoresis cooling must be intensive.

2.2.1. Paper Electrophoresis

Chromatographic paper is the most commonly used support (e.g., Whatman 1, Whatman 4, Schleicher-Schüll 2043b). The paper is dipped into the electrolyte, then put cautiously between two dry filter papers and pressed with a gum roller. This wet paper is fixed in the electrophoresis chamber using one of three variations:

1.   Open-Strip Technique.

The impregnated paper is fixed either vertically, horizontally, or in a V-shape in the electrophoresis apparatus with both ends dipping into the electrolyte troughs with the electrodes. The chamber is closed to maintain a humid atmosphere and to prevent evaporation of the water from the strip. This technique is possible only with low voltage, since with high voltage the heat is too strong and rapid evaporation occurs.

2.   Immersion-Strip Technique.

The strip soaked with the electrolyte is dipped into a liquid immiscible with water which serves as a cooling medium for the heat caused by the current flow. Tetrachloromethane or chlorobenzene are used for this purpose. This technique is suitable for high voltage.

3.   Strip-Inclusion Technique.

The wet filter paper is pressed between two glass plates with both ends dipping into the electrolyte plates. The glass plates are then placed on a cooling block. With this arrangement the application of high voltage is possible.

High-voltage electrophoresis (electric field strength 30–100 V/cm) produces short separation times of the substances and this, in conjunction with a strong reduction in diffusion, yields sharp zones, which leads to better resolution than is possible under low voltage conditions.

Procedure

The wet paper is impregnated as described earlier and the sample is spotted at places which have been marked on a line with pencil. With humid paper only a few µL are possible; for greater volumes, dry paper must be used and, in this case, the electrolyte is applied with a pipet on both sides of the sample spots. If the places with the sample on the paper on are raised by putting glass rods under them, then the front of the electrolyte approaches from both sides and compresses the diffused sample zones.

After this procedure, the ends of the paper are dipped into the electrode vessels, the chamber is closed and the voltage is switched on[10].

2.2.1.1. Paper Electrophoresis of Mono- and Oligosaccharides and Related Compounds

These compounds themselves do not contain any charged groups and, therefore, they do not migrate in the electric field. With suitable reagents it is possible to convert these compounds into charged complexes, which then migrate in the electric field. The migration rate is strongly dependent on the structure of the sugar complexes and these differences enable the separation of sugar mixtures.

The following compounds have been used as complexing agents: borate, phenylborate and its sulfonic acids, molybdate, tungstenate, vanadate, arsenite, bisulfite, alkali and alkali-earth ions.

2.2.1.1.a. Borate Ions

Borate ions give negatively charged complexes in an alkaline medium with polyhydroxy compounds with the following chemical structures.

Structure of Borate–Sugar Complexes

In an alkaline medium the equilibrium is located on the side of the complexes, which migrate in the electric field. The magnitude of the charge is dependent on the strength of the complex, which is influenced by the stereochemical positions of the OH groups relative to one another; it depends particularly on the distances between the oxygen atoms. The maximum interaction between borate and OH groups occurs at an O–O distance of 2.4 Å, which is not present generally in sugar molecules. The migration rates are strongly influenced further by the pH of the borate solution, the maxima being between pH 9 and 10.

In Table 1 the relative mobilities of the most common monosaccharides and related sugar alcohols are listed at various borate concentrations[3,11-13].

Table 1.   Relative Electrophoretic Mobilities M_G of Several Monosaccharides and Sugar Alcohols

Compound	A	B	Compound	A	B
D, L-Glyceraldehyde	0.79	-	D-Idose	-	1.02
1,3-Dihydroxyacetone	0.78	-	D-Talose	-	0.87
D-Erythrose	-	-	D-Fructose	0.90	0.89
L-Threose	-	-	L-Sorbose	0.95	0.97
L-Arabinose	0.96	0.91	D-Tagatose	-	0.95
D-Ribose	0.77	0.75	D-Psicose	-	0.76
D-Xylose	1.00	1.01	meso-Erythritol	-	0.75
D-Lyxose	-	0.71	D-Arabitol	0.90	0.87
2-Desoxy-D-ribose	0.33	-	D-Xylitol	-	0.79
L-Fucose	0.89	0.83	D-Ribitol (Adonitol)	-	-
L-Rhamnose	0.52	0.49	D-Sorbitol (Glucitol)	0.89	0.83
D-Galactose	0.93	0.93	D-Galactitol (Dulcitol)	0.98	0.97
D-Glucose	1.00	1.00	D-Mannitol	0.90	0.91
D-Mannose	0.72	0.69	Allitol	-	0.90
D-Allose	-	0.83	L-Iditol	-	0.81
L-Altrose	-	0.97	D-Talitol	-	0.89
D-Gulose	-	0.82			

A =	Electrolyte:	sodium borate 0.2 M, pH 10
	Paper:	Whatman 3
	Electrophoresis:	500 V (15 V/cm)
	Reference substance:	D-glucose (Bourne, Foster and Grant[3])
B =	Electrolyte:	sodium borate 0.05 M, pH 9.2
	Paper:	Whatman 4
	Electrophoresis:	1300–1400 V (20–25 V/cm), 18–20 °C
	Reference substance:	D-glucose (Frahn and Mills[11])

Comparing the migration rates of the individual monosaccharides, three groups with similar patterns emerge. They are:
1.   Xylose, glucose, idose, altrose.
2.   Arabinose, galactose, talose, fructose, fucose.
3.   Ribose, lyxose, mannose, gulose, allose, rhamnose.
The migration rates of group 3 are remarkably lower than those of groups 1 and 2.

Substitution of the hydrogens of the OH groups by alkyl or carbonylalkyl groups sometimes leads to a dramatic alteration in the migration rate. For methyl glucosides and methyl xylosides the rates for furanoside structures are higher than those of pyranosides as a result of the individual strength of the complexes. For methyl galactosides this effect is not so pronounced. The relative migration rates of methyl glucosides and methyl galactosides are listed in Table 2[12-14].

Table 2.   Relative Electrophoretic Mobilities M_G of Some Methyl Glycosides and Anhydro Sugars

Compound	A	Compound	A
Methyl-α-D-xylopyranoside	0.00	Methyl-α-D-glucofuranoside	0.73
Methyl-β-D-xylopyranoside	0.00	1,6-Anhydro-β-D-glucopyranose	0.00
Methyl-α-D-xylofuranoside	0.30	Methyl-α-D-galactopyranoside	0.38
Methyl-β-D-xylofuranoside	0.30	Methyl-β-D-galactopyranoside	0.38
Methyl-α-D-glucopyranoside	0.11	Methyl-α-D-galactofuranoside	0.41
Methyl-β-D-glucopyranoside	0.19	Methyl-β-D-galactofuranoside	0.31

A =	Electrolyte:	borate 0.3 M, pH 10
	Paper:	Whatman 3
	Electrophoresis:	900–1500 V
	Reference substance:	D-glucose (Foster, Stacey[13,15]; Weigel[12])

With regard to methylation of the primary or a secondary OH group in the sugar molecule, the position of the methyl group is of great importance. This is shown in Table 3 for D-glucose. Methylation of the OH group at C-2 and C-4 of the glucose molecule reduces the migration rate to a quarter of the nonsubstituted molecule, whilst methylation of OH at C-3 and C-6 causes only a 20% reduction. Substitution of the OH at C-2 and C-4, or C-1 and C-4 as well as C-1 and C-6 leads to a migration rate of zero (e.g., methyl 4-O-methyl-β-D-glucopyranoside and 1,6-anhydroglucose).

These results lead to the assumption that for hexoses the formation of complexes between borate and OH groups occurs at two independent positions namely at C-1/C-2 or C-4/C-6, C-4/C-5 and C-5/C-6. The C-4/C-6 complex seems to be weaker than those of C-1/C-2, C-4/C-5 and C-5/C-6. This explains the mobilities of the methyl α- and β-glucopyranosides compared to those of the methyl α- and β-D-methylglucofuranosides[15]. It has also been assumed that a *cis* oriented pair of hydroxyl groups at C-2 and C-4 is a favorable configuration for complex formation with borate ions[52].

Table 3.   Relative Electrophoretic Mobilities M_G of the O-Methyl-D-glucoses

Compound	A		A
2-O-Methyl-D-glucose	0.23	3,4-Di-O-methyl-D-glucose	0.31
3-O-Methyl-D-glucose	0.80	2,3,4-Tri-O-methyl-D-glucose	0.00
4-O-Methyl-D-glucose	0.24	2,3,6-Tri-O-methyl-D-glucose	0.00
6-O-Methyl-D-glucose	0.80	2,4,6-Tri-O-methyl-D-glucose	-
2,3-Di-O-methyl-D-glucose	0.12	3,5,6-Tri-O-methyl-D-glucose	0.71
2,4-Di-O-methyl-D-glucose	<0.05	2,3,4,6-Tetra-O-methyl-D-glucose	0.00

A =	Electrolyte:	borate 0.2 M, pH 10
	Paper:	Whatman 3
	Electrophoresis:	500 V, 1.5 h
	Reference substance:	D-glucose (Weigel[12], Foster[13,14])

For disaccharides the situation is analogous. Those with 1-3 and 1-6 glycosidic linkages such as turanose, nigerose, laminaribiose, melibiose or gentiobiose have distinctly higher electrophoretic mobilities than those with 1-1, 1-2 and 1-4 linkages. The positions of the OH groups relative to each other do not remarkably influence the mobilities as shown in Table 4.

Table 4.   Relative Electrophoretic Mobilities M_G of Some Oligosaccharides

Compound	A	B	Compound	A	B
1-1-Disaccharide			1-4-Disaccharide		
Trehalose α,α'	0.19	-	Maltose	0.32	0.30
Trehalose α,β	0.23	-	Lactose	0.38	0.37
Trehalose β,β	0.19	-	Cellobiose	0.23	0.22
1-2-Disaccharide			1-6-Disaccharide		
Sophorose	0.24	-	Gentiobiose	0.75	-
Saccharose	0.18	0.16	Melibiose	0.80	0.77
1-3-Disaccharide			Isomaltose	0.69	-
Nigerose	0.69	-			
Laminaribiose	0.69	-			
Turanose	-	0.64			

A =	Electrolyte:	borate 0.2 M, pH 10
	Paper:	Whatman 3
	Electrophoresis:	500 V, 1.5 h
	Reference substance:	D-glucose (Foster[13,14])
B =	Electrolyte:	borate 0.05 M, pH 9.2
	Paper:	Whatman 4
	Electrophoresis:	1300–1400 V, 1.5 h
	Reference substance:	D-glucose (Frahn, Mills[11])

Further Complex Forming Reagents

Paper electrophoretic separations have been carried out with the following carbohydrate complexes.

2.2.1.1.b. Phenylboric Acid, Diphenylboric Acid

Phenylboric acid is three times stronger than the nonsubstituted boric acid, but the migration properties of the carbohydrate complexes are similar[16]; the same is true for diphenylboric acid (see Tables 5 and 6)[17].

2.2.1.1.c. Sulfonated Phenylboric Acid

The complexes with monosaccharides and sugar alcohols have, at pH 6.5, very different mobilities, as shown in Table 5. Those for ketoses are 5–8 times higher than those for aldoses[12,16].

2.2.1.1.d. Germanate

With germanates in an alkaline medium carbohydrates yield 1:1, 1:2 and 1:3 complexes; at pH 10.7 the following structures have been assumed[12,18].

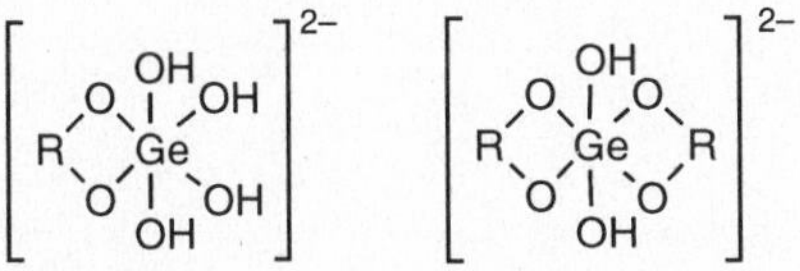

Structure of the Germanate–Sugar Complexes

The strengths of the complexes are dependent on the positions of the OH groups in the sugar molecules.

In an alkaline medium, tin ions similarly give stannate complexes with carbohydrates which also have different mobilities.

2.2.1.1.e. Arsenite

The addition of vicinal polyhydroxy compounds to arsenous acid ($K = 8 \times 10^{-10}$) causes an increase in the strength of this acid, but not to the same extent as that of boric acid. The distances of the oxygen atoms in As_4O_6 is 2.76 Å (they are similar to the [AsO–OH]$^-$ configuration) suggesting complex formation with the vicinal diol groups, and the electric mobilities of carbohydrates in an arsenite medium support this proposal (see Table 5)[11,12].

2.2.1.1.f. Molybdate

In weak acidic solutions many polyhydroxy compounds yield complexes with negative charges. Their migration in the electric field is strongly dependent on the constitution of the relevant compound. Vicinal diols do not give such complexes. For vicinal trihydroxy compounds, the complex formation is strongly dependent on the position of the OH groups relative to each other; those with the highest mobilities are the all *cis* compounds. For the pyranose form of the aldohexoses, the range of the stability of the molybdate complexes is small (pH 5.7–7.8), whilst for sugar alcohols this range is broad (pH 1–8). In a strong acid medium dimolybdate forms complexes, in a weak acid medium a 1:1 ratio of the molybdate and the sugars is assumed. In the latter medium most of the aldoses either do not migrate or migrate at low speed. The mobilities of ketoses and sugar alcohols are higher, since in most cases the above-mentioned configurations can be formed. In such cases a good separation between monosaccharides and the corresponding sugar alcohols is possible[19–21].

Procedure

Electrolyte: dissolve 25 g of sodium molybdate(VI) dihydrate in 1 L of water and adjust to a pH of 5.0 with concd sulfuric acid.

Electrophoresis: 45 V/cm; for common paper with a length of 50–60 cm 2500–3000 V are needed.

Visualization: AgNO$_3$/NaOH according to Trevelyan and Procter (see Paper Chromatography, page 110).

2.2.1.1.g. Vanadate

Sodium vanadate also yields charged complexes with vicinal polyhydroxy compounds whose mobilities are strongly dependent on their individual structures. The complex formation can not be attributed to a single stereochemically defined polyol grouping. In general, replacement of the H of an OH group by a second monosaccharide molecule enhances the mobility. Therefore, sugar alcohols and same disaccharides have remarkably higher migration rates than the common monosaccharides, as seen in Table 5 using an 1.5% solution of sodium metavanadate at pH 8.7[22].

Table 5.   Relative Electrophoretic Mobilities M_G of Several Monosaccharides, Related Sugar Alcohols and Some Oligosaccharides

Compound	A	B	C	D	E
Glyceraldehyde	-	-	-	<0.1	-
Dihydroxyacetone	-	-	-	<0.1	-
D-Erythrose	-	-	-	0.9	0.1–0.2
L-Threose	-	-	-	0.6	0.1–0.2
L-Arabinose	2.4	0.80	0.30	<0.1	<0.10
D-Ribose	4.7	0.75	1.00	0.4	0.40
D-Xylose	1.8	0.95	0.17	<0.1	0.10
D-Lyxose	2.3	0.75	0.42	1.1	0.10
D-Mannose	1.1	0.65	0.35	0.0–0.9[a]	<0.10
D-Glucose	1.0	1.0	0.16	<0.1	0.10
D-Galactose	1.8	0.80	0.28	<0.1	<0.10
D-Allose	-	-	0.75	<0.1	-
D-Altrose	5.8	0.90	0.77	<0.1	-
D-Idose	-	-	1.15	<0.1	-
D-Gulose	-	-	0.53	1.1	0.17
D-Talose	-	-	1.19	0.7	-
D-Fructose	9.3	0.80	0.75	0.5	0.45
L-Sorbose	8.5	1.00	0.73	0.3	0.2–0.8
D-Tagatose	8.6	0.90	1.03	-	0.39
D-Psicose	-	-	1.88	-	-
L-Ramnose	0.5	-		0.0–0.6[a]	-
L-Fucose	-	-		-	-
Glycerol	0.0	0.45	-	<0.1	<0.1
Erythritol	0.1	0.75	0.53	-	0.1–0.2
L-Threitol	0.3	0.80	0.96	-	0.1–0.2
Ribitol (Adonitol)	0.3	1.00	0.76	-	0.32
Arabitol	0.6	0.95	1.24	-	0.31
Xylitol	0.9	0.90	1.55	-	0.60
D-Sorbitol (Glucitol)	1.3	1.00	1.61	1.08	1.00
D-Mannitol	1.0	1.00	1.30	1.00	0.40
L-Iditol	1.4	0.95	1.73	-	1.50
Galactitol (Dulcitol)	1.0	1.05	1.45	1.11	0.48
Allitol	-	-	0.92	-	0.70
Talitol	-	-	1.38	-	-
Rhamnitol	0.8	-	-	-	-
1-1 Disaccharide					
Trehalose	-	-	-	-	-
1-2 Disaccharide					
Sucrose	-	-	0.14	-	-
Sophorose	-	-	-	-	0.13
Kojibiose	-	-	-	-	2.9
1-3 Disaccharide					
Nigerose	-	-	-	-	2.9
Laminaribiose	-	-	-	-	0.1
Turanose	-	-	0.30	-	0.26
1-4 Disaccharide					
Maltose	-	-	0.15	-	0.26
Lactose	-	-	0.24	-	-
Cellobiose	-	-	0.15	-	-
1-6 Disaccharide					
Melibiose	-	-	0.32	-	-
Gentiobiose	-	-	-	-	-
Isomaltose	-	-	-	-	0.10

Table 5.   (Continued)

[a]Formation of stripes.

A =   Electrolyte: sulfonated phenylboric acid 0.05 M, pH 6.5
       Paper: Whatman 1
       Electrophoresis: 500 V (10 V/cm)
       Reference substance: D-glucose (for monosaccharides)
           mannitol (for sugar alcohols and disaccharides) (Garegg and Lindberg[16])
B =   Electrolyte: diphenylboric acid 0.1 M, pH 10
       Paper: Whatman 3 MM
       Electrophoresis: 1000 V (20 V/cm)
       Reference substance: D-glucose (for all compounds) (Garegg and Lindström[16])
C =   Electrolyte: sodium arsenite 0.1 M, pH 9.6
       Paper: Whatman 4
       Electrophoresis: 1300–1400 V (20–25 V/cm), 18–20 °C; 1.5 h
       Reference substance: D-ribose (Frahn, Mills[11])
D =   Electrolyte: 25 g sodium molybdate/1200 mL H_2O, pH 5.0 (1)
           25 g sodium molybdate/1000 mL H_2O, pH 5.0 (2)
       Paper: Whatman 3 MM (1); Schleicher-Schüll 2043 b (2)
       Electrophoresis: 300–800 V (30–80 V/cm) (1); 3000 V (45 V/cm) (2)
       Reference substance: D-sorbitol (1) (Bourne, Hutson and Weigel[19,20]) (sugar)
           D-mannitol (2) (Mayer and Westphal[21]) (sugar alcohols)
E =   Electrolyte: 1.5% sodium metavanadate, pH 8.7
       Paper: Whatman 3 MM
       Electrophoresis: 45 V/cm
       Reference substance: D-sorbitol (Searle and Weigel[22])

Table 6.   Relative Electrophoretic Mobilities M_G of Some Methyl Glycosides and O-Methyl Sugars

Compound	A	B	Compound	A	B
Methyl-α-D-xylofuranoside	2.3	0.0	Methyl-α-D-mannopyranoside	-	0.40
Methyl-β-D-xylopyranoside	0.0	0.0	Methyl-β-D-galactofuranoside	0.4	0.30
Methyl-β-D-glucofuranoside	2.0	0.65	Methyl-β-D-galactopyranoside	0.0	-
Methyl-α-D-glucopyranoside	0.0	0.20	2-O-Methyl-D-glucose	0.0	0.30
Methyl-β-D-glucopyranoside	-	0.30	3-O-Methyl-D-glucose	1.3	0.75
Methyl-α-D-mannofuranoside	16.0	1.00	4-O-Methyl-D-glucose	0.0	-
Methyl-β-D-mannopyranoside	0.0	0.25	6-O-Methyl-D-glucose	0.5	0.65

A =   Electrolyte: sulfonated phenylboric acid 0.05 M, pH 6.5
       Paper: Whatman 1
       Electrophoresis: 500 V (10 V/cm)
       Reference substance: D-glucose (Garegg and Lindberg[16])
B =   Electrolyte: diphenylboric acid 0.1 M, pH 10
       Paper: Whatman 3 MM
       Electrophoresis: 1000 V (20 V/cm)
       Reference substance: D-glucose (for all compounds) (Garegg and Lindström[17])

Conversion of Sugars into Charged Compounds

2.2.1.1.h. *N*-Benzylglycosylamines

Reducing sugars react with benzylamine to give *N*-benzylglycosylamines, which migrate in acid medium as the corresponding ammonium compounds.

Procedure for Paper Electrophoresis

Spot an aliquot of the sample on the paper (e.g., Whatman 3), then add an equal volume of a mixture of 1 mL benzylamine, 9 mL methanol and 5 mL 10 M formic acid. Heat the paper for 5 min at 95 °C and then soak the paper with the formate electrolyte, comprising 600 mL of 5% aqueous NaOH adjusted with 90% formic acid to pH 1.8 and made up to 1 L. Carry out the electrophoresis at 600 V, thereafter visualize the separated carbohydrates with the silver nitrate or with the periodic acid reaction[23].

In this case, the electrophoretic mobilities do not depend on the chemical structures of the individual sugars, but mainly on the number of carbon atoms in the molecules as shown in Table 7.

Table 7.   Relative Electrophoretic Mobilities M_G of Mono- and Oligosaccharides as *N*-Benzylglycosylamines and *N*-(2-Pyridyl)glycamines

Compound	A	B
Pentoses	1.09–1.15	1.10
Hexoses	1.00	1.00
Heptoses	0.91	-
Hexose disaccharides	0.71–0.78	0.72
Hexose trisaccharides	0.59–0.63	0.57
Hexose tetrasaccharides	0.49–0.51	0.52
Hexose pentasaccharides	0.42	0.46
Hexose hexasaccharides	0.33	-
N-Acetylhexosamines	-	0.90

A =   Electrolyte: 600 mL 5% NaOH and 400 mL 90% formic acid
       Paper: Whatman 3
       Electrophoresis: 600 V (12 V/cm)
       Reference substance: ratio of the distance *N*-benzylglycosylamine–unreacted sugars/*N*-benzyl-glucosylamine–unreacted glucose (Barker, Bourne, Grant and Stacey[23])

B =   Electrolyte: pyridine–acetic acid–water 45:30:900 v/v, pH 5.0
       Paper: Whatman 1
       Electrophoresis: 40 V/cm
       Reference substance: *N*-2-D-pyridylglucosamine (Hase, Hara and Matsushima[24])

2.2.1.1.i. 2-*N*-Pyridylglycamine

The trend in modern sugar analysis prior to separation is to convert the substances into compounds which have high absorption in the visible or in the UV region, or which have strong fluorescence. These procedures enable quantitative determinations after separation without further treatment with specific reagents, which are usually sources of inaccuracies. One such method is the conversion of reducing sugars by reaction with 2-pyridylamines and sodium cyanoborohydride into 2-pyridylglycamines. This reaction, the so-called "reductive amination", is described in the section on thin-layer chromatography (see pages 141, 152).

In the case of paper electrophoresis, using a mixture of pyridine–acetic acid–water 45:30:900 v/v and 40 V/cm (0.5–1 h) separation takes place according to the molecular masses of the sugars; using an alkaline borate buffer the separation occurs according to their specific configurations. This property has been used for elucidation of the oligosaccharide parts of glycoproteins and for establishing a fingerprint for these substances.

This reductive amination of reducing sugars also plays an important role in the capillary electrophoresis of these substances[59–63] (see Section 2.2.5.).

2.2.1.1.j. Bisulfite

Aldehydes and ketones react with bisulfite yielding hydroxysulfonic acids according to the following reaction scheme.

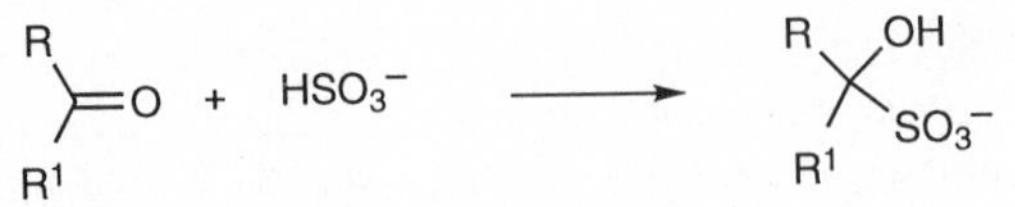

Reaction between Bisulfite and Carbonyl Compounds

These are strong acids which can be considered completely dissociated in the buffers used. The rate of migration of such compounds should depend largely on the migration velocity on the ion itself and on the equilibrium constant; it is, therefore, also dependent on the temperature[25,26,27].

Reducing sugars also give this reaction since the interconversion of the different conformers (α- or β-pyranosides and α- or β-furanosides) takes place through the intermediate formation of the open aldehyde or keto form. The bisulfite ion reacts with the latter two forms according to the equilibrium reaction mentioned above. The sugar sulfonic acids migrate in the electric field. The mobilities are also strongly dependent on the molecular weights. While the sugar sulfonic acids migrate as anions to the anode, the uncharged sugars migrate according to the electro-endoosmosis to the cathode. In the case of 0.4 M sodium bisulfite electrolyte for one aldose two zones are obtained; that of higher mobility refers to the sugar sulfonic acid, the other to the free sugar itself. Ketoses (e.g., fructose, sorbose) do not migrate under these conditions.

Procedure

Electrolyte: 0.4 M sodium bisulfite.

Wet the paper with this electrolyte and wait for equilibrium in the apparatus to be reached with the electrolyte. Then, spot the sugars as their solutions in 0.4 M of sodium bisulfite and carry out the electrophoresis at 8 V/cm for around 2 h. After separation, visualize the separated sugars by dipping the dried papers in aniline picrate in acetone and then heat at 100 °C.

In Table 8 the relative mobilities of some of the common sugars are listed.

Table 8.   Relative Electrophoretic Mobilities M_G of Mono- and Oligosaccharides as Bisulfite Compounds

Compound	A	Compound	A
Arabinose	1.15	Maltose	0.71
Xylose	1.13	Lactose	0.69
Ribose	1.16	Cellobiose	0.70
Glucose	1.00	Melibiose	0.71
Galactose	1.02	Maltotriose	0.55
Mannose	0.97	Maltotetraose	0.45
Rhamnose	1.07		

A =	Electrolyte:	0.4 M sodium bisulfite
	Paper:	Whatman 4
	Electrophoresis:	8 V/cm
	Reference substance:	distance between bisulfite compound–unreacted sugar/glucose bisulfite compound–unreacted glucose (Frahn and Mills[27])

2.2.1.2. Paper Electrophoresis of Sugar Acids

The most important acids of the low molecular weight sugars are the uronic acids, which are units of a great number of natural polysaccharides and which are released upon hydrolysis. For their identification, separation from the neutral sugar units is necessary. For this purpose, electrophoresis is a very suitable method.

Table 9.   Relative Electrophoretic Mobilities M_G of Uronic Acids and Other Sugar Acids

Compound	A	B	C	D	E	G	H
D-Glucuronic acid	1.05	1.44	1.00	1.00	1.00	1.00	1.00
D-Galacturonic acid	0.97	1.18	0.70	0.85	0.98	0.30	0.42
D-Mannuronic acid	1.00	1.00	0.83	-	-	0.75	0.79
L-Guluronic acid	0.93	0.85	0.62	-	-	0.44	0.26
L-Iduronic acid	-	-	0.79	-	-	0.32	0.21
D-Glucose	0.71	1.14	-	-	-	-	-
Erythronic acid	-	-	-	1.12	1.29	-	-
Arabonic acid	-	-	-	0.92	1.09	-	-
Xylonic acid	-	-	-	0.98	1.13	-	-
Galactonic acid	-	-	-	0.90	1.04	-	-
Gluconic acid	-	-	-	0.83	1.02	-	-
2-Ketogluconic acid	-	-	-	1.15	1.06	-	-
5-Ketogluconic acid	-	-	-	1.00	1.04	-	-
α-Saccharinic acid	-	-	-	0.83	0.98	-	-
α-Isosaccharinic acid	-	-	-	0.87	0.99	-	-

A = Electrolyte: 0.01 M borate, pH 9.2
     Paper: Schleicher-Schüll 2043 b
     Electrophoresis: 0.5 mA/cm
     Reference substances: D-mannuronic acid (Haug and Larsen[28])
B = Electrolyte: 0.01 M borate pH 9.2, 0.005 M CaCl$_2$
     Paper: Schleicher-Schüll 2043 b
     Electrophoresis: 0.5 mA/cm
     Reference substances: D-mannuronic acid (Haug and Larsen[28])
C = Eletrolyte: pyridine–acetic acid–water 1:10:89 v/v with
                              98% formic acid adjusted to pH 2.7
     Paper: Tajo filter paper Nr. 51
     Electrophoresis: 100 V/cm
     Reference substance: D-glucuronic acid (Kosakai and Yosizawa[29])
D = Electrolyte: 0.05 M acetate buffer, pH 4.0
     Paper: -
     Electrophoresis: -
     Reference substance: D-glucuronic acid (Theander[31])
E = Electrolyte: 0.05 M phosphate buffer, pH 7.0
     Paper: -
     Electrophoresis: -
     Reference substance: D-glucuronic acid (Theander[31])
G = Electrolyte: 0.1 M zinc acetate, pH 6.6
     Paper: Whatman 1
     Electrophoresis: 600 V; 90 min
     Reference substance: D-glucuronic acid (St. Cyr[30])
H = Electrolyte: 0.1 M barium acetate, pH 7.8
     Paper: Whatman 1
     Electrophoresis: 600 V
     Reference substance: D-glucuronic acid (St. Cyr[30])

Such separations have been described in alkaline as well as in an acid medium. For the first variation a 0.1 M borate buffer pH 9.2 is used. The mobilities of the uronic acids are higher than those of neutral sugars, but the differences are only small[28]. An improvement can be achieved by addition of Ca^{2+} ions (see Table 9)[28].

In acid medium a mixture of pyridine–acetic acid–water has been used; additionally, the pH is adjusted with formic acid (98%) and at pH 2.7 the separation of all the common uronic acids is possible[29].

The acetates of several bivalent metal ions are also suitable for the separation of uronic acids; 0.1 M solutions of Zn^{2+}, Cd^{2+}, Ca^{2+}, Mg^{2+}, Ba^{2+} and Cu^{2+} have been studied. Except for Cu^{2+}, the other compounds migrate to the anode. The best separations have been obtained with barium and zinc acetates[30]. The relative mobilities of uronic acids in borate, pyridine–acetic acid–water, acetate and phosphate buffers, as well as in barium and zinc acetate solutions, are listed in Table 9.

For visualization of the separated uronic acids, the same reagents can be used as described for paper chromatograms (see pages 110–112).

For the separation of other sugar acids such as aldonic acids or ketoaldonic acid, a 0.05 M acetate buffer pH 4.0 and a 0.05 M phosphate buffer pH 7.0 have been described[31,32]. The relative mobilities are also listed in Table 9. It must be recognized that in a neutral medium the differences in the mobilities are distinctly smaller than in an acid medium.

2.2.1.3. Paper Electrophoresis of Sugar Phosphates

Sugar phosphates also migrate without derivatization in the electric field and, hence, can be separated easily by this method from the neutral sugars. For this separation the following electrolytes have been described:

1. 1.0 M Sodium butyrate pH 3.2[33].
2. Pyridine–acetic acid–water pH 3.9 or 6.0[34,35].
3. Cetyltrimethylammonium borate pH 9.6[36].
4. Ammonium acetate[37].
5. Borate[34,38].

For detection, the reagents described for paper chromatography can be used (ammonium molybdate, $FeCl_3$–sulfosalicylic acid, etc.; see Paper Chromatography, pages 118, 119).

In Table 10 the relative mobilities of several sugar phosphates are listed. The differences of the mobilities are small and predominantly determined by the number of charged groups. Only the borate buffer in the pH range 9–10 allows the separation of individual sugar phosphate species with an equal number of phosphate groups (e.g., 0.2 M borate buffer pH 9.5; separation of glucose-1-phosphate, glucose-6-phosphate)[38].

Methylation of the sugar phosphates before electrophoresis leads to an improvement in separation. For this purpose the sugar phosphates are treated in absolute methanol with triethylamine and dicyclohexylcarbodiimide[36]. For the separation of complicated sugar phosphate mixtures a combination of paper chromatography and paper electrophoresis has been used.

The same separation can be obtained by thin-layer electrophoresis on cellulose layers using the following conditions[37].

Support: cellulose powder (for thin-layer chromatography).

Thickness: 0.25 mm.

Electrolyte: 0.28 M ammonium acetate pH 3.6 containing 0.1 g EDTA/L.

The relative mobilites of the sugar phosphates are the same as those on paper; using two-dimensional techniques, however, the separation time can be reduced by a fifth and the sensitivity can be enhanced 20 times more than that on paper.

Table 10. Relative Electrophoretic Mobilities M_G of Sugar Phosphates

Compound	A	B	C
Erythrose-4-phosphate	0.65	-	-
Ribose-5-phosphate	0.69	0.75	0.36
Ribose-1,5-diphosphate	1.04	-	-
Ribulose-5-phosphate	-	-	-
Xylose-1-phosphate	-	-	1.09
Xylulose-5-phosphate	-	-	-
Fructose-1-phosphate	0.63	0.70	0.27
Fructose-6-phosphate	0.63	0.70	0.27
Fructose-1,6-diphosphate	0.95	1.07	-
Glucose-1-phosphate	0.64	0.69	0.85
Glucose-6-phosphate	0.61	0.68	0.36
Galactose-1-phosphate	-	-	0.66
Galactose-6-phosphate	0.63	0.68	0.27
Mannose-1-phosphate	-	-	0.70
Mannose-6-phosphate	0.63	0.70	0.52
Inositol monophosphate	-	1.11	-
Inositol diphosphate	-	1.38	-
Inositol tetraphosphate	-	1.56	-
Inositol hexaphosphate	-	1.70	-
Gluconic acid 6-phosphate	0.91	-	-

A =	Electrolyte:	pyridine–acetic acid–water 20:64:916 v/v, pH 3.9
	Paper:	Whatman 3 washed (L: 57 cm)
	Electrophoresis:	2000 V (35 V/cm), 2 h
	Reference substance:	orthophosphate (Vanderheiden[34])
B =	Electrolyte:	0.25 M ammonium acetate–EDTA (0.05 g/L), pH 3.6
	Paper:	Schleicher-Schüll 589
	Electrophoresis:	600 V
	Reference substance:	orthophosphate (Bieleski and Young[37])
C =	Electrolyte:	0.1 M cetyltrimethylammonium borate pH 9.6
	Paper:	Whatman 1 (L: 50 cm)
	Electrophoresis:	1000 V (20 V/cm)
	Reference substance:	orthophosphate (Piras and Cabib[36])

2.2.1.4. Paper Electrophoresis of Amino Sugars

Amino sugars are important components of natural polysaccharides as free or substituted compounds and their analysis is of practical importance. Besides paper, thin-layer and ion exchange chromatography, paper electrophoresis also enables separation of these compounds, e.g. N-acetylamino sugars, using a borate buffer at pH 10.0[39]. For visualization specific reagents are applied which are described in the section on paper chromatography. The separation of glucosamine, galactosamine, N-acetylglucosamine and N-acetylgalactosamine can be achieved after reduction to the corresponding amino sugar alcohols with sodium borohydride; the separation itself is carried out with 0.06 M borate buffer pH 9.5 and a field strength of 60 V/cm[40]. The relative mobilites of some of the nonreduced amino sugars are listed in Table 11.

Table 11.   Relative Electrophoretic Mobilities M_G of Amino Sugars and Alcohols

Compound	A	Compound	A
N-Acetyl-D-fucosamine	0.14	N-Acetyl-D-allosamine	0.42
N-Acetyl-D-glucosamine	0.23	N-Acetyl-D-talosamine	0.60
N-Acetyl-D-xylosamine	0.23	N-Acetyl-D-gulosamine	0.63
N-Acetyl-D-galactosamine	0.33	N-Acetyl-D-mannosamine	0.65

A =   Electrolyte: borate buffer pH 10; 23.4 g sodium tetraborate and 30 mL 1 M NaOH are
                                        made up to 1L with distilled water
          Paper: Whatman 3
          Electrophoresis: 1300 V
          Reference substance: D-glucose (Crumpton[39])

For quantitative determinations a method using reduction with ^{3}H-sodium borohydride has been described; after electrophoretic separation, the zones are located on the paper with a radioscanner as for paper chromatograms[41] and the radioactivity measured with a liquid scintillation counter.

2.2.2. Cellulose Acetate

This wide porous material is commonly used for the separation of physiologically active polysaccharides (mucopolysaccharides) and rarely for the separation of mono- and oligosaccharides as their derivatives. A special case is the separation of glucuronic acid and iduronic acid from the hydrolysate of glycosaminoglycans on Ti(III)-impregnated cellulose acetate sheets. Using 0.1 M zinc acetate as the electrolyte, a field strength of 50 V/cm and a temperature of 4 °C excellent separations of both compounds has been obtained[42]. For visualization a modified silver nitrate–sodium hydroxide reagent in ethanolic solution is used (see Paper Chromatography, Section 2.1.1.). A semiquantitative evaluation of the electropherogram by densitometry of the patterns is also possible.

2.2.3. Electrophoresis on Glass Fibre Paper

The disadvantage of electrophoresis on normal cellulose filter paper is the limited number of possibilities of visualizing the separated substances since the carbohydrate-specific sulfuric acid and phosphoric acid containing reagents react with the cellulose itself causing a strong background coloration. This disadvantage can be overcome by electrophoresis on glass fibre paper. The practical procedures for carrying out the electrophoresis are nearly the same as on cellulose filter sheets, with the use of 0.2 mol borate buffer pH 10.0 as electrolyte and a field strength of 15 V/cm. For visualization of the sugars all the sulfuric and phosphoric acid containing reagents described in the section on thin-layer chromatography, e.g. naphthoresorcinol–sulfuric acid, can be used. For the visualization of the nonreactive polyols like sugar alcohols or aldonic acids the following reagents are suitable:
1.   Silver nitrate–ammonium hydroxide (see Paper Chromatography, page 110).
2.   Periodate–benzidine (or o-tolidine, o-dianisidine; see Paper Chromatography, page 111).
3.   Potassium permanganate–sodium carbonate (see Thin-Layer Chromatography, page 145).
These react generally with all vicinal polyhydroxy compounds.
The electrophoretic mobilities of these compounds on glass fibre paper is higher than on normal cellulose paper (e.g., D-glucose: Whatman 3:3.66 x 10^{-6}, glass fibre sheets: 4.92 x 10^{-6} cm^2 V^{-1} sec^{-1})

Table 12. Relative Electrophoretic Mobilities M_G of Sugars, Sugar Alcohols and Uronic Acids on Glass Fibre Paper

Compound	A	B	C	D	E
DL-Glyceraldehyde	0.75	-	-	-	-
1,3-Dihydroxyacetone	0.75	-	-	-	-
L-Arabinose	0.94	0.91	- 0.03	-	-
D-Ribose	0.74	-	- 0.26	-	-
D-Xylose	1.00	1.00	-	-	-
2-Deoxyribose	0.28	-	-	-	-
L-Fucose	0.88	-	-	-	-
L-Rhamnose	0.49	0.50	0.00	-	-
D-Galactose	0.91	0.90	- 0.01	-	-
D-Glucose	1.00	1.00	0.00	-	-
D-Mannose	0.67	0.69	- 0.04	-	-
D-Fructose	0.88	0.88	- 0.11	-	-
L-Sorbose	0.95	-	-	-	-
meso-Erythritol	0.70	-	-	-	-
D-Arabitol	0.86	-	-	-	-
D-Sorbitol (D-Glucitol)	0.82	-	-	-	-
D-Mannitol	0.85	-	-	-	-
D-Galactitol (Dulcitol)	0.95	-	-	-	-
myo-Inositol	0.48	-	- 0.10	-	-
Sophorose	0.28	-	-	-	-
Nigerose	0.65	-	-	-	-
Cellobiose	0.26	-	-	-	-
Maltose	0.30	0.34	-	-	-
Gentiobiose	0.69	-	-	-	-
Sucrose	0.15	0.16	-	-	-
Lactose	-	0.42	-	-	-
Raffinose	0.23	-	-	-	-
D-Galacturonic acid	-	1.20	0.24	0.30	0.55
D-Glucuronic acid	-	-	1.00	1.00	1.00
D-Mannuronic acid	-	-	0.74	0.76	0.81
L-Guluronic acid	-	-	-	0.41	0.41

A =	Electrolyte:	0.2 M sodium tetraborate pH 10
	Paper:	glass fibre paper (Reeve Angel & Co) untreated
	Electrophoresis:	15 V/cm
	Reference substance:	D-glucose (Bourne, Foster and Grant[3])
B =	Electrolyte:	0.05 M sodium tetraborate pH 9.2
	Paper:	silanized glass fibre paper
	Electrophoresis:	30 V/cm; 8–10 °C; 45 min
	Reference substance:	D-glucose (Bettler, Amado and Neukom[44])
C =	Electrolyte:	0.1 M barium acetate
	Paper:	silanized glass fibre paper
	Electrophoresis:	30 V/cm; 8–10 °C; 45 min
	Reference substance:	D-glucose for sugar, D-glucuronic acid for uronic acids (Bettler, Amado and Neukom[44])
D =	Electrolyte:	0.1 M zinc acetate
	Paper:	silanized glass fibre paper
	Electrophoresis:	30 V/cm; 8–10 °C; 45 min
	Reference substance:	D-glucuronic acid (Bettler, Amado and Neukom[44])
E =	Electrolyte:	0.1 M calcium acetate
	Paper:	silanized glass fibre paper
	Electrophoresis:	30 V/cm; 8–10 °C; 45 min
	Reference substance:	D-glucuronic acid (Bettler, Amado and Neukom[44])

since the adsorption on the glass fibre material is strongly diminished. A disadvantage is the high electro-endoosmosis and the high diffusion on the glass fibre paper.

For overcoming these effects the surface of the glass fibres is made hydrophobic by treatment with dichlorodimethylsilane.

The silylation of the glass fibre paper is carried out according to the following procedure[43,44].

Procedure

Heat the glass fibre paper to remove organic material for 2 h at 400 °C. After cooling, immerse the sheets in a 5% solution of dichlorodimethylsilane in chloroform and leave at room temperature for 24 h; then rinse the material off with toluene.

For electrophoresis, dip the silylated glass fibre sheets for 1 day in the electrolyte, which contains 0.2% Triton X-100. Take out the sheets, remove the excess of the electrolyte by blotting cautiously with filter paper, place in the electrophoresis apparatus and equilibrate under separation conditions for 15 min; then spot the sample (5–15 μL) and carry out the electrophoresis under the following conditions:

Electrolyte: sugars: 0.05 M borate buffer pH 9.2; uronic acids: 0.1 M zinc acetate, 0.1 M barium acetate, 0.1 M calcium acetate.

Field strength: 30 V/cm.

Temperature: 8–10 °C.

As neutral marker for the strength of electro-endoosmosis either 2,3,4,6-tetramethyl-O-glucose or ω-hydroxymethylfurfurol is used.

The relative mobilities of common sugars and sugar alcohols are listed in Table 12.

2.2.4. Silanized Silica Gel

The separation of sugars and related substances on glass fibre paper has the advantage that aggressive carbohydrate reagents can be used for visualization. For silanized materials, however, production procedures are complicated and mechanical stabilities are poor.

For the electrophoretic separation of polysaccharides a suitable inert support has been described, namely silanized silica gel, whose surface is covered with 1-octanol. On this material the aggressive carbohydrate specific reagents can also be applied, the layers are stabile and the diffusion of the separated substances in this layer only small; hence compact sharp zones can be obtained[45,46].

This support is suitable for the separation of monosaccharides, oligosaccharides and related compounds like sugar alcohols, sugar acids and sugar phosphates[47].

Procedure

Preparation of the Layers

Slurry 15 g of silanized silica gel (for thin-layer chromatography) in 50 mL of dichloromethane or methyl acetate containing 6 g of 1-octanol. Stir the mixture intensively for 40 min (magnetic stirrer) and then evaporate the solvent at 50 °C in a rotatory evaporator until a lump-free powder remains.

Pour the dry powder into a porcelain mortar (∅ 160 mm), add 10 mL of a 2% solution of polyvinyl-pyrrolidone (MW: 360 000 Daltons) in the electrophoresis buffer and then 35 mL of the buffer in portions with intense stirring with a pestle until a homogenous slurry is formed. Spread the slurry on 200 x 200 mm glass plates with Desaga equipment for thin-layer chromatography (thickness of the layer 0.30 mm). After coating, transfer the plates immediately into a dessicator containing, on the bottom, distilled water with a few drops of a neutral detergent (Tween 20) and some crystals of thymol.

Table 13. Relative Electrophoretic Mobilities M_G of Sugar, Sugar Alcohols, Sugar Phosphates and a Variety of Sugar Acids

Compound	A	B	Compound	A	B
L-Arabinose	1.00	-	Isomaltose	0.62	-
D-Ribose	0.70	-	Gentiobiose	0.73	-
D-Xylose	1.01	-	Melibiose	0.79	-
D-Lyxose	0.67	-	Palatinose	0.48	-
D-Glucose	1.00	-	Turanose	0.64	-
D-Galactose	0.92	-	Sophorose	0.35	-
D-Mannose	0.69	-	Sucrose	0.20	-
D-Fructose	0.86	-	Trehalose	0.22	-
L-Sorbose	0.95	-	Raffinose	0.29	-
D-Rhamnose	0.53	-	Melezitose	0.28	-
L-Fucose	0.95	-	Maltotriose	0.26	-
meso-Erythritol	0.68	-	Stachyose	0.37	-
Threitol	0.87	-	Glucose-6-phosphate	1.30	0.77
Adonitol (Ribitol)	0.87	-	Glucose-1-phosphate	1.06	0.86
Xylitol	0.79	-	Fructose-6-phosphate	1.21	0.81
Arabitol	0.75	-	Fructose-1,6-phosphate	1.44	1.38
Mannitol	0.90	-	D-Glucuronic acid	1.39	1.00
Sorbitol	0.80	-	D-Galacturonic acid	1.28	0.26
Galactitol (Dulcitol)	0.94	-	D-Mannuronic acid	1.21	0.74
myo-Inositol	0.48	-	L-Guluronic acid	-	0.06
Maltose	0.39	-	D-Gluconic acid	1.21	-
Lactose	0.42	-	D-Galactonic acid	1.23	-
Cellobiose	0.32	-	L-Arabonic acid	1.30	-
Lactulose	0.61	-	D-Ribonic acid	1.17	-

A =	Electrolyte:	0.3 M sodium borate pH 10
	Support:	silanized silica gel–1-octanol
	Electrophoresis:	400 V (20 V/cm); 16–18 °C
	Reference substance:	D-glucose (Scherz[47])
B =	Electrolyte:	0.07 M barium acetate
	Support:	silanized silica gel–1-octanol
	Electrophoresis:	400 V (20 V/cm); 16–18 °C
	Reference substance:	D-glucuronic acid (Scherz[47])

The detergent is added to prevent the formation of a thin film of 1-octanol on the water surface which would reduce water evaporation and accelerate the desiccation of the plates. The thymol crystals are added to prevent microbial contamination. The side wall of the desiccator is taped with wet filter paper, which establishes a saturated moist atmosphere and, therefore, reduces water evaporation from the surface of the plates. After incubation for 3 h the plates can be used for electrophoresis.

Electrophoresis

Electrolytes: 0.3 M borate buffer pH 10 containing 7.44 g Titriplex III/L; 0.07 M barium acetate.

Field strength: 15–20 V/cm (300–400 V).

Temperature: 16–18 °C.

Remove the plates from the dessicator and put them immediately on the cooling block of the electrophoretic chamber. Connect both ends to the electrode vessels with glass fibre strips and fasten them onto the layer with glass strips (200 x 20 x 1 mm), put on them another glass strip (200 x 20 x 4 mm) and another glass plate (200 x 200 mm). With this equipment the vapor room over the layer becomes small and the evaporation of the water during electrophoresis can be diminished. Equilibrate the layer for 30 min, then take off the upper glass plates and make 10 mm long groves with the tip of a needle into the layer. Fill them with 2–5 µL of the samples, then put the glass plates on again and carry out the electrophoresis.

Visualization

After electrophoresis, heat the plate for 30 min at 90–100 °C to evaporate the water and 1-octanol from the layer. Spray the plate with the appropriate reagent (e.g., sugar, sugar phosphates, uronic acid: naphthoresorcinol–sulfuric acid).

The relative mobilities for common sugar and sugar derivatives are listed in Table 13. Figure 1 shows an electropherogram of sugars and uronic acids.

By comparison of the electrophoretic mobilities of the sugar and its related compounds it can be recognized that hardly any differences can be found between those on cellulose filter paper, glass fibre filter paper and silanized silica gel–1-octanol. It can be stated, therefore, that the nature of the support has no remarkable influence on the relative mobilities of the substances when they are compared to each other.

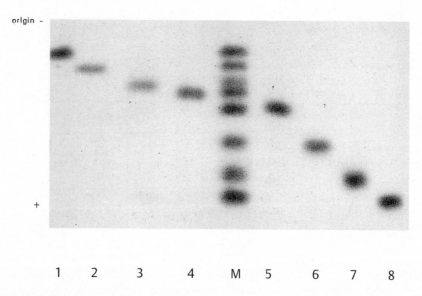

Figure 1a. Thin-Layer Electrophoresis of Mono- and Oligosaccharides on Silanized Silica Gel Covered with 1-Octanol in 0.3 M Borate Buffer pH 10
Electrophoresis: 400 V, 120 min, 18 °C
1: sucrose, **2**: raffinose, **3**: maltose, **4**: lactose, **5**: rhamnose, **6**: mannose, **7**: fructose, **8**: glucose, **M**: mixture of all eight compounds. Sample loading: 10 µg of each compound. Visualization with naphthoresorcinol–sulfuric acid. Origin indicated (Scherz[47]; with permission)

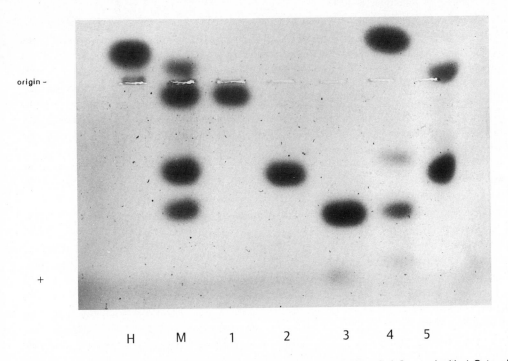

origin –

+

H   M   1   2   3   4   5

Figure 1b. Thin-Layer Electrophoresis of Uronic Acids on Silanized Silica Gel Covered with 1-Octanol in 70 mM Barium Acetate Solution
Electrophoresis: 400 V, 150 min, 16 °C
1: galacturonic acid, **2**: mannuronic acid, **3**: glucuronic acid, **4**: hydrolysate of gum arabic (glucuronic acid), **5**: hydrolysate of sodium alginate (mannuronic and guluronic acid), **M**: mixture of **1–3** and guluronic acid, **H**: ω-hydroxymethylfurfural. Sample loading: 10 µg of each individual compound. Visualization with naphthoresorcinol–sulfuric acid. Origin indicated (Scherz[47]; with permission)

2.2.5. Capillary Zone Electrophoresis

Recently, capillary zone electrophoresis (CZE), developed by Hjerten[48] and Jorgensen[49], has become a powerful method for the separation and quantitative determination of charged compounds[50,51]. Instead of inert supports for stabilizing the system, electrophoresis is carried out in extremely thin capillaries [internal diameter (i.d.): 50–100 µm] made of fused silica (50–100 cm) under a very high applied potential (10–30 kV), which leads to field strengths of 200–400 V/cm. Under these conditions the diffusion of the separated zones is almost prevented; high speeds and high resolutions of the separations are obtained as well as reproducible quantifications of the separated substances. Very small amounts of the substances in the range of nanoliters (nL) are needed. In the field of carbohydrate chemistry, CZE has also emerged as a promising alternative for the analysis of mono- and oligosaccharides as well as for glycoprotein residues.

According to the detection method CZE of carbohydrates can be differentiated into the following groups.

1.   CZE of underivatized carbohydrates by direct photometric detection.
2.   CZE of underivatized carbohydrates by indirect photometric or electrochemical detection.
3.   CZE of derivatized carbohydrates.

2.2.5.1. CZE of Underivatized Carbohydrates

The majority of mono- and oligosaccharides and their derivatives have only a weak absorption in the common UV range because the open-chain configurations, responsible for the UV absorption, are present in only very small amounts in aqueous solutions. Upon addition of borate ions, the absorbances in the low UV range increase significantly (2–20-fold). This enhancement in the absorbances is attributed to (a) the shifts of the equilibria between carbonyl and annular forms from the sugars to the carbonyl forms, and (b) additionally to the formation of oxygen bridges between the carbon and boron atoms. These enhanced absorptions of sugars in the presence of borate allow the UV detection of sugars without derivatization after separation by CZE[52].

Procedure

The following conditions have been described for the separation of sugars:

Capillary: uncoated fused silica tube, L: 94 cm, i.d.: 75 μm.

Potential: 20 kV; field strength: ~250 V/cm.

Detection: UV, 195 nm.

Electrolyte: 50 and 60 mM sodium tetraborate pH 9.3.

Figure 2 shows the separation of some monosaccharides at different temperatures. At elevated temperature ranges (50–60 °C) the separation times become shorter, the peak widths become smaller and the resolutions of the peaks are better than those at room temperature. In Table 14, the absolute and the relative mobilities of several mono- and oligosaccharides are listed (with D-glucose as the reference substance).

Table 14.   Absolute and Relative Mobilities of Several Mono- and Oligosaccharides, Including Uronic Acids and Amino Sugars using 60 mM Borate Buffer pH 9.3 and a Temperature of 60 °C[52]

Compound	Absolute Mobility x 10^{-3} (cm^{2}V^{-1}s^{-1})	Relative Mobility (Reference: D-Glucose)
D-Glucose	0.44	1.00
D-Mannose	0.36	0.82
D-Galactose	0.39	0.89
D-Ribose	0.38	0.86
D-Arabinose	0.40	0.91
D-Xylose	0.47	1.07
L-Fucose	0.36	0.82
D-Fructose	0.37	0.84
D-Glucuronic acid	0.61	1.34
D-Galacturonic acid	0.58	1.32
N-Acetylgalactosamine	0.17	0.39
N-Acetylglucosamine	0.13	0.30
Gentiobiose	0.29	0.66
Lactose	0.17	0.39
Maltose	0.17	0.39
Cellobiose	0.12	0.27
Sucrose	0.06	0.14
Trehalose	0.06	0.14
Maltotriose	0.12	0.27
Raffinose	0.13	0.30
Stachyose	0.16	0.36
Maltotetraose	0.12	0.27

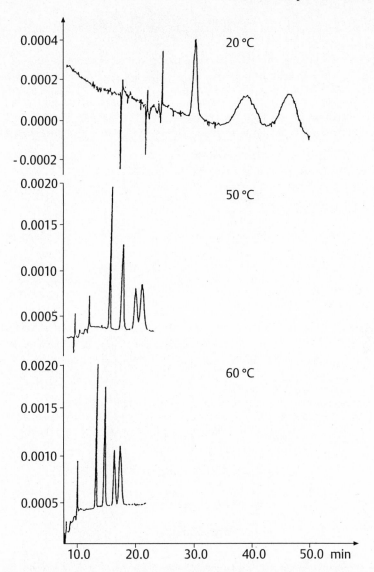

Figure 2.   Electrophoretic Separation of Underivatized Monosaccharides
Capillary: fused silica, L: 94 cm, i.d.: 75 μm; *buffer*: 50 mM borate pH 9.3; *potential*: 20 kV; *detection*: UV, 200 nm; *temperature*: 20 °C, 50 °C, 60 °C; *carbohydrates*: D-mannose, D-galactose, D-glucose, D-xylose (Hoffstetter-Kuhn, Paulus, Grassmann and Widmer[52]; with permission)

CZE is especially suited for the separation of oligosaccharides which are formed by enzymatic digestion of chondroitin sulfate and the hyaluronic acids of connective tissues. The knowledge of the ratio of these compounds to each other is important for medical diagnostic purposes.

These oligosaccharides are highly charged molecules with a characteristic absorbance at 232 nm, as a consequence of an unsaturated group in the molecules, which enables UV detection of the separated zones without difficulty.

Using borate buffer in the range pH 9.0 the disaccharides are separated according to their individual chemical structures, while in acid medium they are separated according to the number of units. In the latter case, UV detection is carried out at 200 nm since at 232 nm the absorbances of these compounds are weak[53].

Procedure

Buffers: (a) 40 mM disodium hydrogen phosphate–10 mM sodium tetraborate–40 mM SDS pH 9.0; (b) 200 mM orthophosphoric acid, adjusted to pH 3.0 with 5 M NaOH.
Capillary: fused silica, L: 72 cm, i. d.: 50 μm.
Potential: 15 kV.

Figure 3 shows the electropherogram of oligosaccharides derived from hyaluronan by digestion with testicular hyaluronidase.

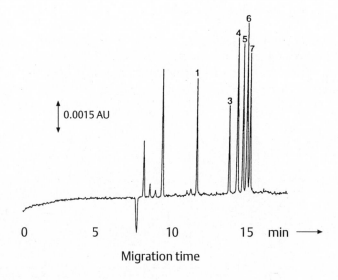

Figure 3. Electrophoretic Separation of Oligosaccharides Derived from Hyaluronan by Digestion with Testicular Hyaluronidase
Capillary: fused silica, L: 72 cm, i.d.: 50 μm; *potential*: 15 kV; *buffer*: 40 mM phosphate–40 mM SDS–10 mM borate; *detection*: 200 nm
1: unsaturated disaccharide of hyaluronan (Δdi-HA), **3**: saturated hexasaccharide of hyaluronan, **4**: saturated octasaccharide of hyaluronan, **5**: saturated decasaccharide of hyaluronan, **6**: saturated dodecasaccharide of hyaluronan, **7**: saturated tetradecasaccharide of hyaluronan. (Carney and Osborne[53]; with permission)

2.2.5.2. CZE of Underivatized Carbohydrates by Indirect UV and Electrochemical Detection

The hydroxyl groups in the molecules of mono- and oligosaccharides, especially the glycosylic OH group at C-1, have slightly acidic characters of different pK. In strong alkaline medium at pH between

11.9–12.5 these substances move with different mobilities in the electric field. The detection of the separated zones is possible either by indirect UV absorption or by amperometric electrode systems.

Indirect UV Photometry
This method is based on the physical displacement of chromophoric, fluorophoric or electroactive compounds by the analytes during the electrophoresis. Negative peaks are observed where the analyte displaces the UV active compounds from the background electrolyte.

This method is particularly useful for the determination of nonreducing sugars since they cannot be derivatized with UV absorbing reagents before the electrophoretic separation. It has been found that sorbic acid is an excellent carrier electrolyte as well as a chromophore with a high absorbance in the middle UV region ($E = 27800$ mol^{-1} cm^{-1} at 276 nm). Furthermore, it does not interact with the analytes and with the capillary surface and it carries a single charge, hence ensuring a good transfer ratio; the latter is defined as the number of chromophore molecules displaced by one analyte molecule[54].

The separation is carried out in the pH range 11.9–12.4 and an exact adjustment of pH is important. A further enhancement of pH causes a sharp decrease in the sensitivity and in the electroosmotic flow; the latter leads to a prolongation of the separation time.

The absolute electrophoretic mobilities of several polyhydroxy compounds are listed in Table 15 which have been determined under the following conditions:
Electrolyte: 6 mM sorbic acid; adjusted with 0.25 M NaOH to pH 12.1.
Capillary: L: 122 cm, i.d.: 50 μm.
Potential: 28 kV.
Temperature: 30 °C.
Detection: UV, 256 nm; vacuum injection.

Table 15.   Mean Electrophoretic Mobilites of Nonderivatized Carbohydrates[54]

Carbohydrates	Mobility x 10^{-5} (cm^{2}V^{-1}s^{-1})	Carbohydrates	Mobility x 10^{-5} (cm^{2}V^{-1}s^{-1})
Raffinose	1.312	N-Acetylgalactosamine	6.287
Sucrose	1.356	D-Lyxose	6.440
2-Deoxy-D-galactose	2.419	L-Sorbose	6.468
2-Deoxy-D-ribose	2.798	L-Rhamnose	6.599
D-Fucose	3.096	Turanose	7.047
Lactose	4.175	D-Fructose	7.140
Maltotriose	4.313	D-Ribose	7.419
D-Galactose	4.358	D-Mannose	7.462
Melibiose	4.623	N-Acetylglucosamine	8.018
D-Galactosamine	4.694	N-Acetylneuraminic acid	19.867
Cellobiose	4.794	D-Galactonic acid	24.950
Maltose	4.813	D-Gluconic acid	25.518
L-Arabinose	5.121	D-Mannonic acid	25.677
D-Glucose	5.135	D-Galacturonic acid	26.789
Lactulose	5.403	D-Arabonic acid	27.593
Palatinose	5.594	D-Glucuronic acid	27.796
D-Glucosamine	5.805	D-Ribonic acid	28.139
D-Xylose	6.268	D-Mannuronic acid	29.315

It has been demonstrated that this method can be applied to the determination of the common sugars such as glucose, fructose and sucrose in foods and in fruit juices (Figure 4)[55].

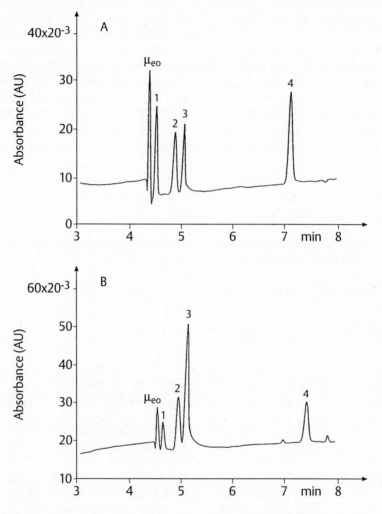

Figure 4.   Electrophoresis of Fruit Juices
A: standard mixture (sugars, 2.22–2.42 mM) and internal standard (GlcAc, 1.15 mM); B: apple juice (diluted 1:50)
Capillary: fused silica, L: 42 cm, i.d.: 50 µm; *potential*: 10 kV; *buffer*: 6 mM sorbate pH 12.2; *detection*: UV, 256 nm
1: sucrose, **2**: glucose, **3**: fructose, **4**: glucuronic acid; μ_{eo}: marking of the electroosmotic flow (Klockow, Paulus, Figueiredo, Amado and Widmer[55]; with permission)

Procedure
Dilute the fruit juices 50–100-fold with highly purified water and in some cases filter it through a 0.22 µm Millipore filter, then adjust the pH to 11.9–12.4 by titration with 1 M NaOH at room temperature.

The range of the sugar concentration should be between 0.1–2 mg/mL; D-glucuronic acid is suggested as an internal standard.

Figure 4 shows the separation of standard mixtures and of apple juices diluted 1:50. The results of some analysis of fruit juices by the CZE and the HPLC method show good agreements and CZE is suitable as a routine method[55].

Electrochemical Detection

An alternative to indirect UV photometry is amperometric detection using the triple-impulse technique. For detection of the separated zones in the CZE capillary, microelectrode systems have been developed. Reference electrodes are Ag/AgCl or calomel electrodes, the working electrodes consist of Cu wires, Cu_2O modified carbon or Au wires, and for the auxiliary electrodes Pt foils or stainless steel tubes[56–58] are used.

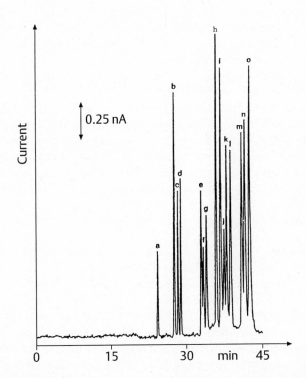

Figure 5.   Electrophoretic Separation of a Mixture Containing 15 Different Carbohydrates (80–150 µM)
Capillary: fused silica, L: 73 cm, i.d.: 50 µm; *potential*: 11 kV; *electrolyte*: 100 mM NaOH
a: trehalose, **b**: stachyose, **c**: raffinose, **d**: sucrose, **e**: lactose, **f**: lactulose, **g**: cellobiose, **h**: galactose, **i**: glucose, **j**: rhamnose, **k**: mannose, **l**: fructose, **m**: xylose, **n**: talose, **o**: ribose. (Colón, Dadoo and Zare[56]; with permission)

Figure 5 shows the separation of a mixture of fifteen different carbohydrates under the following conditions:

Electrolyte: 100 mM NaOH.

Capillary: fused silica, L: 73 cm, i.d.: 50 μm.
Potential: 11 kV.

The mobilities of the sugars appear to be dependent on the pK of the individual compounds. This assumption can be seen by comparison of the absolute mobilities of D-galactose (pK 12.35), D-glucose (pK 12.28) and mannose (pK 12.08) which all have the same molecular weight. During this electrophoresis, the negative charged sugars move against the endoosmotic flow. The weakest acidic sugar of these three, D-galactose, appears first and the strongest one, D-mannose, last as shown in Figure 5.

The mass detection limits are around 20–30 fmol or 1–2 μm/L and are at least two orders of magnitude lower than CZE by indirect UV detection.

2.2.5.3. CZE of Derivatized Carbohydrates

Derivatization of carbohydrates with reagents having strong chromophoric or fluorophoric functions makes UV detection of the separated substances easier and, especially with fluorogenic reagents, the sensitivity of the detection can be increased extensively. One disadvantage is that such derivatization reactions are suitable only for those carbohydrates which have free carbonyl groups such as monosaccharides and reducing oligosaccharides.

The following reactions have been used for derivatization:

(a)   Reductive amination: the free carbonyl group of the sugar reacts with strong UV absorbing amines or aminosulfonic acids to give the so-called "Schiff" bases, which are reduced by the presence of sodium cyanoborohydride or borane–dimethylamine complex to the corresponding glycamines.

A variation of this procedure is the reduction of the sugars with sodium cyanoborohydride in the presence of ammonium ions to 1-amino-1-deoxy sugar alcohols and conversion of the latter compounds to strong UV absorbing or fluorescing derivatives.

(b)   Condensation of the carbonyl group of the reducing sugars with active hydrogens of special heterocyclic compounds affording strong UV absorbing derivatives.

(a) Reductive Amination

The following compounds have been used for this reaction:

2-aminopyridine[59–63], 6-aminoquinoline[63], 4-aminobenzoic acid[64], 4-aminobenzoic acid ethyl ester[65], 4-aminobenzonitrile[66], 8-aminonaphthalene-1,3,6-trisulfonic acid (8-ANTS)[67–69], 7-aminonaphthalene-1,3-disulfonic acid (7-ANDS)[69], 3-aminonaphthalene-2,7-disulfonic acid (3-ANDS)[69], 2-amino-naphthalene-1-sulfonic acid (2-ANMS)[69], 5-aminonaphthalene-2-sulfonic acid (5-ANMS)[69] and 9-aminopyrene-1,4,6-trisulfonic acid (9-APTS)[70]. 2-Aminopyridine was the first substance to be used for this type of reaction according to the scheme on the following page (example: D-glucose).

Procedure[59]

Reagent: 10 mg of sodium cyanoborohydride is dissolved in 1 mL of a methanolic solution of 100 mg of 2-aminopyridine and 100 mg of glacial acetic acid. The reagent must be prepared immediately before use. Another relation: 32 mg cyanoborohydride, 83 mg 2-aminopyridine, 37 μL acetic acid in 0.33 mL methanol.

In a screw-cap vial dissolve a small amount of the sample in distilled water (~20 μL) and add a four- to sixfold volume of the reagent (80–100 μL); the concentration of the sugars should be between 0.2–2

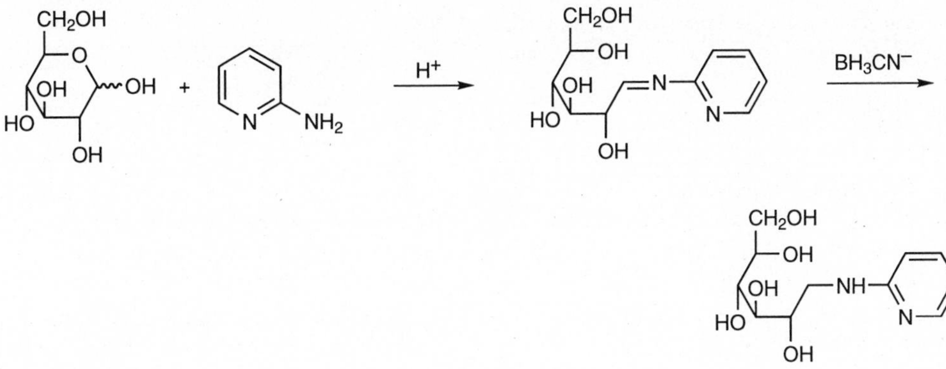

Reaction between D-Glucose and 2-Aminopyridine in the Presence of Cyanoborohydride

mg/mL. Heat the mixture for 2 h at 40–70 °C for the mono- and disaccharides and up to 12 h for higher oligosaccharides. Cool the solution and dilute it with methanol prior to running the electrophoresis. With small modifications this procedure is also valid for the other reagents quoted above[64–70]. In the case of sialic acid containing glycoproteins, the sialyl linkages are hydrolyzed by this procedure to the extent of about 60%. A modification has been developed to prevent the release of sialic acid residues from the carbohydrate moieties[71,72]. This procedure is preferably used for the separation of sugar chains of glycoproteins.

Procedure[71,72]

Reagent A: 550 mg of 2-aminopyridine is dissolved in 200 μL of glacial acetic acid (the pH of the solution should be 6.8 after dilution with 9 volumes of water).

Reagent B: 39 mg of borane–dimethylamine complex is dissolved in 200 μL of glacial acetic acid just before use.

Evaporate the sample with the sugar residues (20–50 μg) in a small, thick-walled vial fitted with a screw cap to dryness and mix the residue with 20–50 μL of reagent A. Heat the mixture at 90 °C for 60 min. After cooling, add 20–50 μL of reagent B and heat the mixture again at 80 °C for 50 min. Dry the reaction mixture with a stream of nitrogen at 40 °C with addition of small amounts of toluene. Then dissolve the residue in the appropriate electrolyte.

In a modification for pyridyl amination of common sugars and acetylated amino sugars, reagent A consists of 1 g of 2-aminopyridine in 0.47 mL of acetic acid and 0.6 mL of methanol, and reagent B of 59 mg borane–dimethylamine complex in 1 mL of acetic acid. Initial heating is carried out at 90 °C for 15 min, followed by evaporation with nitrogen, and the final heating at 90 °C for 30 min.

The electrophoretic separations are carried out under two different conditions:

(a)   Electrophoresis in borate buffer pH around 10. This is used for the separation of individual mono- and oligosaccharides. Their electrophoretic mobilities are dependent on the strengths of their complexes with borate.

(b)   Electrophoresis in neutral or weak acidic medium. Under these conditions, the oligosaccharides are separated mainly according to their molecular size.

(a)   Examples for the Electrophoresis in Alkaline Borate Buffer

Derivatizing Reagent	Capillary	Buffer	Potential/Temperature Detection
2-Aminopyridine[59]	Fused silica L: 65 cm, i.d.: 50 µm coated with polyimine	200 mM pH 10.5	15 kV; - UV: 240 nm
4-Aminobenzoic acid ethyl ester[65]	Fused silica L: 72 cm, i.d.: 50 µm	175 mM pH 10.5	25 kV; 30 °C UV: 305 nm
4-Aminobenzoate[64]	Fused silica L: 72 cm, i.d.: 50 µm	150 mM pH 10.0	28 kV; - UV: 256 nm
9-APTS[70]	Fused silica L: 27 cm, i.d.: 20 µm	100 mM pH 10.2	20 kV; - Fluorescence: Ex: 488 nm, Em: 520 nm

Figure 6 shows the separation of several mono- and oligosaccharides as their 4-aminobenzoic acid ethyl ester derivatives under optimal conditions. Important combinations such as those of D-xylose, D-glucose, L-arabinose and D-galactose, which occur in many common polysaccharides, are separated in a short time. Additionally, uronic acids such as D-glucuronic acid or D-galacturonic acid can be separated and determined in the same run. Unfortunately, the important pair L-arabinose/D-mannose can not be resolved under these conditions.

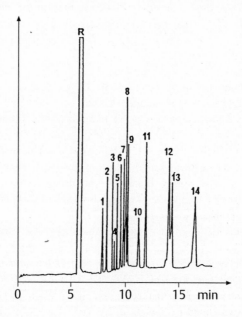

Figure 6.   Electrophoretic Separation of a Mixture of Mono- and Oligosaccharides Derivatized with 4-Amino-benzoic Acid Ethyl Ester
Capillary: fused silica, L: 72 cm, i.d.: 50 µm; *carrier*: 175 mM borate; *potential*: 25 kV; *detection*: UV, 305 nm; *temperature*: 30 °C
R = reagent, **1**: 2-deoxy-D-ribose, **2**: maltotriose, **3**: rhamnose, **4**: cellobiose, **5**: xylose, **6**: ribose, **7**: lactose, **8**: glucose, **9**: arabinose, **10**: fucose, **11**: galactose, **12**: mannuronic acid, **13**: glucuronic acid, **14**: galacturonic acid (Vorndran, Grill, Huber, Oefner and Bonn[65] with permission)

Table 16. Mean Electrophoretic Mobilites of Derivatized Carbohydrates in Borate Buffers[a]

Carbohydrates	Mobility[b] x 10^{-4} (cm^{2}V^{-1}s^{-1})	Mobility[c] x 10^{-4} (cm^{2}V^{-1}s^{-1})	Mobility[d] x 10^{-5} (cm^{2}V^{-1}s^{-1})
Raffinose	n.d.	n.d.	n.d.
Sucrose	n.d.	n.d.	n.d.
2-Deoxy-D-galactose	1.540	1.486	2.912
2-Deoxy-D-ribose	1.230	1.240	2.704
D-Fucose	2.210	2.138	3.393
Lactose	1.830	1.855	2.849
Maltotriose	1.310	1.363	2.178
D-Galactose	2.350	2.257	3.509
Melibiose	1.710	1.705	2.746
Cellobiose	1.560	1.602	2.662
Maltose	1.490	1.540	2.619
L-Arabinose	2.000	1.913	3.274
D-Glucose	1.920	1.871	3.204
Lactulose	n.d.	n.d.	n.d.
D-Xylose	1.740	1.664	3.159
D-Lyxose	1.610	1.522	3.041
L-Sorbose	n.d.	1.731	3.121
L-Rhamnose	1.570	1.490	n.d.
D-Fructose	n.d.	1.898	3.218
D-Ribose	1.770	1.753	2.949
D-Mannose	1.970	1.908	3.204
D-Galactonic acid	n.d.	n.d.	n.d.
D-Gluconic acid	n.d.	n.d.	n.d.
D-Mannonic acid	n.d.	n.d.	n.d.
D-Galacturonic acid	2.990	2.841	4.034
D-Arabonic acid	n.d.	n.d.	n.d.
D-Glucuronic acid	2.820	2.626	3.989
D-Ribonic acid	n.d.	n.d.	n.d.
D-Mannuronic acid	2.800	2.585	3.835

[a] n.d. = not detectable. [b] N-2-Pyridylglycamines, 150 mM borate, pH 10.0, 25 kV, L = 72 cm. [c] Ethyl p-amino-benzoate derivatives, 175 mM borate, pH 10.5, 25 kV, L = 72 cm. [d] p-Aminobenzoate derivatives, 150 mM borate, pH 10.0, 28 kV, L = 72 cm.
(Vorndran, Oefner, Huber, Grill and Bonn[65])

The derivatization with 4-aminobenzoic acid ethyl ester can also be carried out with ketoses, which is not possible with 2-aminopyridine.

In Table 16, the absolute mobilities of the most common sugars and their uronic acids are listed.

In contrast to the underivatized sugars, the glycamines of the monosaccharides have open-chain structures. The strengths of the borate complexes are determined by the OH positions relative to each other in the molecules and it is well-known that cis-oriented OH groups preferentially form borate complexes compared to the trans-disposed OH groups. The OH orientations at the C-3/C-4 positions especially influence and dominate the electrophoretic mobilities[59]. The compounds with cis orientations of the vicinal OH groups generally have higher mobilities than those with trans orientation at this position, as demonstrated for the p-aminobenzoic acid ethyl ester derivatives in Table 17[65].

Hence, it is evident that the glycamines of pentoses and hexoses having the cis configuration at C-3/C-4, such as arabinose, ribose, galactose and fucose, move faster than those with trans orientation, such as lyxose, xylose, glucose and rhamnose; substitution of one of the OH groups as in N-acetylgalactosamine leads to a strong decrease in the electrophoretic mobility of the compound.

Table 17.　Relationship Between Electrophoretic Mobilities and Carbohydrate Structure

Compounds Derivatized with Ethyl p-Aminobenzoate	Mobility x 10^{-4} (cm^{2}V^{-1}s^{-1})	Orientation of the Vicinal Hydroxyl Groups at C-3/C-4
Arabinose	1.917	cis
Lyxose	1.522	trans
Ribose	1.753	cis
Xylose	1.669	trans
Galactose	2.257	cis
Glucose	1.906	trans
Fucose	2.138	cis
Rhamnose	1.490	trans
Galacturonic acid	2.841	cis
Glucuronic acid	2.626	trans
N-Acetylgalactosamine	1.373	cis
N-Acetylglucsoamine	1.833	trans

An improvement to the resolution as well as to the speed of the electrophoretic separation of sugars is the application of the micellar electrokinetic chromatographic separation (MECK). This is generally based on partitioning between the carrier, such as SDS or cyclodextrin of a neutral or ionic substance, and the surrounding medium, and the different migration of the two phases.

By coupling the reducing sugars with 4-aminobenzonitrile via reductive amination the separation of several mono- and oligosaccharides can be carried out in a very short time[66].

(b)　　Examples for the Electrophoresis in Neutral and Weakly Acidic Media

Derivatizing Reagent	Capillary	Buffer	Potential/Temperature Detection
2-Aminopyridine[60]	Fused silica L: 20 cm, i.d.: 25 µm coated with polyacrylamide	100 mM phosphate pH 2.5	8 kV; - UV: 240 nm
2-Aminopyridine[61]	Fused silica L: 30 cm, i.d.: 50 µm coated with polyimide	100 mM phosphate containing 0.1% hydroxypropyl cellulose	20 kV; - Fluorescence: Ex: 320 nm, Em: 390 nm
2-Aminopyridine a. 6-aminoquinoline[63]	Fused silica L: 50 cm, i.d.: 50 µm interlocked coated with polyether	100 mM sodium phosphate containing tetrabutylammonium bromide pH 5.0	10–20 kV; - UV: 240 nm
8-ANTS[67]	Fused silica L: 27 cm, i.d.: 20 µm uncoated	50 mM sodium phosphate pH 2.0–2.5	17–30 kV; 25 °C, 50 °C UV: 214 nm Fluorescence: Ex: 325 nm, Em: 520nm
8-ANTS	Fused silica L: 30–35 cm, i.d.: 50 µm uncoated	50 mM sodium phosphate pH 2.5	20 kV; 25 °C UV: 223 nm
8-ANTS, 7-ANDS, 3-ANDS, 2-ANMS, 5-ANMS[69]	Fused silica L: 72 cm, i.d.: 50 µm	50 mM sodium phosphate pH 2.5	20 kV; 25 °C UV: 235 nm
9-APTS[70]	Fused silica L: 27 cm, i.d.: 20 and 50 µm untreated	50 mM sodium phosphate pH 2.2	20 kV; - Fluorescence: Ex: 448 nm, Em: 520 nm

These electrophoretic procedures have frequently been applied to the separation of isolated sugar residues of degraded microamounts of glycoproteins and to oligosaccharides of degraded polysaccharides, such as starch or pectin[62,68]. Examples in the field of glycoproteins are studies with ovalbumin[60], human α1-acid glycoprotein[73] and human transferrin, human immunoglobulin, calf fetal serum fetuin, bovine spleen invertase, etc.[61]. By running the samples with an acid and with a borate–alkaline buffer, a two-dimensional mapping of the oligosaccharide residues of the individual glycoproteins can be established[61]. The amounts of sugar samples introduced in the capillaries are generally in the order of 100 fmol–1 pmol.

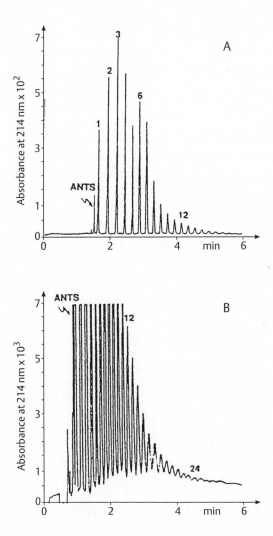

Figure 7.   Electrophoretic Separation of ANTS Derivatized Malto-Oligosaccharides with Acidic pH Conditions at Two Different Temperatures
Capillary: L: 27 cm, i.d.: 50 µm; potential: 15 kV (A), 17 kV (B); buffer: phosphate, 50 mM pH 2.5 (A), 30 mM pH 2.5 (B); detection: fluorescence: Ex: 325 nm, Em: 525 nm; temperature: 25 °C (A), 50 °C (B). Samples were introduced at the cathodic end (Chiesa and Horwarth[67]; with permission)

Using laser-induced fluorescence, the mass detection limit can be decreased to 8–10 fmol with sugar–8-ANTS or sugar–9-APTS derivatives[70]. Figure 7, for example, shows the separation of 8-ANTS derivatized malto-oligosaccharides of a starch hydrolysate.

(b) Reaction with 3-(4-Carboxybenzoyl)-2-quinolinecarboxyaldehyde (CBQCA)

1-Amino-1-deoxyalditols, which are prepared from reducing sugars by reaction with a combination of cyanoborohydride/ammonium ions, react with CBQCA in the presence of cyanide ions to give highly fluorescing isoindole derivatives[74,75], according to the following reaction scheme (example: D-glucose).

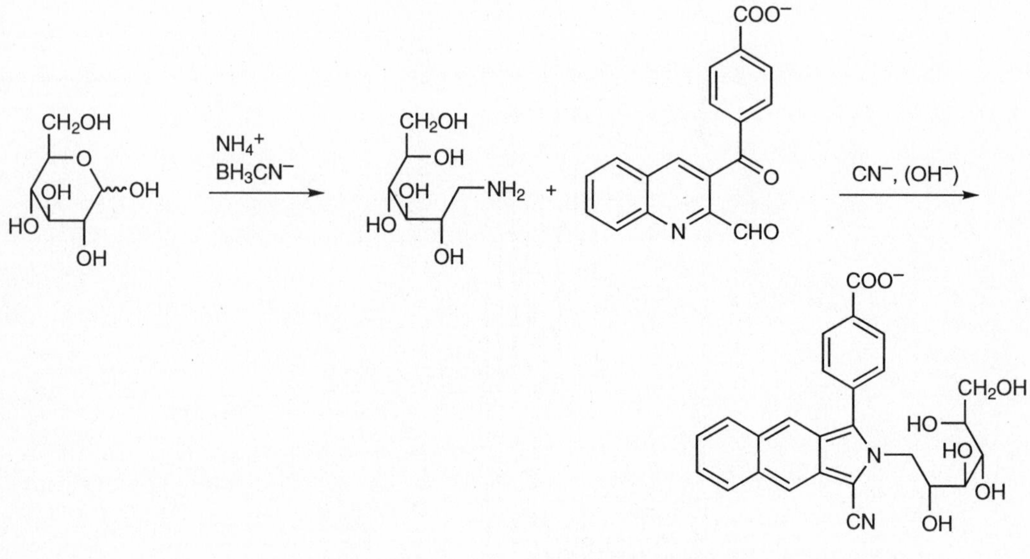

Reaction between D-Glucose and CBQCA in the Presence of Cyanoborohydride and Ammonium Ions

Procedure

Reagent A: 2 mM ammonium sulfate in water or 4 mM ammonium chloride.
Reagent B: 0.4 M sodium cyanoborohydride in water.
Reagent C: 20 mM potassium cyanide in water.
Reagent D: 10 mM CBQCA in methanol.
To an appropriate part of the dry residue of a hydrolysate of a glycoprotein or of a polysaccharide in a thick-walled, screw-cap vial add an excess of reagent A and B. Keep the closed vial for 100–120 min at 100 °C. After cooling, add to ~10 μL of the reaction mixture 10–20 μL of reagent C and 5–10 μL of reagent D and keep the mixture again for 1 h at room temperature. The sample is now ready for the electrophoretic run under the following conditions:
Capillary: fused silica, untreated, L: 90 cm, i.d.: 50 μm.
Buffer: 10 mM disodium phosphate–10 mM tetraborate, pH 9.4.
Potential: 22 kV.
Detection: fluorescence, argon ion laser, Ex: 457 nm, Em: 552 nm.

The mass detection limit of the single sugar derivative is lower here than for the reagents described before and moves in the order of 2–20 attomoles (10^{-18} mol) by introducing 2–5 nL into the capillaries. This reaction has been tested with glycoproteins[74] (fetuin) and with oligosaccharides of degraded starch and polygalacturonic acid samples[75].

Figure 8 shows the separation of some neutral, basic and acidic monosaccharides under the conditions described above.

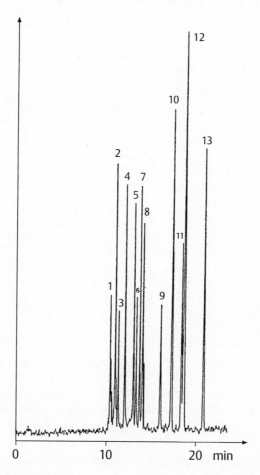

Figure 8.   Electrophoretic Separation of a CBQCA Derivatized Mixture of Monosaccharides and Related Compounds
Capillary: L: 88 cm, i.d.: 50 µm; *potential*: 20 kV; *buffer*: 10 mM phosphate–10 mM borate, pH 9.4; *detection*: fluorescence: Ex: 457 nm, Em: 552 nm; *sample concentrations*: 6.2 µM for glucosamine and galactosamine, 5.5 µM for galacturonic acid, 4.4 µM for other sugars
1: D(+)-glucosamine, **2**: D(+)-galactosamine, **3**: D-erythrose, **4**: D-ribose, **5**: D-talose, **6**: D-mannose, **7**: D-glucose, **8**: D-galactose, **9**: impurity, **10**: D-galacturonic acid, **11**: D-glucuronic acid, **12**: D-glucosaminic acid, **13**: D-glucose-6-phosphate (Liu, Shirota, Wiesler and Novotny[74]; with permission)

(c) Reaction with 3-Methyl-1-phenyl-2-pyrazolin-5-one (MPP)

This reagent condenses in alkaline medium with the aldehyde group of the reducing sugars according to the following reaction scheme (example: D-glucose); ketoses do not react with this reagent[76,77].

Two molecules of MPP react with one molecule of the reducing sugar. The reaction itself is strongly reproducible and almost quantitative (D-glucose 91%, D-xylose 92%, N-acetylglucosamine 89%, L-fucose 88%).

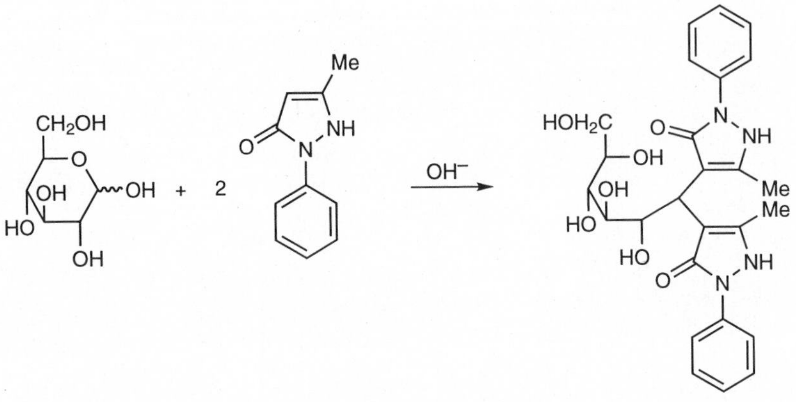

Reaction between D-Glucose and 3-Methyl-1-phenyl-2-pyrazolin-5-one

Procedure[77]

Reagent A: 0.5 M MPP in methanol.

Reagent B: 0.3 M sodium hydroxide in water.

Reagent C: 0.3 M aqueous hydrochloric acid.

To a dry sample of reducing carbohydrates in a screw-cap vial add 50 µL of reagent A and 50 µL of reagent B and keep the solution for 30 min at 70 °C. After cooling, add 50 µL of reagent C and evaporate the mixture to dryness. Add 200 µL of water and 200 µL of chloroform and shake vigorously. Isolate the aqueous phase and evaporate it to dryness; the chloroform phase can be discarded. Dissolve the residue in a small volume of methanol and introduce an aliquot into the CZE capillary for the electrophoretic run under the following conditions:

Capillary: fused silica, coated with polyimine, L: 78 cm, i. d.: 50 µm.

Buffer: 200 mM borate pH 9.5.

Potential: 15 kV.

Detection: UV, 245 nm.

The samples are introduced into the tube at its anodic end and, as in the case of reductive amination, a constant stream of buffer from anode to cathode takes the MPP–aldoses to the detector at the cathodic side. Separation is achieved by the degree of complexation and by the molecular size of the complex.

Figure 9 shows the separation of the individual aldopentoses and aldohexoses; in both groups all the compounds can be separated from each other.

The comparison of these different CZE modifications leads to the following conclusions with regard to advantages, disadvantages and their applicabilities.

(a)   CZE of Nonderivatized Sugars in an Alkaline Medium.

Indirect UV detection or amperometric detection enables the determination of the reducing and also the nonreducing sugars as well as their nonreducing derivatives, such as the sugar alcohols or the aldonic acids. The drawbacks of this method are: (a) its nondiscriminative nature since peptides, amino acids, organic acids, etc., can also yield a positive response; (b) in the case of amperometric detections, different saccharides do not yield a uniform response hence a quantitative standard is required for each of the saccharides; and (c) some sugars like the 3-O-substituents are sensitive against alkali even under the CZE conditions.

(b)   CZE of Derivatized Sugars.

Derivatization of sugars before separation is a discriminating step which enhances the specificity of the determination. Unfortunately, it is only suitable for reducing sugars which contain a reactive carbonyl group.

With regard to the application of these two different CZE approaches, procedure (a) seems to be suitable for determinations of common mono- and oligosaccharides in biological materials such as foods while (b) is useful for the determination of units of polysaccharides formed upon hydrolysis[55,65,78].

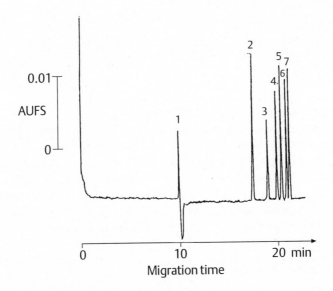

Figure 9a.   Electrophoretic Separation of MPP–Aldopentoses
Capillary: fused silica, L: 78 cm, i.d.: 50 µm; *potential*: 15 kV; *buffer*: 200 mM borate pH 9.5; *detection*: UV, 256 nm
1: methanol, **2**: amobarbital (internal standard), **3**: excess reagent (MPP), **4**: xylose, **5**: arabinose, **6**: ribose, **7**: lyxose

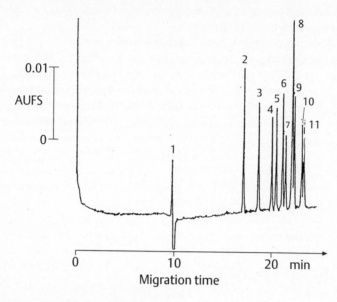

Figure 9b. Electrophoretic Separation of MPP–Aldohexoses
Capillary: fused silica, L: 78 cm, i.d.: 50 µm; *potential*: 15 kV; *buffer*: 200 mM borate pH 9.5; *detection*: UV, 256 nm
1: methanol, **2**: amobarbital (internal standard), **3**: excess reagent (MPP), **4**: glucose, **5**: allose, **6**: altrose, **7**: mannose, **8**: idose, **9**: gulose, **10**: talose, **11**: galactose (Honda, Suzuki, Nose, Yamamoto and Kakehi[77]; with permission)

2.2.5.4. Co-Electroosmotic Separations of Carbohydrates

Capillary electrophoresis of carbohydrates is usually carried out in the counter directional mode where the vectors of the inherent electrophoretic mobility μ_{ep} and the electroosmotic flow (EOF) mobility μ_{eof} are oppositely directed; thus, the sugars are detected after the electroosmotic flow marker. This causes considerably long migration times.

By changing this instrumentation in a way that the anionic analytes co-migrate with the EOF the observed mobilities of the anions may be increased. This is achieved by coating the negatively charged inner surface of the fused silica capillary with cationic surfactants and by reversing the polarity of the power supply.

2.2.5.4.a. Underivatized Carbohydrates

(a) Indirect UV Detection

Underivatized carbohydrates do not exhibit a significant UV absorption nor possess a sufficiently high electrophoretic mobility at moderate pH conditions. Thus, capillary electrophoretic analysis is restricted to certain limits in terms of detection and electrolyte conditions.

Sorbate acts as a suitable background electrolyte as it exhibits several advantages compared to other background electrolytes[54]. First of all it is only singly charged which theoretically enables a transfer ratio of 1 for singly negatively charged analytes. This means that one background electrolyte molecule is replaced by one analyte. Furthermore, it has a high molar absorptivity at the common detection wavelength of 254 nm and it corresponds with the electrophoretic mobilities of the underivatized carbohydrates.

In particular, when detection sensitivity is not a limiting criterion, indirect detection in combination with co-electroosmotic migration results in fast separation of underivatized carbohydrates[79].

Figure 10 illustrates a fast, co-electroosmotic separation of nine underivatized carbohydrates achieved with a separation voltage of 10 kV. Sugar acids exhibit a considerably higher electrophoretic mobility than the other carbohydrates and thus migrate in front of the neutral monosaccharides. This is due to the fact that at under the chosen conditions sugar acids already carry a net charge resulting from their carboxyl group. The order of migration of the carbohydrates is in correspondence to their pK_a values with disaccharides exhibiting a lower electrophoretic mobility than monosaccharides of the same acidity.

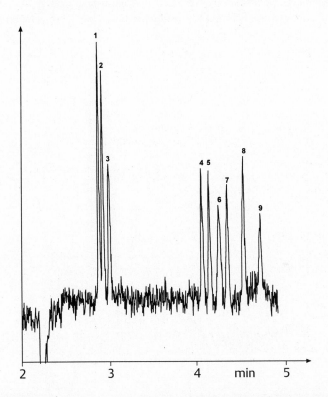

Figure 10. Fast Co-Electroosmotic Separation of a Standard Mixture of Underivatized Carbohydrates
Capillary: fused silica, L: 32 cm, i.d.: 50 µm; *background electrolyte*: 6 mM sorbate, 0.001% hexadimethrine bromide, pH 12.1; *potential*: 10 kV; *injection*: 1 s hydrostatically; *indirect detection*: 254 nm
1: mannuronic acid, **2**: glucuronic acid, **3**: galacturonic acid, **4**: *N*-acetylneuramic acid, **5**: gluconic acid, **6**: fructose, **7**: rhamnose, **8**: glucose, **9**: galactose (Zemann, Nguyen and Bonn[79])

The following conditions are suitable for the separation of sugars with indirect UV detection:
Capillary: uncoated fused silica capillary, L = 32 cm, i.d.: 50 µm.
Potential: 10 kV; field strength approx. 310 V/cm.

Detection: indirect UV, 254 nm.
Electrolyte: 6 mM sodium sorbate, pH 12.1, 0.001% hexadimethrine bromide (w/v) (cationic surfactant).

(b) Direct UV Detection of Borate Complexes

As pointed out previously, the sensitivity of carbohydrates can be significantly increased by forming negatively charged complexes with borate under alkaline conditions. Direct UV detection can be carried out in the low UV range[52]. Figure 11 depicts the fast co-electroosmotic separation of monosaccharide–borate complexes. Glucose and xylose from complexes with the highest electrophoretic mobilities compared to mannose and galactose. The high temperature required for narrow peak zones is due to the fact that the percentage of free carbonyl functionalities is increased when higher temperatures are used. Thus, elevated temperatures accelerate the formation of borate complexes and, as a consequence, shorten the separation time when co-electroosmotic conditions are applied[80].

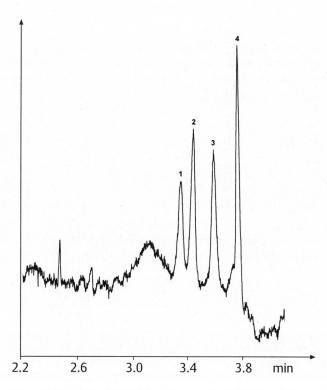

Figure 11. Electropherogram of Underivatized Monosaccharide–Borate Complexes with Direct UV Detection at 185 nm
Capillary: fused silica, L: 50 cm, i.d.: 50 µm; electrolyte: 75 mM borate, 0.0005% hexadimethrine bromide, pH 9.5; potential: 15 kV; injection: 5 s electromigration at 5 kV; direct UV detection: 185 nm
1: glucose, **2**: xylose, **3**: galactose, **4**: mannose (Graffan[80])

The following conditions are suitable for the separation of sugars with direct UV detection:
Capillary: uncoated fused silica capillary, L: 50 cm, i.d.: 50 µm.
Potential: 15 kV; field strength approx. 300 V/cm.
Detection: direct UV, 185 nm.
Electrolyte: 75 mM sodium borate, pH 9.5, 0.0005% hexadimethrine bromide (w/v).

2.2.5.4.b. Derivatized Carbohydrates

The limit of detection for carbohydrates is further improved by the formation of UV absorbing or fluorescent derivatives. However, certain restrictions with specific derivatizing agents may apply. For example, UV absorbing agent 2-aminopyridine reacts only with aldoses and by using p-aminobenzoic acid the reagent itself exhibits an inherent electrophoretic mobility and may interfere with the peak zones of the derivatized carbohydrates in the electropherogram. Also, the high costs of low-wavelength lasers and the difficulties to equip commercial CZE instruments with laser detectors restrict its use to a limited number of laboratories. Nevertheless, derivatization is the method of choice for the sensitive detection of carbohydrates.

(a) Ethyl p-Aminobenzoate Derivatives

Ethyl p-aminobenzoate is a suitable derivatizing agent for carbohydrates[65]. It reacts with both aldoses and ketoses and, after reduction of the imine the derivative, readily forms negatively charged borate complexes which can be separated by capillary electrophoresis. Furthermore, the derivatization reagent does not migrate in the electric field as it carries no charge under the respective separation conditions and thus no interference with the analytes is observed. The sample can be used without prior cleanup because byproducts and impurities possibly present do not interfere with the derivatives as long as they do not exhibit carbohydrate-like structures. However, co-electroosmotic CZE of analytes carrying aromatic moieties always results in broad and retained peak zones. This is explained by strong hydrophobic interactions of the analytes with the aliphatic parts of the electroosmotic flow modifier. This effect can be reduced by either adding organic solvents to the electrolyte or by increasing the ionic strength of the electrolyte[81]. A combination of both organic solvents and high borate concentration results in the electropherogram depicted in Figure 12. Noteworthy, is that the separation is finished within 5 minutes, which is a considerable reduction of the analysis time compared to counter-electroosmotic methods. The optimized electrophoretic mobilities of various carbohydrates derivatized with ethyl p-aminobenzoate are listed in Table 18.

The following conditions are suitable for the separation of sugars with direct UV detection:
Capillary: uncoated fused silica capillary, L: 57.5 cm, i.d.: 50 µm.
Potential: 25 kV; field strength approx. 435 V/cm.
Detection: direct UV, 214 nm.
Electrolyte: 175 mM sodium borate, pH 10.5, 0.001% hexadimethrine bromide (w/v).

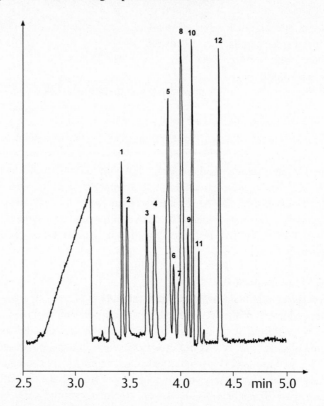

Figure 12. Electropherogram of Carbohydrates Derivatized with Ethyl *p*-aminobenzoate
Capillary: fused silica, L: 57.5 cm, i.d.: 50 μm; *electrolyte:* 175 mM borate, 0.001% hexadimethrine
bromide, 15% methanol, pH 10.5; *potential:* 25 kV; *injection:* 15 s hydrostatically; *direct UV
detection:* 214 nm
1: galacturonic acid, **2:** glucuronic acid, **3:** galactose, **4:** fucose, **5:** glucose, **6:** lactose, **7:** ribose, **8:**
xylose, **9:** cellobiose, **10:** rhamnose, **11:** maltotriose, **12:** 2-deoxy-D-ribose (Nguyen[81])

(b) *p*-Aminobenzonitrile Derivatives

For specific mixtures of carbohydrates it is of advantage to use *p*-aminobenzonitrile as the
derivatization agent for reductive amination[66]. Two strategies may be used to obtain complete
separation of these derivatives with co-electroosmotic CZE methods. The first is the use of high borate
concentrations. One advantage is that only low concentrations of organic modifiers are needed to
obtain baseline separations. However, this is paid for by longer separation times due to the high ionic
strengths of borate, as depicted in Figure 13. The electropherogram shows the separation of
carbohydrates including rhamnose, galactose, and glucose which are often constitutents in plant
materials[82]. Another approach to separate the *p*-aminobenzonitrile derivatives of carbohydrates using
lower concentrations of borate, is to adjust the composition of the electrolyte with regards to the
contents of acetonitrile[81]. Figure 14 shows the fast co-electroosmotic separation of a mixture of
carbohydrates using organic solvents added to the electrolyte. Separation times below three minutes are
achieved with this method. By adjusting the amount of solvent and the composition of the electrolyte it
is possible to set up optimized separation conditions for specific sample mixtures. Table 18 lists the
electrophoretic mobilities of the *p*-aminobenzonitrile derivatives of various carbohydrates.

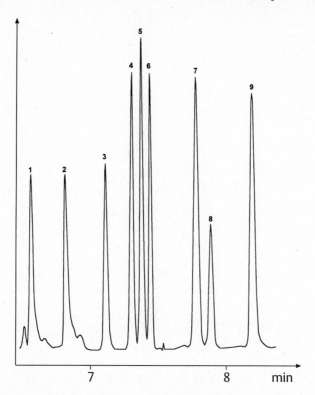

Figure 13. Electropherogram of Carbohydrates Derivatized with *p*-Aminobenzonitrile and High Borate Concentrations
Capillary: fused silica, L: 66 cm, i.d.: 50 μm; *electrolyte:* 500 mM borate, 0.001% hexadimethrine bromide, 5% propanol, 5% methanol, pH 10.5; *potential:* 15 kV; *injection:* 5 s hydrodynamically; *direct UV detection:* 284 nm
1: galacturonic acid, **2**: glucuronic acid, **3**: galactose, **4**: arabinose, **5**: mannose, **6**: glucose, **7**: xylose, **8**: cellobiose, **9**: rhamnose (Lerch[82])

The following conditions are suitable for the separation of *p*-aminobenzonitrile derivatized sugars with direct UV detection:

Capillary: uncoated fused silica capillary, L: 50 cm, i.d.: 50 μm.

Potential: 20 kV; field strength approx. 400 V/cm.

Detection: direct UV, 214 nm.

Electrolyte: 300 mM sodium borate, pH 10.2, 5% acetonitrile, 5% methanol, 0.001% hexadimethrine bromide (w/v).

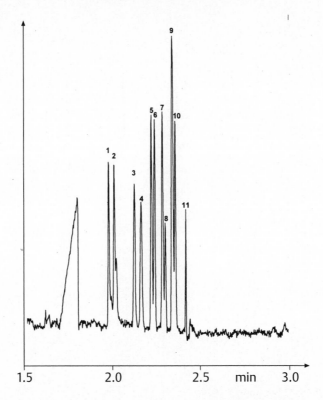

Figure 14. Electropherogram of Carbohydrates Derivatized with *p*-Aminobenzonitrile and Organic Modifiers
Capillary: fused silica, L: 57.5 cm, i.d.: 50 μm; *electrolyte*: 175 mM borate, 0.001% hexadimethrine
bromide, 10% acetonitrile, pH 10.5; *potential*: 25 kV; *injection*: 15 s hydrostatically; *direct UV
detection*: 214 nm;
1: galacturonic acid, **2**: glucuronic acid, **3**: galactose, **4**: fucose, **5**: arabinose, **6**: glucose, **7**: ribose,
8: melibiose, **9**: cellobiose, **10**: rhamnose, **11**: maltotriose (Lerch[82])

Table 18. Electrophoretic Mobilities of Carbohydrates Derivatized with Ethyl *p*-Aminobenzoate and with *p*-
Aminobenzonitrile

Carbohydrates	Mobility[a] $\times 10^{-4}$ $(cm^2V^{-1}s^{-1})$	Mobility[b] $\times 10^{-4}$ $(cm^2V^{-1}s^{-1})$
Galacturonic acid	4.006	5.222
Glucuronic acid	3.850	5.003
N-Acetylneuramic acid	3.825	4.919
Mannuronic acid	3.800	4.887
Galactose	3.644	4.723
Fucose	3.598	4.669
Arabinose	3.483	4.542
Mannose	3.464	4.490
Fructose	3.450	4.480
Glucose	3.434	4.449
Lactose	3.402	4.428
N-Acetyl-D-glucosamine	3.377	4.314
Ribose	3.361	4.298
Sorbose	3.335	4.260

Table 18. (Continued)

Carbohydrates	Mobility[a] x 10^{-4} $(cm^2V^{-1}s^{-1})$	Mobility[b] x 10^{-4} $(cm^2V^{-1}s^{-1})$
Melibiose	3.314	4.157
Xylose	3.283	4.199
Cellobiose	3.214	4.115
Maltose	3.115	4.025
Lyxose	3.113	4.082
Rhamnose	3.110	3.947
Maltotetraose	3.103	3.901
N-Acetyl-D-galactosamine	3.070	3.828
Gentiobiose	3.021	3.810
Maltotriose	2.999	3.794
2-Deoxy-D-ribose	2.956	3.732

[a]Derivatized with ethyl p-aminobenzoate.
[b]Derivatized with p-aminobenzonitrile.

References

1   Haugaard, G.; Kroner, T. D. *J. Am. Chem. Soc.* **1948**, *70*, 2135.
2   Wieland, Th.; Fischer, E. *Naturwissenschaften* **1948**, *35*, 29.
3   Bourne, E.; Foster, A. G.; Grant, P.M. *J. Chem. Soc. (London)* **1956**, 4311.
4   Kohn, J. *Biochem. J.* **1957**, *65*, 9 p.
5   Grassini, G. In *Thin layer Chromatography*; Marini-Bettold, C. B., Ed.; Elsevier: Amsterdam, 1964; p 55.
6   Hannig, K.; Pascher, G. In *Dünnschichtchromatographie, ein Laboratoriumshandbuch*, 2nd ed.; Stahl, E., Ed.; Springer: Berlin, 1967; p 105.
7   Radola, B. J. *Biochim. Biophys. Acta* **1969**, *194*, 335.
8   Wieme, In *Agar-Gel Electrophoresis*; Elsevier: Amsterdam, 1965.
9   Gordon, A. H. In *Laboratory Techniques in Biochemistry and Molecular Biology Vol. 1 - Electrophoresis in Polyacrylamide Gels and Starch gels*; Work, T.S.; Work, E., Eds.; Elsevier: North Holland/America, 1969; p 1.
10 Clotten, R.; Clotten, A. In *Hochspannungselectrophorese*; Thieme: Stuttgart, 1962; p 114.
11 Frahn, J. L.; Mills, A. *Aust. J. Chem.* **1959**, *12*, 65.
12 Weigel, H. *Adv. Carbohydr. Chem. Biochem.* **1963**, *18*, 61.
13 Foster, A. B. *J. Chem. Soc. (London)* **1953**, 982.
14 Foster, A. B. *Adv. Carbohydr. Chem. Biochem.* **1957**, *12*, 81.
15 Foster, A. B.; Stacey, M. *J. Chem. Soc. (London)* **1955**, 1778.
16 Garreg, P. J.; Lindberg, B. *Acta Chem. Scand.* **1961**, *15*, 1913.
17 Garreg, P. J.; Lindström, K. *Acta Chem. Scand.* **1971**, *25*, 1559.
18 Lindberg, B.; Swan, B. *Acta Chem. Scand.* **1960**, *14*, 1043.
19 Bourne, E. J.; Hutson, D. H.; Weigel, H. *J. Chem. Soc. (London)* **1960**, 4252.
20 Bourne, E. J.; Hutson, D. H.; Weigel, H. *J. Chem. Soc. (London)* **1961**, 35.
21 Mayer, H.; Westphal, O. *J. Chromatogr.* **1968**, *33*, 514.
22 Searle, F.; Weigel, H. *Carbohydr. Res.* **1980**, *85*, 51.
23 Barker, S. A.; Bourne, E. J.; Grant, P. M.; Stacey, M. *Nature* **1956**, *177*, 1125.
24 Hase, S.; Hara, S.; Matsushima, Y.; Ikenaka, T. *J. Biochem.* **1979**, *85*, 217, 989.
25 Theander, O. *Acta Chem. Scand.* **1957**, *11*, 717.
26 Sundman, J. Dissertation, University of Helsinki, 1949 .
27 Frahn, J. L.; Mills, J. A. *Chem. Ind.* **1956**, 1137.
28 Haug, A.; Larsen, B. *Acta Chem. Scand.* **1961**, *15*, 1395.
29 Kosakai, M.; Yosizawa, Z. *Anal. Biochem.* **1975**, *69*, 415.
30 St. Cyr, M. J. *J. Chromatogr.* **1970**, *47*, 284.
31 Theander, O. *Svensk. Kem. Tidskr.* **1958**, *70*, 393.

32   Whitacker, J. R. In *Paperchromatography and -electrophoresis*; Zweig, G.; Whitacker, J. R., Eds.; Academic: New York, London, 1967; p 272.
33   Wade, H. F.; Morgan, D. M. *Biochem. J.* **1955**, *60*, 264.
34   Vanderheiden, B. S. *Anal. Biochem.* **1964**, *8*, 1.
35   Runeckles, V. C.; Krotkov, G. *Arch. Biochim. Biophys.* **1957**, *70*, 442.
36   Piras, R.; Cabib, E. *Anal. Chem.* **1963**, *35*, 755.
37   Bieleski, R. L.; Young, R.E. *Anal. Biochem.* **1963**, *6*, 54.
38   Schwimmer, S.; Bevenue, A.; Weston, A. J. *Arch. Biochem. Biophys.* **1956**, *60*, 279.
39   Crumpton, M. J. *Biochem. J.* **1959**, *72*, 481.
40   Takasaki, S.; Kobata, A. *Biochem. J.* **1974**, *76*, 783.
41   Daniel, P. F. *J. Chromatogr.* **1979**, *176*, 260.
42   Miyamoto, I.; Nagase, S. *Anal. Biochem.* **1981**, *115*, 308.
43   Jarvis, M. C.; Threlfall, D. R.; Friend, J. *Phytochemistry* **1977**, *16*, 849.
44   Bettler, B.; Amado, R.; Neukom, H. *J. Chromatogr.* **1990**, *498*, 213, 223.
45   Scherz, H. *Z. Lebensm. Unters. Forsch.* **1985**, *181*, 40.
46   Bonn, G.; Grünwald, M.; Scherz, H.; Bobleter, O. *J. Chromatogr.* **1986**, *370*, 485.
47   Scherz, H. *Electrophoresis* **1990**, *11*, 18.
48   Hjerten, S. *Chromatogr. Rev.* **1967**, *9*, 122.
49   Jorgenson, J. W.; Lukacs, K. D. *Anal. Chem.* **1981**, *53*, 1298.
50   Mikkers, E. F. P.; Everaerts, F. M.; Verheggen, Th. *J. Chromatogr.* **1979**, *169*, 11.
51   Karger, B. L.; Cohen, A.; Guttman, A. *J. Chromatogr.* **1989**, *492*, 585.
52   Hoffstetter-Kuhn, S.; Paulus, A.; Gassmann, E.; Widmer, H. M. *Anal. Chem.* **1991**, *63*, 1541.
53   Carney, S. L.; Osborne, D. J. *Anal. Biochem.* **1991**, *195*, 132.
54   Vorndran. A. E.; Oefner, P. J.; Scherz, H.; Bonn, G. K. *Chromatographia* **1992**, *33*, 163.
55   Klockow, A.; Paulus, A.; Figueiredo, V.; Amado, R.; Widmer, H. M. *J. Chromatogr. A* **1994**, *680*, 187.
56   Colon, L. A.; Dadoo, R.; Zare, R. N. *Anal. Chem.* **1993**, *65*, 476.
57   Xinjian Huang; Kok, W. Th. *J. Chromatogr. A* **1995**, *707*, 335.
58   Wenzhe Lu; Cassidy, R. M. *Anal. Chem.* **1993**, *65*, 2878.
59   Honda, S.; Iwase, S.; Makino, A.; Fujiwara, S. *Anal. Biochem.* **1989**, *176*, 72.
60   Honda, S.; Makino, A.; Suzuki, S.; Kakehi, K. *Anal. Biochem.* **1990**, *191*, 228.
61   Suzuki, S.; Kakehi, K.; Honda, S. *Anal. Biochem.* **1992**, *205*, 227.
62   Nashabeh, W.; Ziad el Rassi *J. Chromatogr.* **1990**, *514*, 57.
63   Nashabeh, W.; Ziad el Rassi *J. Chromatogr.* **1992**, *600*, 279.
64   Grill, E.; Huber, C.; Oefner, P.; Vorndran, A. E.; Bonn, G. *Electrophoresis* **1993**, *14*, 1004.
65   Vorndran, A. E.; Grill, E.; Huber, E.; Oefner, C.; Bonn, G. I. *Chromatographia* **1992**, *34*, 109, 308.
66   Schwaiger, H.; Oefner, P. J.; Huber, C.; Grill, E.; Bonn, G. K. *Electrophoresis* **1994**, *15*, 941.
67   Chiesa, K.; Horvarth, C. *J. Chromatogr.* **1993**, *645*, 337.
68   Klockow, A.; Widmer, M. H.; Amado, R.; Paulus, A. *Fresenius' Z. Anal. Chem.* **1994**, *350*, 415.
69   Chiesa, C.; O'Neill, R. *Electrophoresis* **1994**, *15*, 1132.
70   Evangelista, R. A.; Ming-Sun Liu; Fu-Tai A. Chen *Anal. Chem.* **1995**, *67*, 2239.
71   Kondo, A.; Suzuki, J.; Kuraya, N.; Hase, S.; Kato, I.; Ikenaka, T. *Agric. Biol. Chem.* **1990**, *54*, 2169.
72   Suzuki, J.; Kondo, A.; Kato, I.; Hase, S.; Ikenaka, T. *Agric. Biol. Chem.* **1991**, *55*, 283.
73   Nashabeh, W.; Ziad el Rassi *J. Chromatogr.* **1991**, *536*, 31.
74   Jinping Liu; Shirota, O.; Wiesler, D.; Novotny, M. *Proc. Natl. Acad. Sci.* **1991**, *88*, 2302.
75   Jinping Liu; Shirota, O.; Novotny, M. *Anal. Chem.* **1992**, *64*, 973.
76   Honda, S.; Akao, E.; Suzuki, S.; Okuda, M.; Kakehi, K.; Nakamura, J. *Anal. Biochem.* **1989**, *180*, 351.
77   Honda, S.; Suzuki, S.; Nose, A.; Yamamoto, K.; Kahehi, K. *Carbohydr. Res.* **1991**, *215*, 193.
78   Oefner, P.; Scherz, H. In *Advances in Electrophoresis*, Vol. VII; Chrambach, A.; Dunn, M. J.; Radola, B.J., Eds.; VCH: Weinheim, New York, 1994; p 157.
79   Zemann, A.; Nguyen, D.T.; Bonn, G. *in preparation*.
80   Graffan, A. Thesis, University of Innsbruck, 1995.
81   Nguyen, D. T. Thesis, University of Innsbruck, 1997.
82   Lerch, H. Thesis, University of Innsbruck, 1997.

3. Polysaccharides

The analysis of polysaccharides comprises the group of compounds derived from plant, animal and microbiological sources and can be divided into the following sections.
–   Isolation of pure compounds involving separation procedures when dealing with mixtures of polysaccharides.
–   Identification of single compounds including determination of the chemical structure.
–   Quantitative determination of either a single individual or the sum of the polysaccharides in a sample.

3.1. Isolation of Polysaccharides

Polysaccharides can be classified into three major groups:
A.   Compounds which are totally insoluble in water.
B.   Compounds which are totally soluble in water under neutral conditions.
C.   Compounds which are soluble in water under acid or alkaline conditions.
Polysaccharides are either present in their free form (e.g., cotton fibre), with fat and protein, as in many food raw materials, or linked to protein or polyphenolic molecules (e.g., glycosaminoglycans or polysaccharide–lignin complexes).
The isolation procedures for a sample containing various different groups of compounds such as those in biological systems naturally differ from each other and a great number of variations are in existence depending on whether the isolate belongs to group A, B or C; but in all these cases the procedures should fulfill the following conditions:
(i)   The method should afford the pure polysaccharides without any accompanying substances.
(ii)   The method should be quantitative.
(iii)   The method must avoid alterations or degradations of the polysaccharides during the isolation steps.

3.1.1. Procedures for Isolating Polysaccharides Totally Insoluble in Water (Group A)

These are predominately structural polysaccharides such as cellulose, the mannans of nut shells or the chitin skeletons of insects and crustaceans. The isolation of such substances is carried out according to the following general principles:
–   Removal of water either by evaporation, lyophilization or, together with the lipids, by extraction with water miscible organic solvents.
–   Suspension of the remaining insoluble part in water and removing the proteins, if they are present, by treatment with proteases or by dissolving them by addition of chaotropic agents such as urea (8 M), guanidinium salts or lithium isothiocyanate[1].

–   The remaining residues can contain, in addition to the polysaccharides, small amounts of non-degraded or nondissolved proteins, polymeric phenolic substances (like lignin) and insoluble inorganic substances.

Further purification operations depend on the nature of the material itself; it is, therefore, not possible to claim a general procedure.

As an example, the isolation of polysaccharides from plant materials is described here in detail. The procedure is similar to that for the determination of dietary fibre or thickening agents[2–4].

Procedure

Reagent A: petroleum ether (analytical grade) or dioxane.

Reagent B: protease, type VIII (from *Bacillus subtilis*, e.g., Fa. Sigma N. 5380).

Reagent C: 0.05 M phosphate buffer pH 7.5.

For removal of the lipid fraction extract the fine-powdered dry sample with petroleum ether (reagent A) (preferably in a Soxhlet apparatus) after removal of water either by lyophilization or by drying in vacuo. Alternatively, slurry the wet sample in an excess of dioxane or in a mixture of acetone–dichloromethane where the lipids as well as water are taken up into the liquid phase. Stir vigorously and after centrifugation and removal of the supernatant repeat the procedures two to three times.

For removal of the main part of proteins, if present, slurry approximately 0.5 g of the organic solvent free sample in 50 mL of reagent C and add 100 μL of reagent B. Maintain the sample at 60 °C under slight stirring. Separate the insoluble part either by centrifugation or by filtration.

The residue contains, in most cases, further small amounts of proteins, and sometimes greater amounts of polyphenolic substances or inorganic materials. The parts can by determined by appropriate methods (e.g., protein: Kjeldahl N x 6.25; inorganic materials: ashing).

In several cases, the samples contain remarkable amounts of starch which can interfere with the isolation and determination of other polysaccharides. It can be removed by a two-step enzymatic degradation as follows:

–   Treat the defatted suspension of the samples with preferably heat-stabile amylase at 100 °C before protein degradation.

–   Cleave the formed malto-oligosaccharides and short-chain dextrins after the amylase treatment with amyloglucosidase to D-glucose after protein degradation.

It is unlikely that application of this procedure to the isolation of insoluble polysaccharides will alter the original structure because of the mild conditions employed. Care must be taken that the enzymes used are free of carbohydrate-altering activities.

For further purifications of the raw polysaccharides the procedures must be adapted to the individual compounds. In several cases extraction procedures are successful as demonstrated by the following examples.

–   Starch, glycogen and some hemicelluloses, e.g. *O*-acetylgalactoglucomannan, can be dissolved in pure dimethyl sulfoxide or in dimethyl sulfoxide containing 10% water[5,6]. Bacterial polysaccharides of the amylopectin type as well as starch are extracted easily with 30% aqueous chloral hydrate[7]. Bacterial polysaccharides from gram-negative organisms are successfully extracted with diethylene glycol[8,9].

–   Cellulose and other insoluble polysaccharides are soluble in *N*-methylmorpholine *N*-oxide; plant cell walls can be dissolved completely without degradation or formation of derivatives[10,11].

–   Pectins are present as insoluble calcium compounds in plant cells. Extracting such materials with a monovalent salt solution can replace the bivalent metal ions with sodium or potassium ions and solubilize them[1]. Alternatively, effective chelating reagents such as sodium hexametaphosphate or EDTA can be used. These reagents form strong complexes with bivalent metal ions at pH 4.0; under these conditions the base-catalyzed β-elimination of esterified galacturonic units does not occur[12]. Extraction procedures with acid or alkaline reagents should be avoided when the polysaccharide is to be isolated in its native form.

Another method of purifying the raw polysaccharide is to use reagents which react with accompanying substances yielding water-soluble substances. One such reagent is chlorous acid which is used for the solubilization of lignin and proteins[13–15]. However, in several cases a modification of the polysaccharides can be expected with this reagent.

3.1.2. Procedures for Isolation of Water-Soluble Polysaccharides (Group B)

These compounds are predominately either reserve compounds for energy metabolism of plants such as starch, guar, etc., or protection substances against infection of wounded parts of plants such as gum arabic, etc. (plant excudates).

The main procedure for isolating these compounds from plant materials of foods is almost identical to that used for the determination of water-soluble dietary fibre or thickening agents[2,3,16]. The procedure for the latter case is described here in detail[16].

Procedure
Reagent A: dioxane (analytical grade).
Reagent B: 25 g of 5-sulfosalicylic acid dihydrate dissolved in 100 mL of distilled water.
Reagent C: saturated aqueous sodium chloride.
The weight of the applied sample depends on the amount of soluble polysaccharide expected. For the determination of polysaccharides in foods, suspend 25 g of the wet sample in 40 mL of dioxane (reagent A) and heat the mixture at 60 °C for 10 min in a water bath. After 10 min, centrifuge the slurry, discharge the supernatant and extract the residue twice with 10 mL of 70% ethanol, centrifuging inbetween and discharging the supernatants. Suspend the residue from the last centrifugation in 75 mL distilled water and heat the suspension for 30 min at 90 °C. The latter operation dissolves the major part of the soluble polysaccharides together with soluble proteins.

For deproteination by denaturation and insolubilization, add an appropriate part of reagents B and C (e.g., for a volume of 75 mL use 15 mL of B and 10 mL of C) and, after maintaining at room temperature for 15–30 min, remove the precipitate by centrifugation. To the supernatant add a four- to sixfold volume of ethanol and, after keeping for at least 2 h at room temperature, centrifuge the precipitate which contains mainly the water-soluble polysaccharides.

Instead of ethanol the polysaccharide can be precipitated with a 10–15-fold volume of methanol saturated with ammonium carbonate[17].

Several polysaccharides, especially those with acid units, are precipitated together with the denatured protein; they are lost for further treatments. In the presence of such substances and a great amount of accompanying proteins, an enzymatic degradation of the latter is necessary; for this purpose the following procedure has been developed[18].

Procedure

Reagent: 200 mg of Pronase E (lyophilized; from *Streptomyces Griseus*, ca. 7000 U/g) are dissolved in 5 mL of H_2O.

After the defatting operation, adjust the aqueous suspension of the samples with sodium hydroxide to pH 8 (controlled with a glass electrode) and add 2.5 mL of the reagent. Keep the solution in a covered beaker at 40 °C for approx. 3 h and then for 15 min in a boiling water bath; then carry out the deproteinization operation with sulfosalicylic acid.

As another example, the isolation of animal polysaccharides such as glycosaminoglycans from animal tissues are described here in detail[19,48].

Procedure[48]

Reagent A: 0.5 M sodium acetate buffer pH 7.5.

Reagent B: 10 mM aqueous calcium chloride.

Reagent C: Pronase E (protease of *Streptomyces Griseus*) in acetate buffer (reagent A).

Reagent D: trichloroacetic acid.

Reagent E: 5% potassium acetate in ethanol.

Remove the water from 0.25–1.0 g of the tissue samples by lyophilization and extract the dry materials twice for 24 h with chloroform–methanol 2:1. Freeze the fat-free material with liquid nitrogen, grind it to a fine powder and dry it in a vacuum dessicator at 20–25 °C. Suspend ~50 mg of this material in 2 mL of reagent A and place it for 20 min in a boiling water bath. Cool to room temperature and add reagent C at 12 h intervals, providing a total amount of 10 µg Pronase E/mg of dry defatted material, and reagent B to a final concentration of 1 mmol calcium ions. Incubate the solution for 24 h at 50 °C, then cool and add reagent D for a final concentration of 5%. Keep it at 4 °C for 30 min, centrifuge and save the supernatant. Treat the precipitate again with a 5% aqueous solution of reagent D and repeat the centrifugation. Then discard the precipitate, combine the two supernatants and precipitate the glycosaminoglycans with three volumes of reagent E to one volume of the total supernatant. After storing for 12 h at 4 °C, centrifuge the precipitate, discard the supernatant, and wash the residue successively with 2 mL of absolute ethanol, 2 mL of ethanol–diethyl ether 1:1 and 2 mL of diethyl ether. Finally, remove the diethyl ether by storing under reduced pressure at 20–25 °C.

If one polysaccharide is present in a great excess and interferes with the isolation of the other compounds it must be removed by enzymatic degradation. Starch is one such example and the procedure for its removal is described in the previous section. The operations are carried out before deproteinization.

The precipitation of water-soluble polysaccharides with alcohols or other water-soluble organic solvents is an universal nonspecific method in this field. Fractional precipitation by stepwise addition of the organic solvent to the aqueous solution of a mixture of polysaccharides can achieve the separation of the components, but this method is not effective. Salting out procedures which are generally used for proteins are not effective here as a result of the lack of hydrophobic groups in most of the molecules.

Fractional precipitation of polysaccharides by formation of water-insoluble salts or complexes is effective in several cases and can be the first separation step when mixtures of such compounds are present. The native polysaccharide is regenerated from the precipitated compound by a second operation. Some of the reagents used for this purpose are listed in Table 1.

Table 1.   Reagents for Precipitating Polysaccharides

Reagent	Polysaccharide
Potassium chloride	*kappa*-Carrageenan[20], alginate[21]
Calcium chloride	Chondroitin 4-sulfate[22], alginate, pectin[12,23,24]
Manganese(II) sulfate	Alginate[21]
Neutral cupric acetate	Separation of acidic from neutral polysaccharides[25]
Fehling's solution	Polysaccharides containing D-mannose and D-xylose units[26]
Cetyltrimethylammonium bromide (Cetavlon); Cetylpyridinium bromide	Acidic polysaccharides; neutral polysaccharides as borate complexes: regeneration by treatment of the precipitate with an aqueous solution of monovalent ions[23,27–29]
Uranyl acetate	Carboxymethyl cellulose[30]
Methylene Blue	Carrageenan[31]
Congo Red	β-D-Glucans[32,33]

An example of a successful application is the procedure for purifying the bacterial capsular polysaccharide antigen from *Pneumococcus type 26*, which is reported here in detail[34].

Procedure

Reagent A: aqueous solution of 5% cetyltrimethylammonium bromide.

Reagent B: 10% aqueous sodium chloride.

Dissolve ~500 mg of the isolated polymer in 250 mL of purified water and add 60 mL of reagent A. The precipitate is isolated by centrifugation and, after discharging the supernatant, dissolve it by adding 100 mL of reagent B, and precipitate the polysaccharide by adding 500 mL of ethanol (analytical grade).

3.1.3. Procedures for Isolating Polysaccharides Partially Soluble in Water (Acid or Alkaline Conditions) (Group C)

The procedures are dependent on the individual polysaccharide. It is not posible to claim general methods for this group.

3.2. Separation Methods for Soluble Polysaccharides

After selective precipitation as a first operation, further separation can be achieved by the following methods: ion exchange chromatography, affinity chromatography, molecular sieve chromatography and electrophoresis.

3.2.1. Ion Exchange Chromatography

This method is successful in many cases and is used for the separation of acid polysaccharides from neutral materials and for the separation of acid polysaccharides of different acidity grades. The method was introduced by Neukom and co-workers who used DEAE-cellulose (diethylaminoethyl cellulose) as the ion exchange material[35,36].

Under neutral conditions, DEAE-cellulose (chloride or phosphate form) scarcely absorbs neutral polysaccharides which can be eluted readily and quantitatively with dilute salt solutions[37].

DEAE ion exchange chromatography has also been used for separation of polysaccharide mixtures in the field of analysis of thickening food hydrocolloids; the procedure is described here in detail[38].

Procedure

Preparation of the DEAE-Cellulose Ion Exchanger

Approx. 10 g DEAE-cellulose is suspended twice in aqueous 0.5 M hydrochloric acid, then in water, then in 0.5 M sodium hydroxide, and then in water again. Discharge the supernatant each time. Add 500 mL aqueous 0.5 M phosphate buffer pH 7.2 with stirring and, after 2 min, pour the suspension into

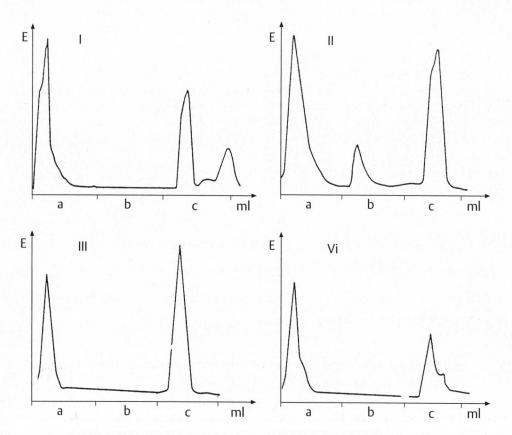

Figure 1.   Separation of Polysaccharide Containing Hydrocolloids on DEAE Columns
Eluant a: 0.067 M phosphate buffer pH 7.2; *eluant b*: 0.067 M phosphate buffer pH 9.2; *eluant c*: 0.1 M sodium hydroxide; photometric determination of the carbohydrates in the single fractions by the carbazole sulfuric acid reaction
I: methyl cellulose + sodium carboxymethyl cellulose, II: guar + tragacanth, III: agar + carrageenan, IV: agar + sodium carboxymethyl cellulose (Mergenthaler and Schmolck[38]; with permission)

a graduated measuring glass and keep for 45 min. The turbid supernatant is removed under vacuo and again 500 mL of phosphate buffer is added. The supernatant is again removed and the suspension is poured into a chromatographic column (L: 150 mm, i.d.: 14 mm). For removal of the excess phosphate ions, the material is washed with 100 mL of distilled water.

Separation Procedure

Eluant A: 0.067 M phosphate buffer pH 7.2.

Eluant B: 0.067 M phosphate buffer pH 9.2.

Eluant C: 0.1 M sodium hydroxide.

Dissolve 20–50 mg of the isolated polysaccharide in 5 mL of eluant A and put it on top of the DEAE-cellulose column. Then fractionate with 100 mL of eluant A, 100 mL of eluant B and 100 mL of eluant C, and collect eluants in portions of 2–5 mL. In 0.2–0.5 mL of each fraction, the carbohydrate content is determined by photometry using, preferably, one of the color reactions with concentrated sulfuric acid [e.g., phenol, anthrone, carbazole (see Section 1.4.)].

Figure 1 shows examples of the separation of mixtures of neutral and acidic thickening agents.

Other materials are DEAE-Sephadex (based on cross-linked dextrans) and DEAE-Sepharose (based on cross-linked agarose). These materials are preferably used for separation and purification of bacterial capsular polysaccharide antigens, such as those of *Pneumococcus*[39] and *Haemophilus influenca*[40], and for fractionating pectin[41]. For the first case[39], the materials are used as their acetate forms and equilibrated with 0.1 M acetate buffer pH 6.05. The polysaccharides isolated by precipitation are dissolved in the same buffer, washed onto the DEAE-Sephadex or -Sepharose column and fractionated with a 0–2 M sodium chloride gradient.

After the chromatographic separation, the polysaccharides in the single fractions are generally released from the electrolytes by dialysis.

3.2.2. Affinity Chromatography

This method is mainly used in the field of glycoproteins and is discussed in Chapter 4.

3.2.3. Molecular Sieve Chromatography

Unlike the majority of proteins, the molecular mass of an individual polysaccharide is not a constant characteristic, but variable and depends on various parameters such as, for example, method of isolation, source of the material or treatment with heat or chemicals.

Molecular sieve chromatography is used mostly for the fractionation of a single polysaccharide for studying, for example, its rheological properties.

3.2.4. Electrophoresis

Like monosaccharides, polysaccharides can be also separated by electrophoretic methods. Compounds with charge-carrying units move in an electric field and neutral polysaccharides can be converted into charged compounds by, for example, reaction with borate ions to give charged borate complexes.

The support materials, which are suitable for the electrophoresis of polysaccharides, must have the following properties:

(a)   The porosity of the support must be large enough so that a molecular sieve effect does not occur.

(b)   The adsorption of polysaccharides on the surface of the support should be minimal.

(c)   The material should be as inert as possible to allow application of the aggressive acid containing, carbohydrate specific visualizing reagents.

The best materials for the separation of polysaccharides have been found to be cellulose acetate sheets, glass fibre paper, silanized silica gel and, for some special purposes, agarose gels.

3.2.4.1. Cellulose Acetate

Cellulose acetate is a neutral, wide porous material bearing only minimal ionic charges on its surface and only a small proportion of free OH groups. Hence, no molecular sieve effects have been found and the electro-endoosmosis as well as the adsorption of polysaccharides on the surface of the support has been found to be small.

Cellulose acetate membranes are commercial products and a widespread tool in the field of clinical chemistry for the separation and characterization of blood serum proteins.

In the field of polysaccharides, this material is used for the separation of polysaccharide-containing hydrocolloids and for glycosaminoglycans. The separation of such hydrocolloids has been described by several authors[42–45] using procedures such as the following one, which is described here in detail[45].

Procedure

Reagent A: borate buffer, pH 10: 12.4 g of boric acid and 4 g of sodium hydroxide are dissolved in distilled water and made up to 1 L.

Reagent B: Methylene Blue, Toluidine Blue O; 0.2 g of each are dissolved in 100 mL of distilled water.

Reagent C: fuchsin–sulfurous acid: 1 g of fuchsin is dissolved in ca. 80 mL of refluxing water by stirring vigorously for 90 min; after cooling, 1 mL of concd hydrochloric acid and 2 g of potassium metabisulfite ($K_2S_2O_5$) are added. The solution is diluted with water to 100 mL and allowed to stand for 12 h. About 25 min before use, the reagent is cleared with 1.2 g of powdered charcoal.

Reagent D: 2 g of periodic acid is dissolved in 10 mL of water and 90 mL of ethanol (analytical grade) is added. The reagent has to be prepared freshly before use.

Wet the cellulose acetate membranes along both sides with the holes for fixing in the electrophoretic cell bridge or on a wooden model with the same parameter. Wet approximately eight spots on the membrane with 0.25 µL of reagent A using a microzone applicator, then apply 0.5 µL of the sample and dry under a stream of cool air to avoid spot broadening. Wet the membrane with reagent A and start the electrophoresis at 300 V for 20 min.

Visualization

(a) With Toluidine Blue or Methylene Blue

After electrophoresis, dip the cellulose acetate membranes into 50 mL of the staining solutions (reagents B) and shake gently for 20 s. Remove the excess of dye by rinsing with 5 portions of 40 mL of tap water, each for ca. 20 s. For staining with Toluidine Blue, add 10 mL of ethanol to the last portion and rinse for 20–30 s.

Only the acid polysaccharides are stained by the reagents; carrageenan and furcellaran change from blue to violet.

(b) With Fuchsin–Sulfurous Acid

Incubate the membrane after electrophoresis for exactly 5 min in reagent D, then cautiously remove the excess with 4 portions of 40 mL of tap water, shaking for 30–60 s each time. Plunge the sheet into 30 mL of colorless filtered reagent C and hold until the whole sheet is deep red (30 min to 10 h). Remove the excess of the dye with 15-mL portions of 90% ethanol until the membranes become white or pink; the separated polysaccharides appear as red zones. The reagent reacts with acidic and neutral polysaccharides.

As a variation regarding the detection, the polysaccharides are stained prior to separation. This can be performed with reactive dyes which form covalent linkages with polysaccharide molecules. One group of reactive dyes suitable for this purpose are the Procion dyes possessing a dichlorotriazinyl group, which reacts with hydroxyl, amino or amido groups. A great number of such dyes have been tested with several polysaccharides[46,47] and the following procedure has been found to be the most suitable[47].

Procedure

Reagent: freshly prepared aqueous solution of 50 mg of Procion Brilliant Red M-2B in 5 mL of distilled water.

Dissolve 50 mg of the polysaccharide sample in 5 mL of distilled water and add the reagent. After 5 min add 0.2 g of sodium chloride and after 30 min 0.02 g of sodium carbonate. Leave the mixture at 25 °C for 12 h, then clarify the solution by centrifugation, apply it directly to a molecular sieve column (e.g., Sephadex G-25) and elute with distilled water. The colored polysaccharide appears at the void volume, whilst the low molecular weight substances, including the excess of the dye, follow later. After freeze-drying the dyed polysaccharide, fraction dissolve the dry residue either in 0.1 M ammonium carbonate buffer pH 8.9, 0.1 M ammonium acetate buffer pH 4.7, or 0.05 M sodium tetraborate buffer pH 9.2.

As the support, porous cellulose acetate sheets (called Phoroslide) were found to be better than the common cellulose acetate membranes.

Immerse the sheets for 5 min in the relevant buffer and press them tightly between filter sheets. Apply the polysaccharides with microsyringes as thin bands at a short distance from the cathode. After a short drying period, carry out the electrophoresis using one of the buffers named above at a potential of approx. 40 V/cm for a running time of 10 min.

Procion dyes react much more readily with primary than with secondary hydroxyl groups and polysaccharides containing uronic acid units require an enhanced alkaline medium to get intensively dyed products (optimum conditions: 0.4–0.5% sodium carbonate). With this method several plant polysaccharides such as those from different acacia varieties were separated and characterized[47].

Cellulose acetate membranes have also been described for the separation of glycosaminoglycans by application of the following electrolytes[48–51]:

–   0.05 M Lithium chloride in dilute 0.01 M hydrochloric acid pH 2.0[48].
–   0.05 M Sodium phosphate buffer pH 7.2[48].
–   0.06 M Sodium barbital buffer pH 8.6[49].
–   0.05 M Barium acetate pH 5.8[50] or 0.10 M barium acetate pH 8.0[49,51].
–   0.2 M Zinc acetate pH 5.1[48] or 0.1 M zinc acetate pH 6.3[51].
–   0.1 M Pyridine–0.47 M formic acid pH 3.0[51]

The most important glycosylaminoglycans are substances found in human skin and connective tissue such as chondroitin 4-sulfate, chondroitin 6-sulfate, keratan sulfate, dermatan, and heparin. The total separation of these compounds can only be achieved by a two-dimensional technique using the following buffer systems[51]:

1st Direction: 0.1 M pyridine–0.47 M formic acid pH 3.0.

2nd Direction: 0.1 M barium acetate pH 8.0.

The visualization of the separated glycosaminoglycans is the same for all electrophoretic procedures, namely, staining with Alcian Blue in 0.1% aqueous solutions[52] or dissolving in 0.1% aqueous acetic acid[51] or in 95% ethanol saturated with sodium chloride[49]. A two-step staining has also been described using firstly 0.1% aqueous Toluidine Blue and in the second step 0.1% aqueous Alcian Blue[52]. In another procedure the cellulose acetate strip, after the electrophoretic separation, is immersed in a solution of bovine serum albumin in 50 mM acetate buffer pH 4.0, washed three times with 150 mM acetate buffer for removing the unbound albumin, and then the strip is stained with 0.2% Amido Black in 50% aqueous methanol containing 10% acetic acid[50].

The destaining of the colored membranes is carried out generally with 1–5% acetic acid in water or water/alcohol mixtures.

Another technique uses as a support specially prepared Ti(III)–cellulose acetate sheets and between two and three electrophoretic runs the membranes are immersed in mixtures of barium acetate solutions with various amounts of ethanol[52]. This method was used for the separation of the urinary glycosaminoglycans[53].

3.2.4.2. Agarose

Agarose gel has been found to be an alternative to cellulose acetate for the separation and identification of several glycosaminoglycans. The sizes of the pores of the gel are large enough so that the glycosaminoglycan molecules can move without hindrance; the molecular sieve effect can be excluded for this support[54–56].

In the classical study on this subject the following procedure has been proposed[54,55].

Procedure

Reagent A: aqueous 0.06 M barbital buffer pH 8.6.

Reagent B: aqueous 0.025–0.05 M 1,3-diaminopropane adjusted with acetic acid to a pH of 8.5.

Reagent C: 0.1% aqueous solution of cetyltrimethylammonium bromide.

Reagent D: 0.1% solution of Toluidine Blue in acetic acid–ethanol–water 0.1:5:5 v/v.

Preparation of the Agarose Gel

For preparation of a 1% solution suspend the dry agarose powder in the relevant buffer and heat the suspension at 98 °C in a water bath. Pour the clear solution into a flat glass gel trough with a comb at one side for the slots for the samples and cool it rapidly; the agarose solution changes to a gel. Remove the side frame and put the gel, which is fixed to the glass plate, into the electrophoresis apparatus.

Electrophoresis

Apply the samples (3–5 μL) containing 1–10 μg of the substances into the slots of the gel slide and subject them to electrophoresis at 5 °C at a potential of 150 V for 60 min. The zone of cresol red which is applied additionally as indicator should migrate 30 mm from the origin in the barbital buffer (reagent A) and 45 mm in the diamine buffer (reagent B).

After electrophoresis, immerse the slides in reagent C for 3 h, cover the gel sides with filter paper which has been previously wetted with reagent C and then place them under a 250 V infrared lamp for

2 h in a current of air from a fan. Immerse the slides in reagent D for staining and remove the excess of the dye with the same solvent mixture.

For special purposes when a complete separation of a complex mixture of glycosylaminoglycans (heparin, heparitin sulfate, hyaluronic acid, etc.) is required a two-dimensional technique has been described using the two aforementioned buffer systems. The best results are obtained by using the barbital buffer for the first dimension and the diamine buffer for the second dimension.

Agarose gel as a support for the electrophoresis has the following advantages over cellulose acetate: (a) it can be used for preparative purposes by enhancing the thickness of the agarose blocks, (b) the loading capacity and the resolution is better (small individual compounds can be identified in a mixture), and (c) contamination of the samples does not affect the electrophoretic separation. The drawback of this support compared to cellulose acetate is the longer time of development and the more complex preparation procedure of the agarose gel.

3.2.4.3. Glass Fibre Paper

This inert material is suitable for the separation of polysaccharides due to its lower adsorption capacity and the possibility of visualizing the separated compounds with acid containing, carbohydrate specific reagents. The drawback of untreated products is the high electro-endoosmosis caused by free silanol groups on the surface. This effect can be diminished by treatment with silylating agents according to the following procedure[57–59].

Procedure[58,59]

Preparation of Silanized Glass Fibre Papers

Cut strips of glass fibre approx. 16 cm long and heat them for 2 h at 400 °C to remove the organic materials. After cooling, immerse the strips in a solution of 2–5% dichlorodimethylsilane in chloroform or carbon tetrachloride for 18–24 h at room temperature; then rinse the strips with toluene and dry. Put the strips into the relevant buffer containing 0.2% Tween 20 to complete the wetting. The glass fibre strips must be handled carefully since they tear readily.

Electrophoresis

Blot the wet silanized glass fibre strips with filter paper to remove the excess of the relevant buffer and place into the electrophoresis chamber; equilibrate the strips with a potential of ~6 V/cm. Apply the samples onto the support with a microsyringe and carry out the electrophoretic run with the equilibration potential (6 V/cm) for 90 min at room temperature.

Visualization

Reagent A: naphthoresorcinol–sulfuric acid: 0.2 g of naphthoresorcinol are dissolved in a mixture of 100 mL of ethanol and 4 mL of concd sulfuric acid.

After the electrophoretic run spray the dried strips extensively with reagent A and heat them for 10–15 min at 110 °C. The polysaccharide containing zones appear as deep blue zones on a gray background. The reaction is suitable for acidic and neutral polysaccharides.

Reagent B: Toluidine Blue: 0.2 g dissolved in 100 mL of ethanol, analytical grade.

Immerse the dried strips after the electrophoretic run for 10 min in reagent B and remove the excess of the dye by rinsing with tap water. Only the acid polysaccharides appear as blue zones since they yield insoluble complexes with this dye.

An alternative dipping procedure has been described using either the naphthoresorcinol reagent, the orcinol–Fe(II)–sulfuric acid reagent or aniline–diphenylamine–phosphoric acid[58].

3.2.4.4. Silanized Silica Gel

The preparation of silanized glass fibre papers is complicated and the obtained material is sensitive towards mechanical effects.

These drawbacks can be overcome with silanized silica gel whose surface is coated with a thin film of a higher saturated alcohol such as 1-octanol or 1-decanol. The slurry of this material is spread on glass plates and, using polyvinylpyrrolidone as the binder, mechanically stabile layers are obtained. Polysaccharides separated on silanized silica gel can be visualized by sensitive and specific acid containing carbohydrate reagents. This method had been developed for the separation of polysaccharide containing hydrocolloids[60] and soluble hemicellulose[61] and is also suitable for the detection of polysaccharides in biological materials, such as glycogen or hydroxyethyl starches[62] in blood plasma or in liver samples. Since this method is also suitable for the separation of mono- and oligosaccharides as well as their derivatives a detailed description of the preparation of the layers and the performance of the electrophoresis is given in Section 2.2.4.

For the separation of polysaccharides, the conditions for the electrophoretic separation are nearly the same.

For the polysaccharide samples the viscosities of their solutions are important. If they are high, stripes instead of sharp zones are obtained and visualization is sometimes difficult.

These negative effects can be diminished by reducing the molecular mass. This can be achieved by (a) careful treatment of the solution with ultrasound as in the case of carrageenan or (b) by treatment with hydrochloric acid.

Table 2.   Relative Mobility M_E of Several Polysaccharide Containing Thickening Agents[a]

Compound	M_E	
	Native Product	**Partial Degradation**
Wheat starch	0; 0.10; 0.30	-
Starch monophosphate	0.40	-
Oxydized starch	0.35	-
Guar	-	0.90[b]
Locust bean gum	-	0.80[b]
Tara gum	-	0.85[b]
Agar	0.60	-
Pectin[d]	1.80	-
Alginate	1.45	1.95[b]
kappa-Carrageenan	-	1.60[c]
lambda-Carrageenan	-	2.30[c]
Gum tragacanth	0.30; 0.40; 0.90	0.45; 1.00; 1.20[b]
Gum arabic	1.00	1.30[b]
Gum ghatti	0.55[e]	1.20; 1.35[b]
Gum karaya	-	1.40[b]
Xanthan	-	1.30[b]

[a]0.3 M Borate buffer, pH 10; 250 V; M_E relating to gum arabic; mean values obtained from five measurements. [b]Partially degraded with concd HCl. [c]Partially degraded with ultrasound. [d]Mobility after total hydrolysis of the ester groups. [e]Mobility of the main constituent; M_E of accompanying substances 0.90; 1.20.

Treatment with acid (b) leads, in several cases, to an alteration of the electrophoretic properties. For the identification of polysaccharides by this method the test materials must be treated in the same manner.

Procedure

Suspend approximately 150–200 mg of the polysaccharide sample in 1 mL of concd hydrochloric acid and keep it for 30 min at room temperature. Add 1 mL of distilled water and 20 mL of ethanol, isolate the precipitate by centrifugation and discard the supernatant. Wash the precipitate until it is free of acid with 80% methanol, then with acetone and finally with diethyl ether. Dry it at room temperature and dissolve an adequate amount in the relevant buffer. This method had been used for identification of hydrocolloids and their relative mobilities are listed in Table 2.

The visualization procedure must also be altered for this purpose as follows.

Dry the thin-layer plates after the electrophoretic run at 80 °C to remove water and 1-octanol (or 1-decanol). After cooling, spray the plate with the naphthoresorcinol reagent described in Section 2.1.2.1.e. and keep it for at least 30 min at room temperature for prehydrolysis of the polysaccharides; then heat the plates for 30 min at 110 °C. The separated polysaccharides appear as blue or bluish-gray zones. Figure 2 shows the separation of some polysaccharide containing hydrocolloids.

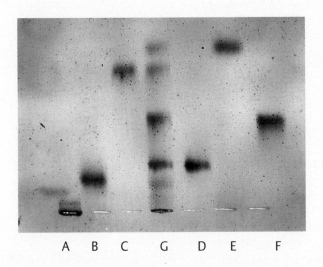

Figure 2.   Electropherogram of Some Polysaccharide Containing Hydrocolloids
Support: silanized silica gel–1-octanol; 0.3 M borate buffer pH 10; *electrophoresis:* 250 V, 25 mA, 180 min, 23 °C
A: wheat starch, **B:** starch monophosphate, **C:** *kappa*-carrageenan, **D:** agar, **E:** pectin, totally deesterified, **F:** gum arabic, **G:** mixture **A–F**; 0.2% solutions in 0.3 M borate buffer pH 10 (Scherz[60]; with permission).

3.3. Identification of Polysaccharides

The identification of polysaccharides after their isolation and purification is achieved by several chemical and physical methods. It is beyond the scope of this book to describe all the different techniques necessary for elucidating complete structures. This section restricts itself only to those methods which are preferably suitable for the identification of polysaccharides with known structures. They can be classified into the following operations:

1.   Degradation of the polysaccharide to its monomer units including their identification.
2.   Determination of the position of the external ether bridges.
3.   Special reactions to modify the polysaccharide for structural elucidation, e.g. periodate scission.

3.3.1. Degradation of the Polysaccharides to Their Monomeric Units

The common method for this purpose is hydrolysis with acids. The conditions required to achieve a total depolymerization varies strongly from one material to another. The acid treatment of polysaccharides leads to the loss of hydrolysis products by additional side reactions such as conversion to furan derivatives by internal dehydration.

The optimal conditions afford maximum depolymerization and minimum destruction of the sugars. In general, aldose containing polysaccharides can be hydrolyzed completely with a minimum loss of units by treatment with sulfuric acid or with trifluoroacetic acid according to the following procedures[63,64].

Procedures
(a) Hydrolysis with Sulfuric Acid[63]
Reagent A: 1 M aqueous sulfuric acid.
Reagent B: 0.25 M aqueous barium hydroxide.
Suspend 10–15 mg of the weighed polysaccharide in 5–10 mL of reagent A and heat it under reflux at 100 °C for 5 h. After cooling, dilute the solution to 25 mL with water and titrate with reagent B to a pH of 5–6 to remove sulfate ions. After standing for 1 h, filter the solution and wash the barium sulfate precipitate with distilled water. Evaporate the whole filtrate under vacuo at 35 °C to dryness.

(b) Hydrolysis with Trifluoroacetic Acid (TFA)[64]
Reagent: 2 M aqueous trifluoroacetic acid.
Suspend 5 mg of the weighed polysaccharide in 2 mL of the reagent and heat the mixture in a sealed tube at 120 °C for 2 h. After cooling, dilute with water and evaporate the hydrolysate on a rotary evaporator under vacuo to dryness.

Both methods are suitable for noncellulosic polysaccharides containing neutral sugars. The use of TFA has the advantage in that it is easily removed. Under the aforementioned conditions, the release of monosaccharides was found to be optimal with a minimum loss of the single substances. Further heating leads to remarkable degradation of aldoses. Table 3 shows the results of a test where mixtures of equimolar quantities of sugars were heated with 1 M aqueous sulfuric acid at 100 °C for 2, 5, 8, 12 and 18 h, and were then analyzed by gas chromatography as the alditol acetates (see Section 2.1.4.)[63].

Table 3.   Recovery of Sugars after Heating with 1 M Sulfuric Acid[63]

Compound	Recovery[a] (%)				
	Heating Time				
	2 h	5 h	8 h	12 h	18 h
Rhamnose	100	100	98	92.2	81
Arabinose	99	96.3	95	90.8	75.3
Xylose	99	98	94	84	67
Mannose	100	98	97	92	81
Galactose	98	95	93.7	92	81
Glucose	100	98	91.4	87.4	82

[a]Percentage recovery is based on values obtained on treatment with 1 M sulfuric acid (0 h).

Polysaccharides containing glycosiduronic acids or 2-amino-2-deoxyglycosidic units are more resistant to acidic depolymerization. In the case of uronic acids, a complete hydrolysis is rarely possible as a result of parallel degradation reactions such as decarboxylation. A moderate destruction can be achieved by treating the polysaccharide with 72% sulfuric acid at room temperature according to the following procedure[63].

Procedure

Wet 10 mg of the weighed polysaccharide with 0.82 g of 72% sulfuric acid and leave the mixture for 3 h at 20 °C. Dilute the mixture with distilled water and carry out the previously described neutralization of the solution and removal of the sulfate ions with barium hydroxide. In the filtrate, the uronic acids are determined by a photometric method.

In the case of cellulose containing materials, additional heating of the diluted (1M) solution at 100 °C for 2 h is necessary; uronic acids undergo partial degradation with this treatment and should be determined without heating.

Another possibility for obtaining complete hydrolysis of glycosiduronic glycans is to reduce the uronic acid residues to neutral units of the polymer which can be hydrolyzed more readily. This is possible in an aqueous medium by reaction of the carboxylic groups with water-soluble carbodiimides; the activated carboxylic groups are reduced with sodium borohydride to the corresponding alcohols[65] according to the following reaction scheme.

Procedure

Reagent A: 1-ethyl-3-(3-aminopropyl)carbodiimide (EDC) or 1-cyclohexyl-3-(2-morpholinoethyl) carbodiimide (CMC), both as solid materials.

Reagent B: 2 M aqueous solution of sodium borohydride.

To 10 mL of a solution containing 100 µequiv. carboxylic acid (monomer or polymer) add 1 mmol EDC or CMC (reagent A) and maintain the pH of the solution at 4.75 using an automatic titrator with 0.1 M hydrochloric acid. Plot the consumption of acid versus time; the uptake of hydrogen ions should not last more than 2 h. After the uptake of hydrogen ions has ceased, slowly add reagent B which causes a rapid increase in pH. Maintain the pH at a value around 7 by automatic titration in a pH-stat with 4 M hydrochloric acid; a total volume of 15–25 mL of reagent B is usually required for complete reduction which needs approximately 60 min. Purify the reduced polysaccharide by dialysis.

Reduction of Carboxylic Groups of Uronic Acid Units in a Polysaccharide According to the Method of Taylor and Conrad

It is important that a pH 7 is maintained during the reduction with borohydride. Under this condition polyuronides are reduced with maximum yields (e.g., 95% of the uronic acid residues of capsular polysaccharides and 88% of polygalacturonic acid). An increase in pH above 8 decreases the reduction yield to less than 50%.

Methanolysis
Depolymerization of polysaccharides can also be carried out with methanolic hydrochloric acid yielding the corresponding methyl glycosides. This reaction causes less destruction of the monomeric units, especially uronic acids. A detailed description of the procedure is given in Section 2.1.4. (Gas Chromatography) and in the literature[66,67].

Enzymatic Degradation of Polysaccharides
Enzymes catalyze the specific depolymerization of polysaccharides under very mild conditions, such as almost neutral aqueous solutions and at room temperature. The specificity of the enzyme towards the individual units is further controlled by the position of the glycosidic linkages between the single monomeric units and their anomeric forms (α or β). They can further be classified into "endoenzymes" which depolymerize the polysaccharides by a random splitting of the interior glycosidic bonds and "exoenzymes" which cut the sugar units mainly from the nonreducing end of the polymers.

Table 4 gives a compilation of the most important enzymes for polysaccharide depolymerization; this is a short outline of the excellent review on this subject by McCleary et al[68].

Table 4. Survey of the Most Important Carbohydrate Splitting Enzymes

Enzyme EC Number	Specificity	Degradation Products	Source of the Enzyme
Endoenzymes			
α-Amylase; 1,4-α-glucan-D-glucanase EC 3.2.1.1	Random cleavage of the α-1,4-glycosidic D-glucose bonds at the internal parts of amylose and amylopectin	D-Glucose and oligomers of D-glucose with 6–7 units; the anomeric form of the cleaved bond is retained	Pork pancreas; barley malt; *Aspergillus oryzae*; *Bacillus subtilis*
Dextranase; Endo-1,6-α-D-glucanase EC 3.2.1.11	Cleavage of the α-1,6-glycosidic D-glucose bonds	Dependent on the enzyme itself; α-1,6-oligomers of D-glucose with 5–9 units	*Arthrobacter spp*; *Penicillium lilacinum*
Endo-1,4-β-D-mannase EC 3.2.1.78	Cleavage of the β-1,4-glycosidic D-mannose bonds as with β-mannans or β-galactomannans (e.g., guar)	Oligomers of mannose up to 9 units; for galactomannans, dependence on the content of D-galactose	Several plants, such as those which contain galactomannans; *Aspergillus niger*
Cellulase; Endo-1,4-β-D-glucanase EC 3.2.1.4	Cleavage of the β-1,4-glycosidic D-glucose bonds (e.g., cellulose)	Cellobiose and cellotriose	*Trichoderma viride*
Polygalacturonase; Endo-1,4-α-D-galacturonase EC 3.2.1.15	Random cleavage of glycosidic α-1,4 galacturonic acid bonds with nonesterified galacturonans	α-1,4-Oligomers of D-galacturonic acid	Many plants; several microorganisms such as *Aspergillus niger, Bacillus subtilis*
Endo-1,3-β-glucanase EC 3.2.1.39	Random cleavage of the glycosidic β-1,3-D-glucose bonds of β-1,3-glucans such as laminaran, scleroglucan or yeast cell glucans	For laminaran: laminaribiose and further oligomers	Microorganisms such as *Arthrobacter spp.*
Endo-1,3-α-glucanase EC 3.2.1.59	Cleavage of the glycosidic α-1,3-D-glucose bonds of α-1,3-glucans such as nigeran	D-glucose; nigerose; oligomers of nigerose	Microorganisms such as *Trichoderma viride*; *Streptomyces spp.*; *Flavobacterium EK-14*; *Cladosporium resinae*
Endo-1,6-β-glucanase EC 3.2.1.75	Cleavage of the glycosidic β-1,6-D-glucose bonds of β-1,6-D-glucans such as pustulan	D-glucose, gentiobiose	Microorganisms such as *Bacillus circulans*; *Penicillium brefeldianum*
Pullulanase; Endo-1,6-α-glucanase EC 3.2.1.41	Random cleavage of the glycosidic α-1,6-D-glucose bonds such as of pullulan or amylopectin	Multiple oligomers of maltotriose	Microorganism: *Enterobacter aerogenes*

Table 4.   (Continued)

Enzyme EC Number	Specificity	Degradation Products	Source of the Enzyme
Levanase; Endo-2,6-β-fructanase EC 3.2.1.65 Inulanase; Endo-2,1-β-fructanase EC 3.2.1.7	Random cleavage of the glydosidic β-2,6- resp. β-2,1-D-fructose bonds with fructans such as levan or inulin	D-Fructose; levanbiose and oligomers (levanase) β-2,1 oligomers of D-fructan (Inulanase)	Micoorganisms; Levanase: *Athrobacter spp.* Inulanase: *Aspergillus niger*
Chitinase EC 3.2.1.14	Cleavage of the glycosidic β-1,4-D-N-acetylglucosaminic bonds of chitin	Chitobiose and chitotriose	Microorganisms: *Aspergillus niger, Staphylococcus aureus*
Exoenzymes			
β-Amylase; Exo-α-1,4-glucanase EC 3.2.1.2	Splitting of the glycosidic α-1,4-D-glucose bonds from the nonreducing end of α-1,4-glucans such as amylose or amylopectin; stops at the branching points of the molecule	Maltose, maltotriose; inversion of the anomery since β-maltose is liberated	Higher plants like batata, soja, barley; some microorganisms such as *Bacillus polymyxa, Bacillus meganterium, Pseudomonas spp.*
Glucoamylase, amyloglucosidase; Exo-α-1,4; 1,6-glucanase EC 3.2.1.3	Splitting of the glycosidic α-1,4-D-glucose bonds from the nonreducing end of amylose and amylopectin; α-1,6-bonds are also hydrolyzed but noticeably slower.	D-Glucose	Microorganisms such as *Aspergillus niger, Cladosporium resinae*
Glucodextranase; Exo-α-1,6-glucanase EC 3.2.1.70	Splitting of the glycosidic α-1,6-D-glucose bonds from the reducing end of dextran	D-Glucose	Microorganisms such as *Athrobacter globiformis* or *Streptococcus mitis*
Expolygalacturonase; Exo-α-1,4-galacturonase EC 3.2.1.67	Splitting of glycosidic α-1,4-bonds of galacturonic acid from the nonreducing end of polygalacturonic acid chains such as deesterified pectin, which is incompletely depolymerized due to substitions with neutral sugar units; highly esterified pectins are not degraded	D-Glacturonic acid	*Aspergillus niger*

3.3.2. Determination of the Position of the External Glycosidic Bonds

The location of the external glycosidic linkage between two monomeric units in a polysaccharide is carried out by the so-called "methylation analysis". It is based on the principle of the total methylation of the polysaccharide, hydrolysis of the product to monomeric units and identification of the resulting O-methyl sugars, preferably by gas chromatography–mass spectrometry of the acetylated O-methyl alditols (see Section 2.1.4.).

Several methods exist for the methylation of polysaccharides and they can be classified into two main groups, namely, (a) methylation in the presence of alkali and (b) methylation in a neutral medium. Methods for the methylation of polysaccharides in an alkaline medium are listed in Table 5.

Table 5.   Methylation of Polysaccharides in an Alkaline Medium[a]

Base	Methylation Agent	Solvent	Literature
NaOH	Dimethyl sulfate	Water	Haworth[69]
BaO + Ba(OH)$_2$; Ag$_2$O	Methyl iodide or dimethyl sulfate	N,N-Dimethylformamide or DMSO	Kuhn et al.[70,71]
NaH + DMSO	Methyl iodide	DMSO	Hakomori[72]
KH + DMSO	Methyl iodide	DMSO	Phillips, Fraser[73]
Solid NaOH	Methyl iodide	DMSO	Ciucanu, Kerek[74]

[a]DMSO = dimethyl sulfoxide.

Methylation in neutral medium can be carried out with, for example, trifluoromethanesulfuric acid methyl ester with 2,6-di-*tert.*-butylpyridine as catalyst in dichloromethane[75] or trimethyl phosphate[76] as solvents. In these media the polysaccharides are less soluble and the yields of the methylated products are, therefore, poor. The procedure of Hakomori is the usual method for the methylation of polysaccharides since it generally needs only one step for complete reaction. Here, the dry sample of the polysaccharide is dissolved in absolute dimethyl sulfoxide (DMSO) and reacted with methylsulfinylmethylsodium (prepared by adding sodium hydride to dry DMSO) to give polyalkoxy anions, which react further with methyl iodide to give the corresponding poly-O-methyl ethers. A drawback is the time-consuming reaction between DMSO and NaH which has to be carried out under nitrogen. A detailed description of the procedure is given in Section 2.1.4.

An alternative is the procedure of Ciucanu and Kerek[74], where the polyalkoxide anions are formed in DMSO with solid, fine-powdered sodium hydroxide according to the following scheme.

$$R(OH)_n + OH^- \; \rightleftarrows \; R(OH)_{n-1}\, O^- + H_2O$$

The water is bound by the solid sodium hydroxide which is present in excess (3–5 equivalents per exchangeable H atom and 1–3 mmol base/mL DMSO). For a detailed procedure see Section 2.1.4. The advantage of this method is its simplicity and the excellent yields of the methylation products[77,78]. The methylation reaction leads to esterification of the uronic acid residues. In such an alkaline medium a β-elimination reaction can occur, according to the following scheme, which leads to a partial depolymerization of the molecule. Such an elimination reaction takes place when the reaction in alkaline medium lasts a long time. Therefore, the complete methylation should be done in one step as it

Degradation of Esterized Pectin by Base-Catalyzed β-Elimination

is possible with the two forenamed procedures. In the case of the procedure of Ciucano and Kerek, for complete methylation of uronic acids and uronic acid polysaccharides a decrease in the concentration of sodium hydroxide to 2.5 moles per 1 mol exchangeable hydrogen and its reduction with 1 mmol/mL DMSO[79] is necessary.

After isolation of the completely methylated polysaccharide by extraction from the reaction mixture with chloroform, the solvent of the extract is evaporated and the residue dried in vacuo for several hours. The completeness of the methylation can be checked by IR spectroscopy. In the case of complete methylation, the OH vibrations at 3400–3600 cm^{-1} should be absent, although traces of water due to the hygroscopicity of the methylated polysaccharide can lead to misinterpretations. Another method is gas chromatography of the methylated, reduced and acetylated units; traces of higher acetylated products indicate incomplete methylation.

The methylated polysaccharides are hydrolyzed with acid to the methylated monomers. The procedure with 2 M trifluoroacetic acid (TFA) at 120 °C according to Albersheim et al.[64] was found to be most suitable due to the slight volatility of the TFA, which can be removed easily by simple evaporation. A detailed procedure is given in Section 3.3.1. The hydrolysis time depends on the polysaccharide and varies from 30 min to 12 h. Usually the methylated sugars obtained are reduced with borohydride to the corresponding partially methylated sugar alcohols and acetylated to the O-methyl-O-acetyl alditols, which are identified by GS–MS (see Section 2.1.4.).

For location of the glycosidic carbon atom of the original methylated sugar in the methylated alditols, the reduction must be carried out with sodium borodeuteride in deuterium oxide before acetylation and GC–MS.

Reduction of the Carboxylic Group

Acid hydrolysis causes loss of uronic acid residues. This can be prevented by reduction of the esterified carboxylic group of the methylated polysaccharide into a primary alcohol group with lithium aluminum hydride in dry tetrahydrofuran using the following procedure[80,81,86].

Procedure[81,86]

Reagent: a suspension of 100 mg of lithium aluminum hydride in 10 mL of dry tetrahydrofuran under nitrogen.

Dissolve 20 mg of the dry permethylated uronic acid containing the polysaccharide in 5 mL of tetrahydrofuran in a 50-mL flask equipped with a drying tube containing anhydrous calcium chloride. Add the reagent and stir for 48 h at room temperature. Add 1 mL of ethyl acetate, then carefully add 20 mL of deionized water and 10 mL of 0.5 M aqueous sulfuric acid. Pour the mixture into a separating funnel and extract the aqueous phase three times with chloroform. Collect the organic phase, wash with 0.5 M sulfuric acid and water, dry with anhydrous sodium sulfate and evaporate the chloroform. The dry residue is the reduced methylated polysaccharide containing a free OH group which can be recognized by IR spectroscopy.

Reductive Cleavage of the Glycosidic Bond

Simple acid hydrolysis of a permethylated polysaccharide has the disadvantage that no information about the ring structure of the monomers in the polysaccharide molecules is available as a result of the furanose–pyranose equilibrium. Reduction of the methylated sugars to the corresponding alditols completes the loss of the ring structure and open-chain compounds are formed. This drawback can be overcome by the so-called "reductive cleavage". This method includes depolymerization of the methylated polysaccharide with a Lewis acid in the presence of a hydride donor yielding the cyclic anhydroalditols according to the following scheme[82–84].

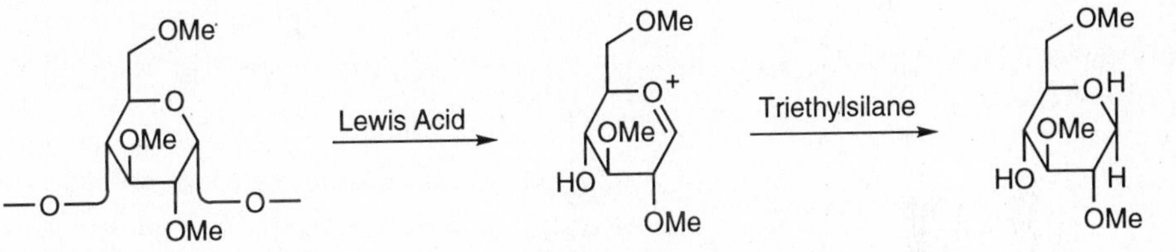

Reductive Cleavage of the Glycosidic Bonds of Permethylated Polysaccharides

Several Lewis acids such as boron trifluoride–diethyl ether complex, trifluoroacetic acid, etc., have been tested as cleavage reagents. The most suitable compounds have been found to be trimethylsilyl trifluoromethanesulfonate (TMS-O-triflate)[84] and a mixture of trimethylsilyl methanesulfonate (TMS-O-mesylate) and boron trifluoride–diethyl ether complex (BF$_3$•Et$_2$O)[85]. Reductive cleavage of the methylated polysaccharide is carried out either with 5 equivalents of TMS-O-triflate per equivalent glycosidic bond or 5 equivalents TMS-O-mesylate and 1 equivalent BF$_3$•Et$_2$O. The reduction itself is carried out with 5 equivalents of triethylsilane per equivalent of glycosidic bond.

After the reaction, the O-methyl anhydroalditols are acetylated immediately with either acetic anhydride/N-methylimidazole (TMS-O-triflate) or acetic anhydride/trifluoroacetic acid (TMS-O-mesylate/BF$_3$•Et$_2$O).

Procedure

(a) Reaction with TMS-O-Triflate[84,86]

Reagent A: triethylsilane (TES), pure.

Reagent B: O-trimethylsilyl trifluoromethanesulfonate (TMS-O-triflate), pure.

Reagent C: acetic anhydride.

Reagent D: N-methylimidazole.

Evaporate 0.5–2 mg of the permethylated polysaccharide, in a closeable vial, with a nitrogen stream to dryness and dissolve the residue in ~200 µL dry CH$_2$Cl$_2$ [dried with CaH$_2$ (0.2 g/9 mL)]. Add 5 equiv. of reagent A and reagent B per equiv. glycosidic bond and stir for 3–20 h at room temperature in a closed vessel. Add 20 µL reagent C and 10 µL reagent D and stir for 1 h at room temperature. Then add 200 µL dichloromethane and several mL of saturated aqueous sodium bicarbonate solution to the reaction mixture and stir for 30 min. Separate the organic phase and extract the aqueous phase again with 200 µL chloroform. Collect the organic phase, wash 3 times with deionized water, dry with anhydrous sodium sulfate or anhydrous calcium chloride, and evaporate the organic phase again to dryness. Then, add 200 μL dichloromethane containing tetra-*O*-acetyl-*meso*-erythritol as internal standard for subsequent GC–MS studies.

(b) Reaction with TMS-O-Mesylate–BF₃•Et₂O[85,86]

Reagent A: a mixture of *O*-trimethylsilyl methanesulfonate (TMS-*O*-mesylate), BF₃•Et₂O and triethyl-silane (TES): 5 equiv. TMS-*O*-mesylate, 1 equiv. BF₃•Et₂O and 5 equiv. TES per 1 equiv. glycosidic bond prepared immediately before use.

Reagent B: a mixture of acetic anhydride–trifluoroacetic acid 10:1 v/v.

Dissolve 0.5–2.0 mg of the permethylated polysaccharide, in a 2 mL vial, under nitrogen in 200 μL dry dichloromethane and add reagent A. Stir for 3–14 h at room temperature, add 30 μL of reagent B and stir for another 15 min at 50 °C. Isolate the acetylated compounds according to the procedure above (a).

Figure 3 shows the GC of the partially methylated anhydroalditol acetates by the reductive cleavage (using the TES–TMS-*O*-mesylate–BF₃•Et₂O reagent) of methylated agarose[86].

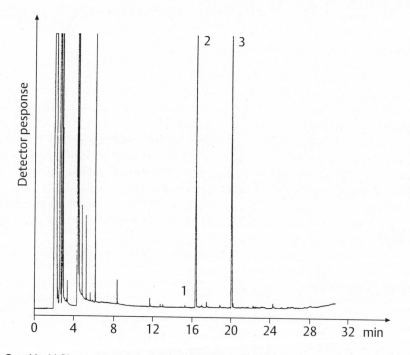

Figure 3.   Gas–Liquid Chromatogram of Partially Methylated Anhydroalditol Acetates Derived from TES–TMS-*O*-Mesylate–BF₃•Et₂O Cleavage of Per-*O*-methylated Agarose
Column: JW-DB-5 fused silica column, L: 30 m, i.d.: 0.25 mm; *temperature/time:* 70 °C, 3 min; 70–120 °C/20 °C min⁻¹; 120 °C, 5 min; 120–300 °C/5 °C min⁻¹
1: 1,5-anhydro-2,3,4,6-tetra-*O*-methyl-D-galactitol, **2:** 4-*O*-acetyl-1,5:3,6-dianhydro-2-*O*-methyl-L-galactitol, **3:** 3-*O*-acetyl-1,5-anhydro-2,4,6-tri-*O*-methyl-D-galactitol (Kiwitt-Haschemie, Heims, Steinhart and Mischnik[86]; with permission)

3.3.3. Other Methods for the Determination of Polysaccharide Structures

Other methods for the elucidation of the chemical structures of polysaccharides must determine (a) the stereochemical structure of the single units including their linkages and (b) the sequences of the single units in the chains of the polysaccharide molecules.

In the first case, a specific oxidation reaction with periodate and sodium borohydride ("Smith degradation") can be used in addition to the methylation procedure (Section 3.3.2.).

3.3.3.1. Cleavage of the Polysaccharide with Periodate (Smith Degradation)

The reaction of vicinal polyhydroxy compounds with periodate ions, the so-called "Malaprade reaction" (see Section 1.3.1.), leads to cleavage of the C–C bond between two vicinal primary or secondary H–C–OH groups yielding two aldehyde groups, which can be reduced in aqueous solution with sodium borohydride to the corresponding alcohols. Application of this reaction to polysaccharides depends on the position of the glycosidic linkages between the single units and whether the single unit itself has, for example, a furanose or pyranose structure, as demonstrated in the following schemes[87–89] (see page 328).

Different practical procedures exist for carrying out this reaction[90–92]; the micromethod developed for the elucidation of the structures of soluble bacterial polysaccharides is described here in detail[92].

Procedure

Reagent A: 0.1 M aqueous sodium acetate buffer pH 3.9.

Reagent B: 0.2 M aqueous solution of sodium metaperiodate.

Reagent C: ethylene glycol, pure.

Reagent D: sodium borohydride pure.

Reagent E: 50% aqueous acetic acid.

Dissolve approximately 30 mg of the polysaccharide in a small amount of reagent A and add 5 mL of reagent B. Keep the solution in the dark at 4 °C for 120 h, then add 1 mL of reagent C to destroy the excess of reagent and dialyze the solution against tap water overnight. Concentrate the solution of the oxidized polysaccharide in vacuo to approx. 50 mL, add 300 mg of reagent D and keep the solution for 8 h at room temperature. Decompose reagent D by addition of reagent E and dialyze the solution overnight; concentrate the solution or lyophilize it to obtain a dry product.

The obtained altered polysaccharide can be treated in different ways:

– Hydrolysis with weak acid (0.05–0.1 M sulfuric acid) for cleaving only the acetal groups, liberating low molecular sugar alcohols like glycerol, erythritol, etc.

– Hydrolysis with medium strength acids like 0.25 M sulfuric acid giving lower polyalcohols as well as nonaltered sugar units.

– Methylation of the periodate-treated polysaccharide followed by hydrolysis.

The low molecular weight degradation products are analyzed qualitatively and quantitatively, preferably by GC–MS, as their *O*-acetyl derivatives (see Section 2.1.4.). In several cases, polysaccharides show so-called "nonideal behavior" which can be divided into two groups, namely, (a) overoxidation and (b) incomplete oxidation. Overoxidation[93,94] is encountered when the reaction with periodate yields either tartronaldehyde derivatives or tartronic acid–half aldehyde derivatives (e.g., from hexafuranosides or hexuronic acid end groups). Such nonspecific oxidation reactions are limited when the reaction is carried out at a pH range of 2.2–4.0, but preferably at pH 3.6.

Reactions of Periodate with Differently Linked Sugar Units

Incomplete oxidation has two main causes:

(a)   Hemiacetal formation between aldehyde fragments in oxidatively cleaved residues and hydroxyl groups in adjacent but not yet oxidized residues; the latter units are then protected against further oxidation[95]. The highest degree of such incomplete oxidation occurs in 4-linked polysaccharides without primary hydroxyl groups at C-6[96].

(b)   Hydrogen bonds between one of the pair of hydroxyl groups resp. carboxyl groups and a suitably disposed acetamino group on a neighboring sugar residue[97,98].

3.3.3.2. Further Reactions

These are applied for particular purposes, especially for the determination of the frequencies of the single units in a polysaccharide chain. They are only briefly mentioned here.

(a) Base-Catalyzed β-Fragmentation

This can be initiated by strongly electron-withdrawing functional groups leading to the cleavage of the glycosidic linkage[99]. The best known is the degradation of esterified polyuronic acid units containing polysaccharides, such as highly esterified pectins[100,101] or 3-O-polyglycan laminaran[101–103] (see pages 323, 324). Another modification is the conversion of the free primary hydroxyl group of a methylated polysaccharide unit into a sulfone group with 4-toluenesulfonyl chloride (Ts–Cl) according to the following scheme[104–106].

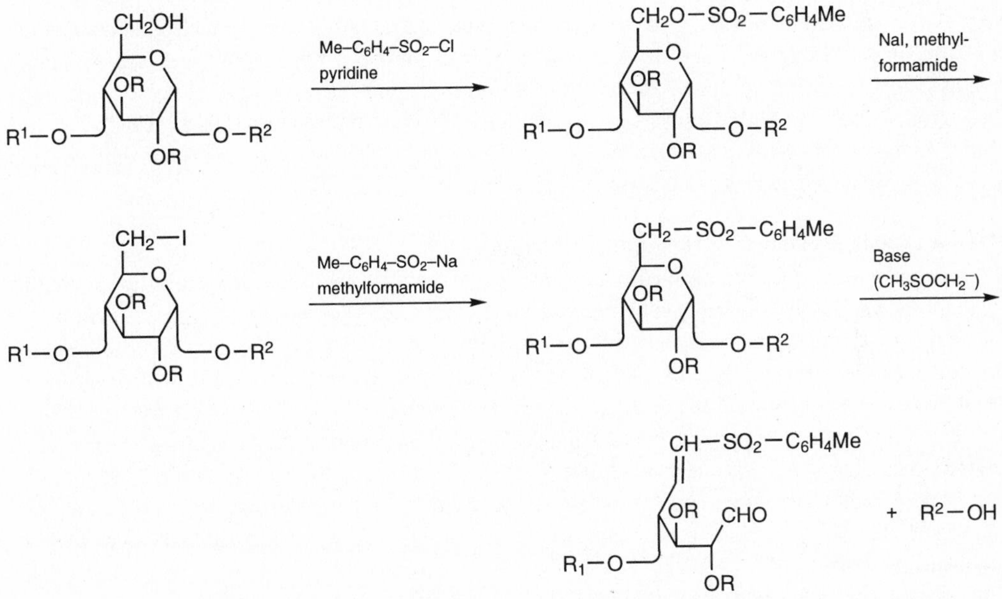

Base-Catalyzed β-Elimination of 6-Toluene Sulfone Units of a Methylated Polysaccharide

The methylated polysaccharide possessing a free primary alcohol group (e.g., obtained by reduction of a methylated uronic acid containing polysaccharide with lithium aluminum hydride) is treated with 4-toluenesulfonyl chloride in pyridine yielding the respective tosyl ester. These units are converted with sodium iodide in methylformamide into the 6-deoxy-6-iodo residues. Further reaction with sodium 4-toluenesulfinate in methylformamide introduces the toluene sulfone group into the 6-position of these units of the methylated polysaccharides. Treatment of the sulfone units with base cleaves the glycosidic bond to the next unit. The residue R–OH can be identified by further methylation.

The latter reaction has been used for the elucidation of the structures of bacterial polysaccharides[104, 105] as well as galactomannans[106].

(b) Partial Depolymerization in an Aqueous Medium[107,108]

The method of partial depolymerization results in a mixture of different oligosaccharides. After separation, the structure of each compound is determined by methods described in the previous chapters and additionally by ^{13}C and ^{1}H NMR and GC–MS. The structure of the polysaccharide can be elucidated by piecing the structure of the single oligosaccharides together like a puzzle.

As remarked in Section 3.3.1., differences in the cleavage rates of the glycosidic linkages exist among sugar units of the same type and ring sizes, but linked at different positions as well as among the single units themselves[107]. α-D-Glycopyranosyl linkages are generally more easily hydrolyzed than β-D-bonds; linkages of the (1→6)-type are more resistant than those at other positions and furanoside rings hydrolyze $10–10^3$ faster than pyranoside rings. Glycosiduronic acid[109] and particularly 2-amino-2-deoxy glycosidic linkages are resistant towards acid hydrolysis. The property of the latter compounds is caused by the inductive effect of the totally protonated 2-amino group which prevents further protonation of the glycosidic bond[110]. Acetylation of such amino groups diminishes the resistance against hydrolysis. In native polysaccharides these amino sugar units are acetylated in most cases. After deacetylation, either by hydrazinolysis[111] or treatment with sodium hydroxide in DMSO[112], the resulting 2-amino-2-deoxy sugar linkages are almost resistant to hydrolysis and such amino sugar disaccharides can be isolated as products of partial hydrolysis.

(c) Partial Depolymerization in Nonaqueous Medium

Acetolysis: polysaccharides are cleaved with a mixture of acetic anhydride and sulfuric acid yielding acetylated oligosaccharides, which can be separated and identified. The rates of cleavage depend on the nature of the glycosidic linkages and are different from those in an aqueous medium. For example, the (1→6)-linkages are more easily cleaved than the others[113] (in aqueous medium it is inverted). The terminal 6-deoxyhexapyranosyl bonds (e.g., those of L-rhamnose[115] or L-fucose[114]) are remarkably stable. The sialic acid linkages also remain intact during this treatment[116], in contrast to that in an aqueous acid medium where they are very labile.

It must be remarked further, that anomerization occurs during acetolysis particularly at (1–6)-linkages; the energetically mote stable α-anomers are formed also from β-linkages of the original polysaccharides[117].

Trifluoroacetolysis: this method uses mixtures of trifluoroacetic anhydride and trifluoroacetic acid and is frequently applied in the field of glycoconjugates[118–120].

References

1   Aspinall, G. In *The Polysaccharides*, Vol. 1; Aspinall, G., Ed.; Academic: New York, 1982; p 26.

2   Southgate, D. A. T.; Hudson, G. J.; Englyst, H. *J. Sci. Food Agric.* **1978**, *29*, 979.

3   Rabe, E. *Getreide, Mehl, Brot* **1987**, *41*, 302.

4   Scherz, H.; Mergenthaler, E. *Z. Lebensm. Unters. Forsch.* **1980**, *170*, 280.

5   Leach, H. W.; Schoch, T. J. *Cereal Chem.* **1962**, *39*, 318.

6   Bouveng, H. O.; Lindberg, B. In *Methods in Carbohydrate Chem.*, Vol. 5; Wistler, R. L., Ed.; Academic: New York, 1965; p 147.

7   Bourne, E. J.; Weigel, H. In *Methods in Carbohydrate Chem.*, Vol. 5; Wistler, R. L., Ed.; Academic: New York, 1965; p 78.

8   Davies, D. A. L. *Adv. Carbohydr. Chem.* **1960**, *15*, 271.

9   Morgan, W. T. J. In *Methods in Carbohydrate Chem.*, Vol. 5; Wistler, R. L., Ed.; Academic: New York, 1965; p 80.

10   Joseleau, J. P.; Chambat, G.; Chumpitazi-Hermoza, B. *Carbohydr. Res.* **1981**, *90*, 339.

11   Chanzy, H.; Dube, M.; Marchessault, R. H. *J. Polymer. Sci. Polymer. Lett. Ed.* **1979**, *18*, 219.

12   Stoddart, R. W.; Barrett, A. J. Northcote, D. H. *Biochem. J.* **1967**, *102*, 194.

13   Green, J. W. In *Methods in Carbohydrate Chem.*, Vol. 3; Wistler, R. L., Ed.; Academic: New York, 1963; p 21.

14   Painter, T .J. In *Methods in Carbohydrate Chem.*, Vol. 5; Wistler, R. L. Ed.; Academic: New York, 1965, p 98.

15   Selvendran, R. R.; Davies, A. M. C.; Tidder, E. *Phytochemistry* **1975**, *14*, 2169.

16   Preuss, A.; Thier, H. P. *Z. Lebensm. Unters. Forsch.* **1983**, *176*, 5.

17   Crössmann, F.; Kupfer, H.; Souci, S .W.; Mergenthaler, E. *Z. Lebensm. Unters. Forsch.* **1970**, *142*, 410.

18   Schwendig, P.; Camps, G.; Thier, H. P. *Z. Lebensm. Unters. Forsch.* **1989**, *188*, 545.

19   Antonopoulos, C. A.; Gardell, S.; Szirmai, J. A.; Tyssonsk, E. R. *Biochim. Biophys. Acta* **1964**, *83*, 1.

20   Painter, T. J. *J. Chem. Soc. (London)* **1964**, 1396; In *Methods in Carbohydrate Chem.*, Vol. 5; Wistler, R.L., Ed.; Academic: New York, 1965; p 100.

21   Haug, A.; Smidsrød, O. *Acta Chem. Scand.* **1962**, *16*, 1569; In *Methods in Carbohydrate Chem.*, Vol. 5; Wistler, R. L., Ed.; Academic: New York, 1965; p 71.

22   Jeanloz, R. W. In *Methods in Carbohydrate Chem.*, Vol. 5; Wistler, R. L., Ed.; Academic: New York, 1965; p 110.

23   Graham, H. *J. Food Sci.* **1971**, *36*, 1052.

24   Diemair, W.; Weichel, H. H. *Dtsch. Lebensm. Rdsch.* **1958**, *54*, 151.

25   Erskine, A. J.; Jones, J.K.N. *Can. J. Chem.* **1956**, *34*, 821.

26   Jones, J. K. N.; Stoodley, R. J. In *Methods in Carbohydrate Chem.*, Vol. 5; Wistler, R. L., Ed.; Academic:, New York, 1965; p 36.

27   Jones, A. S. *Biochim. Biophys. Acta* **1953**, *10*, 607.

28   Bouveng, H. O.; Lindberg, B. *Acta Chem. Scand.* **1958**, *12*, 1977.

29   Scott, E. *Methods Biochem. Anal.* **1960**, *8*, 145; In *Methods in Carbohydrate Chem.*, Vol. 5; Wistler, R. L., Ed.; Academic: New York, 1965; p 38.

30   Crössmann, F.; Klaus, W.; Mergenthaler, E.; Souci, S. W. *Z. Lebensm. Unters. Forsch.* **1964**, *125*, 413.

31   Graham, H. D. *Food Res.* **1960**, *25*, 720.

32   Wood, P. J.; Fulcher, R. G. *Cereal Chem.* **1978**, *55*, 952.

33   Wood, P. J. *Carbohydr. Res.* **1980**, *85*, 271.

34   Kenne, L.; Lindberg, B.; Madden, J. K. *Carbohydr. Res.* **1979**, *73*, 175.

35   Neukom, H.; Deuel, H.; Heri, W. J.; Kündig, W. *Helv. Chim. Acta* **1960**, *43*, 64.

36   Heri, W.; Neukom, H.; Deuel, H. *Helv. Chim. Acta* **1961**, *44*, 1939.

37   Hirst, E. L.; Rees, D. A.; Richardson, N. G. *Biochem. J.* **1965**, *95*, 453.

38   Mergenthaler, E.; Schmolck, W. *Z. Lebensm. Unters. Forsch.* **1974**, *155*, 193.

39   Guy, R. C. E.; How, M. J.; Stacey, M. *J. Biol. Chem.* **1967**, *242*, 5106.

40   Branefors-Helander, P.; Classon, B.; Kenne, L.; Lindberg, B. *Carbohydr. Res.* **1979**, *76*, 197.

41   Aspinall, G.; Craig, J. W. T.; Whyte, J. L. *Carbohydr. Res.* **1968**, *7*, 442.

42   Padmoyo, M.; Miserez, A. *Mitt. Geb. Lebensmittelunters. Hyg.* **1967**, *58*, 31.

43   Chang, J. C.; Renoll, M. V.; Hansen, P. M. T. *J. Food Sci.* **1974**, *39*, 97.

44   Marrs, W. M.; Kilcast, D.; Fry, J. C. In *Gums and Stabilisers for the Food Industrie*, Vol. 2; Phillips, G.O.; Wedlock, P. J.; Williams, P. A., Eds.; Pergamon: Oxford, 1984; p 507.

45   Pechanek, U.; Blaicher, G.; Pfannhauser, W.; Woidich, H. *J. Ass. Off. Anal. Chem.* **1982**, *65*, 745.

46   Dudman, W. F.; Bishop, C. T. *Can. J. Chem.* **1968**, *46*, 3079.

47   Anderson, D. M. W.; Hendrie, A.; Millar, J. R. A.; Munro, A. C. *Analyst* **1971**, *96*, 870.

48 Breen, M.; Weinstein, H. G.; Blacik, J. J.; Borcherding, M. S.; Sittig, R. A. In *Methods in Carbohydrate Chem.*, Vol. 7; Wistler, R. L., Ed.; Academic: New York, 1976; p 101.
49 Wessler, E. *Anal. Biochem.* **1968**, *26*, 439.
50 Szewczyk, B. *Anal. Biochem.* **1983**, *130*, 60.
51 Hata, R.; Nagai, Y. *Anal. Biochem.* **1972**, *45*, 462.
52 Cappelleti, R.; Del Rosso, M.; Chiarugi, V. P. *Anal. Biochem.* **1979**, *93*, 37; *99*, 311.
53 Hopwood, J. J.; Harrison, J. R. *Anal. Biochem.* **1982**, *119*, 120.
54 Dietrich, C.; Mc Duffie, N. M.; Sampaio, L. O. *J. Chromatogr.* **1977**, *130*, 299.
55 Dietrich, C.; Dietrich, S. *Anal. Biochem.* **1976**, *70*, 645.
56 Biachini, P.; Nader, H. B.; Takahashie, H. K.; Osima, B.; Straus, A. H.; Dietrich, C. P. *J. Chromatogr.* **1980**, *196*, 455.
57 Jarvis, M. C.; Threlfall, D. R.; Friend, J. *Phytochemistry* **1977**, *16*, 849.
58 Schäfer, H.; Scherz, H. *Z. Lebensm. Unters. Forsch.* **1983**, *177*, 193.
59 Bettler, B.; Amado, R.; Neukom, H. *Mitt. Geb. Lebensmittelunters. Hyg.* **1985**, *76*, 69; *J. Chromatogr.* **1990**, *498*, 213.
60 Scherz, H. *Z. Lebensm. Unters. Forsch.* **1985**, *181*, 40.
61 Bonn, G.; Grünwald, M.; Scherz, H.; Bobleter, O. *J. Chromatogr.* **1986**, *370*, 485.
62 Haidacher, D.; Bonn, G. K.; Scherz, H.; Nitsch, E.; Wutka, R. *J. Chromatogr.* **1992**, *591*, 351.
63 Selvendran, R. R.; March, J. F.; Ring S. G. *Anal. Biochem.* **1979**, *96*, 282.
64 Albersheim, P.; Nevins, D. J.; English, P. D.; Karr, A. *Carbohydr. Res.* **1967**, *5*, 340.
65 Taylor, R. L.; Conrad, H. E. *Biochemistry* **1972**, *11*, 1383.
66 Chambers, R. E.; Clamp, J. R. *Biochem. J.* **1971**, *125*, 1009.
67 Preuß, A.; Thier, H. P. *Z. Lebensm. Unters. Forsch.* **1982**, *175*, 93.
68 Matheson, N. K.; Mc Cleary, B. V. In *The Polysccharides*, Vol. 3; Aspinall, G., Ed.; Academic:, New York, 1985; p 1ff.
69 Haworth, W. N.; Hirst, E. L.; Thomas H. A. *J. Chem. Soc.* **1931**, 821.
70 Wallenfels, K.; Bechtler, G.; Kuhn, R.; Trischmann, H.; Egge, H. *Angew. Chem., Int. Ed. Engl.* **1963**, *2*, 515.
71 Kuhn, R.; Egge, H. *Chem. Ber.* **1963**, *96*, 3338.
72 Hakomori, S. *J. Biochem.* **1964**, *55*, 205.
73 Phillips, L. R.; Fraser, A. *Carbohydr. Res.* **1981**, *90*, 149.
74 Ciucanu, I.; Kerek, F. *Carbohydr. Res.* **1984**, *131*, 209.
75 Arnap, J.; Kenne, L.; Lindberg, B.; Lönngren, J. *Carbohydr. Res.* **1975**, *44*, C5–C7.
76 Prem, P. *Carbohydr. Res.* **1980**, *78*, 372.
77 Anumula, K. R.; Taylor, P. B. *Anal. Biochem.* **1992**, *203*, 101.
78 Heims, H.; Steinhart, H. *Carbohydr. Polym.* **1991**, *15*, 207.
79 Ciucanu, I.; Luca, C. *Carbohydr. Res.* **1990**, *206*, 71.
80 Björndahl, H.; Lindberg, B. *Carbohydr. Res.* **1970**, *12*, 29.
81 Lindberg, B.; Ginsburg, V. (Ed.) *Methods Enzymol.* **1972**, *28*, 178.
82 Rolf, D.; Gray, G. R. *J. Am. Chem. Soc.* **1982**, *104*, 3539.
83 Rolf, D.; Gray, G. R. *Carbohydr. Res.* **1986**, *152*, 343.
84 Rolf, D.; Bennek, J. A.; Gray, G. R. *J. Carbohydr. Chem.* **1983**, *2*, 373.
85 Jun, J. G.; Gray, G. R. *Carbohydr. Res.* **1987**, *163*, 247.
86 Kiwitt-Haschemie, K.; Heims, H.; Steinhart, H.; Mischnik, P. *Carbohydr. Res.* **1989**, *191*, 343; **1993**, *248*, 267; Kiwtt-Haschemie, K. Thesis, University of Hamburg, 1994.
87 Aspinall, G. In *The Polysaccharides*, Vol. 1; Aspinall, G., Ed.; Academic:, New York, 1982; p 81.
88 Hay, G. W.; Lewis, B.A.; Smith, F. In *Methods in Carbohydrate Chem.*, Vol. 5; Wistler, R. L., Ed.; Academic: New York, 1965; pp 357, 377.
89 Goldstein, I. J.; Hay, G. W.; Lewis, B. A.; Smith, F. In *Methods in Carbohydrate Chem.*, Vol. 5; Wistler, R. L., Ed.; Academic: New York, 1965; p 361.
90 Churms, S. C.; Stephen, A. M. *Carbohydr. Res.* **1971**, *19*, 211.
91 Dutton, G. S.; Gibney, K. B. *Carbohydr. Res.* **1972**, *25*, 99.
92 Lindberg, B.; Lönngren, J.; Nimmich, W.; Ruden, U. *Acta Chem. Scand.* **1973**, *27*, 3787.
93 Hough, L. In *Methods in Carbohydrate Chem.*, Vol. 5; Wistler, R. L., Ed.; Academic: New York, 1965; p 370.
94 Perlin, A. S. In *The Carbohydrates*, Vol. 1B; Pigman, W.; Horton, D., Eds.; Academic:, New York, 1980; p 1167.
95 Painter, T. J.; Larsen, B. *Acta Chem. Scand.* **1970**, *24*, 813, 2366, 2724.
96 Fahmy Ishak, M.; Painter, T. M. *Acta Chem. Scand.* **1971**, *25*, 3875.
97 Scott, J. E.; Tigwell, M. J. *Biochem. J.* **1978**, *173*, 103.

98   Fransson, L. A. *Carbohydr. Res.* **1978**, *62*, 235.
99   Lindberg, B.; Lönngren, J.; Thompson, J. L. *Carbohydr. Res.* **1973**, *28*, 351.
100   Barrett, A. J.; Northcote, D. H. *Biochem. J.* **1965**, *94*, 617.
101   Anet, E. F. L. J. *Adv. Carbohydr. Chem.* **1964**, *19*, 181.
102   Sowden, J. C. *Adv. Carbohydr. Chem.* **1957**, *12*, 35.
103   Aspinall, G. O.; Khan, R.; King, R. R.; Pawlak, Z. *Can. J. Chem.* **1973**, *51*, 1359.
104   Björndahl, H.; Wägström, B. *Acta Chem. Scand.* **1969**, *23*, 3313.
105   Kenne, L.; Lindberg, B.; Svensson, S. *Carbohydr. Res.* **1975**, *40*, 69.
106   Baker, C. W.; Whistler, R. L. *Carbohydr. Res.* **1975**, *45*, 237.
107   Szeitly, J. *Säurehydrolyse glykosidischer Bindungen*; VEB Fachbuchverlag: Leipzig, 1976.
108   Aspinall, G. In *The Polysaccharides*, Vol. 1; Aspinall, G., Ed.; Academic: New York, 1982; p 62.
109   BeMiller, J. N. *Adv. Carbohydr. Chem.* **1967**, *22*, 25.
110   Graham, E. R. B.; Neuberger, A. *J. Chem. Soc. C.* **1968**, 1638.
111   Yosizawa, Z.; Sato, T.; Schmid, K. *Biochim. Biophys. Acta* **1966**, *121*, 417.
112   Kenne, L.; Lindberg, B. In *Methods in Carbohydrate Chem.*, Vol. 8; BeMiller, J. N., Ed.; Academic: New York, 1980; p 295.
113   Lee, Y. C.; Ballou, C.E. *Biochemistry* **1965**, *4*, 257.
114   Aspinall, G. O.; Baillie, J. *J. Chem. Soc.* **1963**, 1702.
115   Aspinall, G. O.; Charlson, A. J.; Hirst, E. L.; Young, R. *J. Chem. Soc.* **1963**, 1696.
116   Bayard, B.; Montreul, J. *Carbohydr. Res.* **1972**, *24*, 427.
117   Aspinall, G. O.; Mc Kenna, J. P. *Carbohydr. Res.* **1968**, *7*, 244.
118   Nilsson, B.; Svensson, S. *Carbohydr. Res.* **1979**, *72*, 183.
119   Lundblad, A.; Svensson, S.; Löw, B.; Messeter, L.; Cedergren, B. *Eur. J. Biochem.* **1980**, *104*, 323.
120   Nilsson, B.; Svensson, S. *Carbohydr. Res.* **1978**, *65*, 169.

General References

The Polysaccharides; Aspinall, G., Ed.; Academic: New York, 1982–1985.
Methods in Carbohydrate Chem.; Wistler, R. L., Ed.; Academic: New York, 1965–1980.

4. Glycoproteins

4.1. Introduction

Glycoproteins are combinations of proteins and carbohydrates connected by covalent linkages. They are present in all living systems. This group of proteins includes nearly all serum proteins, many plasma proteins, blood group substances, antibodies (immunoglobulins), lectins, and several food proteins such as the caseins. Other glycoproteins are compounds of cell membrane and cell surface proteins; the latter are responsible for the recognition of viruses. At the molecular level, the carbohydrate moieties of the glycoproteins protect the peptide chains against proteolytic attack, reduce the tendency for denaturation, stabilize the protein conformation, and determine the specificity of the blood group proteins.

As remarked previously (see Introduction, page 22), two classes of glycoproteins exist according to the type of covalent linkage between the glycan and the peptide chain, namely, the N-glycosyl and the O-glycosyl connections.

In the case of the N-glycosyl glycoproteins only one type has been found, namely, the linkage between β-N-acetyl-D-glucosamine (N-acetyl-2-amino-2-deoxy-D-glucose) and the L-asparagine residue of the protein chain. In contrast, several variations are found for the O-glycosyl proteins, namely:

−　　　Linkage between α-N-acetyl-D-galactosamine (N-acetyl-2-amino-2-deoxy-D-galactose) and L-serine or L-threonine (alkali labile) (Figure 1, A).

−　　　Linkage between β-D-xylose and L-serine which occurs predominately in proteoglycans (alkali labile) (Figure 1, B).

−　　　Linkage between β-D-galactose and 5-hydroxy-L-lysine which occurs predominately in collagen (alkali stable).

−　　　Linkage between β-L-arabinose and 4-hydroxy-L-proline (alkali stable).

The primary chemical structure of the carbohydrate part of the glycoprotein is divided into two sections, namely, (a) the inner core and (b) the antenna. The inner cores are the nonspecific invariant oligosaccharide sequences which are linked to the protein chain and which determine the type of glycoprotein. In contrast, the outer part of the glycan, the so-called "antenna", connected to the inner core, shows great variation with many cases of branched structures as shown in the following example (Figure 1, C).

Gal(β1-3)GalNAc(α1-3)Ser or Thr　　**A**

Gal(β1-3)Gal(β1-4)Xyl(β1-3)Ser　　**B**

Man(α1-3)
＼
　　Man(β1-4)GlcNAc(β1-4)GlcNAc(β1-N)Asn　　**C**
／
Man(α1-6)

Figure 1

The analytical chemistry of glycoproteins is complex as it comprises the analytical methods for proteins as well as those for carbohydrates. The treatment of both exceeds the scope of this book, therefore, only the methods dealing with the glycan parts of the glycoproteins will be discussed here.

4.2. Isolation of the Carbohydrate Parts of Glycoproteins

The important point here is distinction between O-glycosyl and N-glycosyl linkages.

4.2.1. Cleavage of the O-Glycosidic Linkages

4.2.1.1. Treatment with Dilute Alkali

The O-glycosyl linkages between the glycan and β-hydroxyamino acids are cleaved easily with dilute aqueous alkali solution under mild conditions by a β-elimination according to the following scheme (Figure 2).

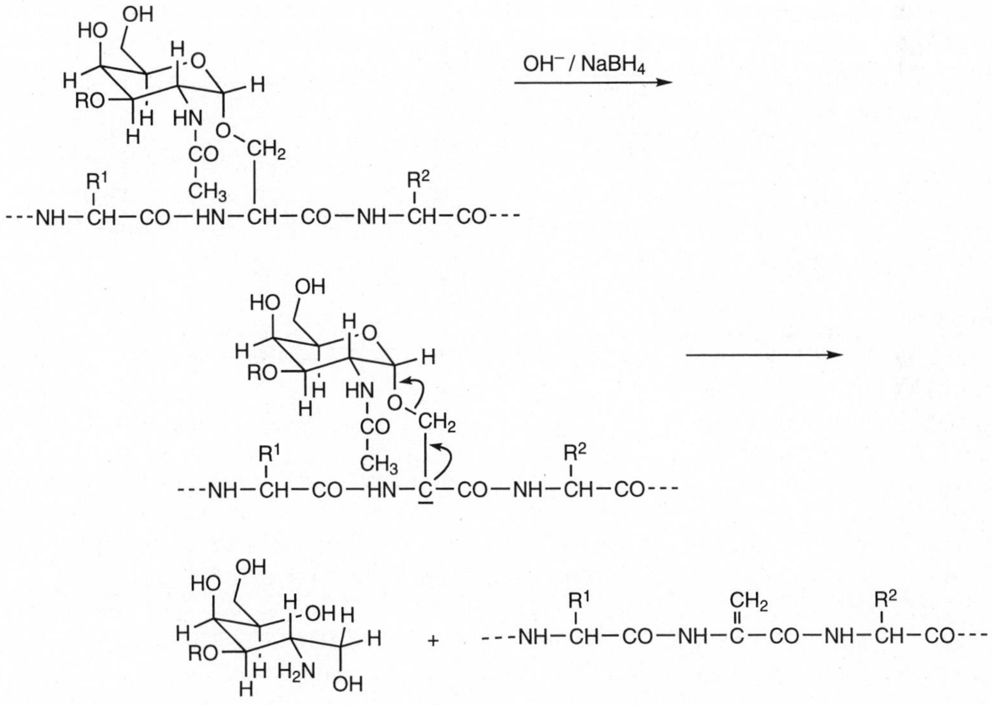

Figure 2

For minimizing the destruction of the carbohydrate parts, the reaction is carried out in the presence of sodium borohydride and the glycan moieties are obtained as their corresponding alditols. The serine and threonine residues are converted to dehydroamino acids[1-3].

Procedure

Reagent A: a freshly prepared solution of 1 M sodium borohydride in 0.1 M dilute aqueous sodium hydroxide.

Reagent B: 50% aqueous acetic acid.

Dissolve approximately 10–20 mg of the glycoprotein or the isolated glycopeptides, after digestion of the glycoprotein with proteases such as Pronase E[4], in 1 mL of water. Adjust the solution with dilute sodium hydroxide to pH 10, add 1 mL of reagent A and heat the solution for 16–24 h at 45 °C. After cooling, add reagent B to destroy the excess of borohydride; a pH of 6 should be maintained. Evaporate the solution to dryness, remove the boric acid by repeated evaporation with methanol, and isolate the glycans, which are present as oligosaccharide alditols, by gel chromatography, ion exchange chromatography, HPLC and capillary electrophoresis. Their structures can be determined by methods described in the preceding chapters of this book.

This reaction is not as selective as described. It has been shown that mild alkali borohydride treatment also cleaves *N*-glycosidically linked oligosaccharides to a lesser extent[5].

4.2.2. Cleavage of the *N*-Glycosidic Linkage

4.2.2.1. Treatment with Strong Alkali

In the presence of strong alkali, *N*-glycosidic linkages are cleaved quantitatively yielding D-glucosamine containing the free glycan. Since the reaction must be carried out in the presence of sodium borohydride to prevent destruction of the glycan by alkali, the *N*-glycosidically linked D-glucosamine residue appears as D-glucosaminitol[1,6].

Procedure

Reagent A: a freshly prepared solution of 1 M sodium borohydride in 1 M aqueous sodium hydroxide.

Reagent B: 50% aqueous acetic acid.

Reagent C: saturated solution of sodium bicarbonate.

Reagent D: acetic anhydride, pure.

Dissolve approximately 50 mg of the glycoprotein in 1 mL reagent A and heat the mixture under reflux for 6 h. Cool the solution in an ice bath and keep it at pH 6 during the neutralization with reagent B when the excess of borohydride is destroyed. Isolate the glycan fraction by gel chromatography on a Biogel P-2 column (L: 400 mm, i.d.: 20 mm) or Sephadex G-15 column (L: 1000mm, i.d.: 18 mm), collect the carbohydrate containing fraction and lyophilize. Stabilize the amino sugars by reacetylating the amino groups by dissolving the dry lyophilic compound in reagent C (app. 1 mL per mg glycan), and add aliquot parts of reagent D (~10 µL) to the solution at 5 min intervals keeping at 4 °C for the first 20 min and then at room temperature for 30 min. Remove the sodium ions by passing the solution through a cation exchange column in the H⁺-form (e.g., Dowex 50WX8) and lyophilize the effluent again. Separate the oligosaccharides as their alditols either by preparative HPLC, affinity chromatography on immobilized lectins or, in exceptional cases, by paper chromatography or paper electrophoresis.

Traces of boric acid can be removed from the residue of the lyophilized effluent after the cation exchange passage by evaporation 2–3 times with absolute methanol.

4.2.2.2. Hydrazinolysis

The reaction of glycoproteins or glycopeptides with anhydrous hydrazine leads to an unspecific cleavage of the O-glycosidic and N-glycosidic linkages of the glycan moieties[1,7–11]. In the case of the N-glycosidic bonds between N-acetyl-D-glucosamine residues of the glycan and the asparagine group, the following reaction proceeds (Figure 3).

Steps 1 and 2 of the reaction occur quickly, the last one rather slowly. The reaction product of step 3, the D-glucosamine hydrazone, is degraded to several products; therefore, prolonged hydrazinolysis should be avoided.

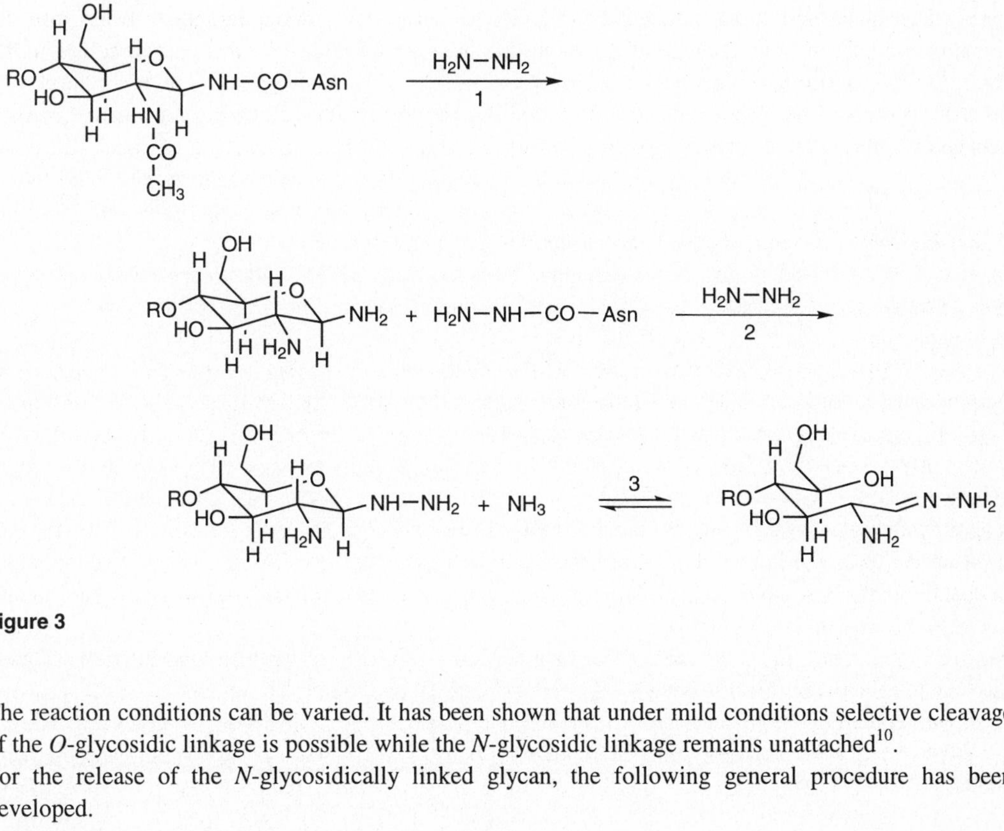

Figure 3

The reaction conditions can be varied. It has been shown that under mild conditions selective cleavage of the O-glycosidic linkage is possible while the N-glycosidic linkage remains unattached[10].

For the release of the N-glycosidically linked glycan, the following general procedure has been developed.

Procedure[1,7,8]

Reagent A: anhydrous hydrazine (freshly distilled), pure.
Reagent B: saturated sodium bicarbonate.
Reagent C: acetic anhydride, pure.
Reagent D: sodium borohydride.

Suspend the dry, salt-free glycoprotein or glycopeptide containing 0.2–1 mg of the asparagine linked sugar chain in 0.5–1 mL of reagent A in a thick-walled glass tube with a screw cap and equipped with a Teflon–silicon disk. Close the vessel immediately, gently shake and heat for 4–8 h at 95 °C. Cool the reaction mixture and remove the excess hydrazine either by repeated evaporation under a stream of dry nitrogen in the presence of anhydrous toluene[1] or under high vacuum at a temperature not exceeding 25 °C[10]. Remove the last traces by storage in a vacuum dessicator over concentrated sulfuric acid. Dissolve the dry residue in a small amount of water, fractionate the material on a Biogel-P-2 column, collect the carbohydrate containing fractions and lyophilize. For reacetylation, dissolve the dry residue in reagent B (approx. 1 mL per 2 mg glycan) and add reagent C as is described in Section 4.2.2.1. After acetylation, pass the reaction mixture through a column of cation ion exchanger in the H$^+$-form (e.g., Dowex 50W), wash the column five times with distilled water, combine the effluent and washings and evaporate to dryness, preferably by lyophilization.

Reacetylation of the D-glycosamine hydrazone affords the corresponding 1-acetylhydrazo-N-acetylglucosamine, which is converted by treatment with the cation exchange resin into N-acetylglucosamine (60%) which yields N-acetyl-D-glucosaminitol after the borohydride reduction; the remaining 40% of the glycan hydrazone is reduced to 1-N-acetylhydrazino-N-acetylglucosaminitol.

For the reduction of the N-acetylglucosamine, dissolve the dry residue after the ion exchange column passage in dilute 0.05 M sodium hydroxide and add 5 mg of reagent D per mg of sugar. Keep the mixture for 16 h at 20 °C and stop the reaction by adding 1 M acetic acid or Dowex 50W H$^+$ which also removes the sodium ions. Then, remove the boric acid by repeated evaporation with absolute methanol. The residue contains the oligosaccharides as the corresponding alditols.

These can be separated by gel chromatography, anion exchange HPLC with amperometric detection, ion exchange chromatography on DEAE-cellulose, high voltage paper electrophoresis or paper chromatography.

As an alternative procedure, reacetylate the dry hydrazine-free residue of the hydrazinolysis as described in Section 4.2.2.1. without separation of the carbohydrate fraction by gel filtration. After the reaction, pass the mixture through a cation exchange column (H$^+$-form) (e.g., Dowex AG 50X12 or Dowex 50W), wash with distilled water, combine the effluent and washings, and evaporate the whole fraction to dryness. Dissolve the dry residue in a small volume of distilled water and spot it on a sheet of chromatographic paper (e.g., Whatman 3MM). Develop the paper with 1-butanol–ethanol–water 4:1:1 v/v for 2 days. The area 0–50 mm from the baseline contains the oligosaccharide fraction while the peptide moieties move faster. This section of paper is cut out and the oligosaccharides are recovered by elution with water[9,10].

For structure determination, the methods of permethylation, GC–MS of the methylated alditol acetates, cleavage to subunits by specific enzymes as well as ^{1}H NMR and ^{13}C NMR are used.

As remarked before, the procedure can be modified so that only the O-glycosidic linkages are cleaved. For this purpose the hydrazinolysis is carried out at 60–65 °C for 5 hours as shown, for example, with fetuin[10,11].

4.2.3. Deglycosylation with Trifluoromethanesulfonic Acid

Treatment of the dry glycoproteins with trifluoromethanesulfonic acid leads to the release of the O-glycosidically and N-glycosidically linked glycans together with the remaining carbohydrate-free intact protein chains. The procedure can be modified so that only the O-glycosyl bonds are cleaved while the

N-glycosidically linked glycan remains attached to the protein. The latter procedure is described in detail and is the method of choice for obtaining intact protein chains[12,13].

Procedure[12]
Reagent A: 2 mL of trifluoromethanesulfonic acid mixed with 1 mL of anisol in a glass tube with a Teflon lined screw cap and cooled to 0 °C.
Reagent B: aqueous pyridine 50% (v/v).
Dissolve 5–10 mg of the dry glycoprotein in 1 mL of reagent A in a dry glass vial with a Teflon screw cap, bubble a stream of dry nitrogen for 30 min through the solution and stir magnetically at 0 °C for 3 h. Dilute the solution with an excess of diethyl ether which has been cooled to –40 °C and carefully add an equal volume of the chilled reagent B. Stir the mixture intensively to redissolve the precipitate and remove the diethyl ether phase. Repeat the diethyl ether extraction and dialyze the aqueous phase against approximately 4 L of 2 mM pyridine acetate buffer pH 5.5; lyophilize the dialysate.
This method has been tested on fetuin, human erythrocyte sialoglycoprotein and bovine nasal septum proteoglycan.

4.2.4. Enzymatic Deglycosylation

Several enzymes exist for the deglycosylation of glycoproteins. To obtain the whole unaltered part of the glycan, endoglycosidases are necessary which specifically cleave the glycan part from the protein or peptide moiety[1,14].
The enzymatic deglycosylation of the asparagine linked glycans can be achieved with the enzyme peptide -N^4-(*N*-acetyl-β-D-glucosaminyl)asparagine amidase (PNG-ase, F; EC 3.5.1.52), which hydrolyzes the glycosaminic linkages of many glycoproteins or glycopeptides yielding aspartic acid residues in the protein/peptide chain and free 1-amino oligosaccharides. This enzyme has been isolated from cultures of *Flavobacterium meningosepticum*[15,16] in addition to other endoglycosidases such as endo-β-*N*-acetyl-D-glucosaminidase F (ENDO-F; EC 3.2.1.96), which cleaves the β-(1→4)-linked di-*N*-acetylchitobiose core of the *N*-glycosidic glycans[17]. One *N*-acetylglucosamine molecule remains linked with the asparagine residue of the protein/peptide, the other becomes the reducing terminus of the liberated oligosaccharide. A similar enzyme, the "ENDO-H" has been isolated from *Streptomyces griseus (plicatus)*[18] and a PNG-ase A from almond emulsin[19].
The cleavage of the *N*-glycosidic bond of the glycoprotein/peptide by the PNG-ases is nearly independent of the structure of the glycan, which remains unaltered in contrast to the ENDO-H(F) enzymes.
These enzymes are used mainly for the preparation of the intact glycan-free protein/peptide chains of the glycoproteins/glycopeptides, which are then submitted for determination of the amino acid sequence and for localization of the glycan parts in the intact glycoproteins/glycopeptides[20]. The procedures deal, therefore, mainly with operations for isolating the protein/peptide part and not the glycans.
Another group of enzymes, preferably specific exoenzymes, which remove terminal monosaccharides from the glycans, which consist partly of several branches, are used for the determination of their structures[21]. Some of them are listed in Section 3.3.1. of this book (pages 321, 322). The description of the whole spectrum of the enzymes used for this purpose is beyond the scope of this monograph.

4.2.5. Isolation of the Glycopeptides, the Free Oligosaccharides and Glycoproteins by Affinity Chromatography

Lectins are proteins or glycoproteins of non-immunogen origin which are able to bind sugars with various affinities. They possess at least two sugar-binding regions which enables them to precipitate polysaccharides and glycoproteins. Such precipitates can be redissolved in solutions containing specific monosaccharides or their glycosides, so-called "sugar haptens", which are also active for lectin molecules. In several cases, the low molecular weight "sugar haptens" must be applied in high concentrations to verify the exchange of sugar moieties in the sugar–lectin complex since the binding affinities of the lectins toward higher oligosaccharides are much greater than toward monosaccharides[1,22].

Today, a great spectrum of such lectins has been identified from a lot of different sources and many of them have several regions of sugar-binding properties in their molecules. In such cases, the interaction of the glycoconjugate (glycoprotein, glycopeptide) to the polysaccharide or free oligosaccharide is determined by the binding affinity of each of the regions. When a glycoconjugate contains several carbohydrate determinants, the binding strength of such a complex rises dramatically and splitting of it sometimes requires drastic conditions such as treatment with alkaline borate buffer which denatures the lectin irreversibly.

The affinity of such a lectin to the glycoconjugate is selective and it must be determined empirically in each case. For example, it is possible that a lectin binds the glycoconjugate with the specific oligosaccharide, but the released free oligosaccharide is not retained.

Today lectins are valuable tools for the elucidation of complex carbohydrate structures in the field of glycoproteins, glycolipids, glycosaminoglycans and polysaccharides. A few of them, their sources and their properties are listed (see Table); this data is taken from excellent reviews[1,22].

For the application of the lectins for the isolation of the glycoconjugates and the free oligosaccharides by affinity chromatography they must be immobilized on a water-insoluble inert matrix.

The following materials have been used for this purpose.

(a) Cross-Linked Agarose = Sepharose 4B (Pharmacia)

The material is activated with CNBr in a suspension of 2 M aqueous sodium carbonate according to the procedure of March[35].

For coupling, the appropriate lectin is dissolved in a solution of 0.2 M sodium hydrogen carbonate and 0.5 M NaCl containing 0.2 M hapten sugar, and the solution is chilled to 4 °C. The wet activated Sepharose is added and the reaction mixture is stirred gently for 24 h at 4 °C. Then, the reaction mixture is filtered and the remaining gel with the immobilized lectin is washed with ice-cold distilled water, 0.1 M NaHCO₃ and water again. The nonreacted active Sepharose 4B is then inactivated by addition of glycine at room temperature.

(b) Polyacrylhydrazido-Sepharose (PAHS)[1,36]

This commercially available material is first gently stirred with a 10% (w/v) aqueous solution of glutaric dialdehyde for 4 h at 4 °C. The lectin is dissolved in a solution of 0.1 M sodium hydrogen carbonate and 0.9 M NaCl containing 0.1 M hapten monosaccharide, and stirred gently with the activated PAHS overnight. After filtering and washing, the lectin–PAHS is treated further with a sodium chloride–sodium borohydride solution for reducing the excess free aldehyde groups and the imino groups of the Schiff bases.

(c) Wide-Porous Activated Silica Gel[37,38]

The silica gel is activated by suspending in an aqueous solution of γ-glycidoxypropyltrimethoxysilane and then heated for 2 h at 90 °C at pH 5.5 and then for 1 h at pH 3.0; under the latter conditions the oxirane ring of the gel is converted into glycol groups. This material is treated with periodate to obtain reactive aldehyde groups, which are then coupled with the free amino groups of the lectin molecules in 0.1 M NaHCO$_3$ solution at pH 7.9. The last step is the reduction of the aldimine groups with NaBH$_4$ to secondary amino groups.

In several cases the immobilized lectins are commercially available.

For the immobilization of the lectin and the subsequent affinity chromatographic separation, the following guidelines should be considered.
(a)   During the coupling operations the binding regions of the lectin molecules must be protected by hapten sugars.
(b)   To limit the nonspecific reactions between glycoproteins and the immobilized lectins the buffers must contain an electrolyte (0.1–1.0 M NaCl).
(c)   Some lectins possess metal ion complexing sites which stabilize the conformation of the molecule and which are necessary for their binding activities to the glycans. The eluting buffers must contain such metal ions to maintain the activity of the immobilized lectin.
(d)   Immobilized lectins must be stored at 4 °C in buffers containing 0.02% sodium azide as bacteriostatic agent.
The most common technique of lectin affinity chromatography is the preparation of a small column (L: 10–15 cm, i.d.: 0.7–2.5 cm) containing the immobilized lectin. The reservoir containing the eluting buffer is connected to the top of the column and the effluent is collected by a fraction collector. The chromatographic separation is carried out mainly by gravity flow. The glycoconjugates and oligosaccharides should be labeled either by radioactive (^{14}C, ^{3}H) or fluorescing groups for improved detection in the effluent, which contains the hapten monosaccharides. The isolated glycoconjugate or oligosaccharide is separated from the buffer electrolytes and from the hapten monosaccharides by gel chromatography (e.g., Biogel P-2).

The lectin affinity chromatography is a valuable method for the separation and isolation of specific glycopeptides and oligosaccharides. It is important to know precisely the specificity of the immobilized lectin which makes it possible to predict the structures of the glycan or glycopeptide.
In contrast, the separation of glycoproteins by lectin affinity chromatography is less successful since, in several cases, nonspecific interactions between the glycoprotein molecules and the immobilized lectin are observed.
The conditions for the practical performance of the lectin affinity chromatography are strongly dependent on the individual separation problem and what kind of lectin is used for that purpose. A common procedure is given here.

Procedure
(a)   Preparation of the column: pour the suspension of the immobilized lectin in the eluting buffer containing 0.02% of sodium azide into the chromatographic column and equilibrate with the same buffer.
(b)   Dissolve the glycopeptide or the oligosaccharide in a small volume of the same buffer and apply it to the top of the column.

Table. Small Survey of Lectins Used for Affinity Chromatography of Glycopeptides and Oligosaccharides

Trivial Abbreviation	Source of the Lectin	Monosaccharide Specificity	Hapten Sugars	Comments on Binding
Concanavalin A[23,24] ConA	Canavalia ensiformis	α-D-Mannose α-D-Glucose	α-Methyl-D-mannoside α-Methyl-D-glucoside	Low binding to biantennal glycopeptide; strong to high mannose/hybride type
Pea lectin[23,25]	Pisum sativum	α-D-Mannose	α-Methyl-D-mannoside	Binding to bi- and triantennal glycopeptide containing L-fucose in the glycan core
Lentil lectin[23,25]	Lens culinaris	α-D-Mannose	α-Methyl-D-mannoside	Binding to bi- and triantennal glycopeptide containing fucose in the glycan core
RCA-1[26]	Ricinus communis agglutinin	β-D-Galactose N-acetyl-β-D-galactosamine	Lactose	Binding glycopeptides with β-(1→4)-linked terminal galactosyl residues
L₄PHA[27]	Phaseolus vulgaris leukoagglutinin	β-D-Galactose	N-Acetyl-D-galactosamine	Binding to tri- and tetraantennal types of glycopeptides
E₄PHA[27-29]	Phaseolus vulgaris erythroagglutinin	β-D-Galactose	N-Acetyl-D-galactosamine	Binding to bi- and triantennal types of glycopeptides and oligosaccharides
SNA[30]	Sambucus nigra agglutinin	β-D-Galactose	Lactose	Binding to sialic acid containing glycopeptides with NeuAc-α(2→6)Gal terminal sequences
LTA[31] Lotus lectin	Tetragonolobus purpureas agglutinin	α-L-Fucose	Fucose	Binding to glycopeptides with fucose residues at the outer part of the glycan
GS-I-B₄[32]	Griffonia (Bandeiraea) Simplicifolia 1-34	α-D-Galactose N-acetyl-α-D-galactosamine	Raffinose	Binding to oligosaccharides with α-galactosyl residues at one terminal
UEA-I[22]	Ulex europaeus I	α-L-Fucose	Fucose	Binding to oligosaccharides with outer α-L-fucose residues
HPA[22,33]	Helix pomatia agglutinin	N-α-Acetyl-D-galactosamine	N-Acetyl-D-galactosamine	Binding to N- and O-linked oligosaccharides with terminal α-GalNAc with high affinity
Jacalin[34], lectin (Jackfruit lectin)	Artocarpus integrifolia	α-D-Galactose α-2-N-acetylgalactosamine	α-D-Methylgalactoside	Binding to glycopeptides and oligosaccharides with O-α-galactosyl linked terminals

(c)   Elute the column at first with a fixed volume of the equilibrium buffer and collect small fractions of the effluent with a fraction collector; change the eluent to solutions of the same buffer containing the hapten sugars with enhancing concentrations.

By this procedure glycopeptides and oligosaccharides can be differentiated into three classes, namely:
(a)   Glycopeptides and oligosaccharides with no affinities to the immobilized lectin; they are eluted with the void volume of the column with the equilibrium buffer.
(b)   Glycopeptides and oligosaccharides with weak affinities to the immobilized lectin; they are eluted with the equilibrium buffer at volumes greater than the void volume.
(c)   Glycopeptides and oligosaccharides with medium and strong affinities to the immobilized lectin; they are eluted with the equilibrium buffer containing related concentrations of the hapten sugar. After these operations, the fractions containing the glycopeptides and the free oligosaccharides are collected and both substances are isolated by individual procedures. Their structures as well as their quantities are determined predominately with methods described in Chapters 2–4 of this book.

References

1   Montreuil, J.; Bouquelet, S.; Debray, H.; Fournet, B.; Spik, G.; Strecker, G. In *Carbohydrate Analysis*; Chaplin, M. F.; Kennedy, J. F., Eds.; Irl-Press: Oxford, Washington, 1986; p 143.
2   Iyer, R. N.; Carlson, D. M. *Arch. Biochem. Biophys.* **1971**, *142*, 101.
3   Carlson, D. M: *J. Biol. Chem.* **1968**, *243*, 616.
4   Fukuda, M. *Methods Enzymol.* **1989**, *179*, 17.
5   Ogata, S. I.; Lloyd, K. *Anal. Biochem.* **1982**, *119*, 351.
6   Yuan Chuan Lee; Scocca, J. R. *J. Biol. Chem.* **1972**, *247*, 5753.
7   Yosizawa, Z.; Sato, T.; Schmid, K. *Biochim. Biophys. Acta* **1966**, *121*, 417.
8   Fukuda, M.; Kondo, T.; Osawa, T. *J. Biochem.* **1976**, *80*, 1223.
9   Takasaki, S.; Mizuochi, T.; Kobata, A. *Methods Enzymol.* **1982**, *83*, 263.
10   Patel, T.; Bruce, J.; Merry, A.; Bigge, C.; Wormald, M.; Jaques, A.; Parekh, R. *Biochemistry* **1993**, *32*, 679.
11   Patel, T. P.; Parekh, R. B. In *Guide to Techniques in Glycobiology; Methods in Enzymology 230*; Lennarz, W. J.; Hart, G. W., Eds.; Academic: San Diego, 1994; p 57.
12   Edge, A.; Faltynek, C. R.; Hof, L.; Reichert, L. E.; Weber, P. *Anal. Biochem.* **1981**, *118*, 131.
13   Sojar, H. T.; Bahl, O. M. *Methods Enzymol.* **1987**, *138*, 341.
14   Thotakura, N.; Bahl, O. P. *Methods Enzymol.* **1987**, *138*, 350.
15   Tarentino, A. L.; Plummer, T. H., Jr. In *Guide to Techniques in Glycobiology; Methods in Enzymology 230*; Lennarz, W. J.; Hart, G. W., Eds.; Academic: San Diego, 1994; p 44.
16   Maley, F.; Trimble, R. B.; Tarentino, A. L.; Plummer, T. H., Jr. *Anal. Biochem.* **1989**, *180*, 195.
17   Plummer, T. H., Jr.; Elder, J. H.; Alexander, S.; Phelan, A. W.; Tarentino, A. L. *J. Biol. Chem.* **1984**, *259*, 10700.
18   Tarentino, A. L.; Maley, F. *J. Biol. Chem.* **1974**, *249*, 811.
19   Taga, E. M.; Waheed, A.; van Etten, R. L. *Biochemistry* **1984**, *23*, 815.
20   Parekh, R. B. In *Guide to Techniques in Glycobiology; Methods in Enzymology 230*; Lennarz, W. J.; Hart, G. W., Eds.; Academic: San Diego, 1994; p 340.
21   Jacob, S. G.; Scudder, P. In *Guide to Techniques in Glycobiology; Methods in Enzymology 230*; Lennarz, W. J.; Hart, G. W., Eds.; Academic: San Diego, 1994; p 280.
22   Cummings, R. In *Guide to Techniques in Glycobiology; Methods in Enzymology 230*; Lennarz, W. J.; Hart, G. W., Eds.; Academic: San Diego, 1994; p 66.
23   Kornfeld, K.; Reitman, M. L.; Kornfeld, R. *J. Biol. Chem.* **1981**, *256*, 6633.
24   Baenzinger, J. U.; Fiete, D. *J. Biol. Chem.* **1979**, *254*, 2400.
25   Cummings, R. D.; Kornfeld, S. *J. Biol. Chem.* **1982**, *257*, 11235.
26   Baenzinger, J. K.; Fiete, D. *J. Biol. Chem.* **1979**, *254*, 9795.

27   Cummings, R. D.; Kornfeld, S. *J. Biol. Chem.* **1982**, *257*, 11230.
28   Yamashita, K.; Hitoi, A.; Kobata, A. *J. Biol. Chem.* **1983**, *258*, 14753.
29   Kobata, A.; Yamashita, K. *Methods Enzymol.* **1989**, *179*, 46.
30   Shibuya, N.; Goldstein, I. J.; Broekaert, W. F.; Nsimba-Lubaki, M.; Peeters, B.; Peumans, W. J. *Arch. Biochem. Biophys.* **1987**, *254*, 1.
31   Srivatsan, S.; Smith, F.; Cummings, R. D. *J. Biol. Chem.* **1992**, *267*, 20196.
32   Goldstein, I. J.; Blake, D. A.; Ebisu, S.; Williams, T. S.; Murphy, L. A. *J. Biol. Chem.* **1981**, *256*, 3890.
33   Do, I. S.; Cummings, R. D. *J. Biochem. Biophys. Methods* **1992**, *24*, 153.
34   Hortin, G. L. *Anal. Biochem.* **1990**, *191*, 262.
35   March, S. C.; Parikh, I.; Cuatrecasas, P. *Anal. Biochem.* **1974**, *60*, 149.
36   Wilchek, M.; Miron, T. *Mol. Cell. Biol.* **1974**, *4*, 181.
37   Ohlson, S.; Hansson, L.; Larsson, P. O.; Mosbach, K. *FEBS-Letters* **1978**, *93*, 5.
38   Zopf, D.; Ohlson, S.; Dakour, J. Wang, W.; Lundblad, A. *Methods Enzymol.* **1989**, *179*, 55.

General References

Montreuil, J.; Bouquelet, S.; Debray, H.; Fournet, B.; Spik, G.; Strecker, G. In *Carbohydrate Analysis*; Chaplin, M. F.; Kennedy, J. F., Eds.; Irl-Press: Oxford, Washington, 1986.
Guide to Techniques in Glycobiology; Methods in Enzymology 230; Lennarz, W.J.; Hart, G. W., Eds.; Academic: San Diego, London, 1994.

Index